IoT Security Paradigms and Applications

Computational Intelligence in Engineering Problem Solving

Series Editor:
Nilanjan Dey

Computational intelligence (CI) can be framed as a heterogeneous domain that harmonized and coordinated several technologies, such as probabilistic reasoning, artificial life, multiagent systems, neurocomputing, fuzzy systems, and evolutionary algorithms. Integrating several disciplines such as machine learning (ML), artificial intelligence (AI), decision support systems (DSS), and database management systems (DBMS) increases the CI power and impact in several engineering applications. This book series provides a well-standing forum to discuss the characteristics of CI systems in engineering. It emphasizes on the development of CI techniques, their role as well as the state-of-the-art solutions in different real-world engineering applications. This book series is proposed for researchers, academics, scientists, engineers, and professionals who are involved in the new techniques of CI. CI techniques, including artificial fuzzy logic and neural networks, are presented for biomedical image processing, power systems, and reactor applications.

Applied Machine Learning for Smart Data Analysis
Nilanjan Dey, Sanjeev Wagh, Parikshit N. Mahalle, and Mohd. Shafi Pathan

IoT Security Paradigms and Applications: Research and Practices
Sudhir Kumar Sharma, Bharat Bhushan, and Narayan C. Debnath

For more information about this series, please visit: https://www.crcpress.com/Computational-Intelligence-in-Engineering-Problem-Solving/book-series/CIEPS

IoT Security Paradigms and Applications
Research and Practices

Edited by
Sudhir Kumar Sharma, Bharat Bhushan
and Narayan C. Debnath

CRC Press
Taylor & Francis Group
Boca Raton London New York

CRC Press is an imprint of the
Taylor & Francis Group, an **informa** business

First edition published 2021
by CRC Press
6000 Broken Sound Parkway NW, Suite 300, Boca Raton, FL 33487-2742

and by CRC Press
2 Park Square, Milton Park, Abingdon, Oxon, OX14 4RN

© 2021 Taylor & Francis Group, LLC

First edition published by CRC Press 2021

CRC Press is an imprint of Taylor & Francis Group, LLC

ISBN: 978-0-367-51496-9 (hbk)
ISBN: 978-1-003-05411-5 (ebk)

Typeset in Times
by codeMantra

Contents

Preface

The Internet of Things (IoT), the next Internet-related revolution, is an emerging technology that integrates technologies and aspects coming from varied approaches. Pervasive computing, ubiquitous computing, communication technologies, sensing technologies, Internet Protocol, and embedded devices are integrated together to form a system where the digital and real worlds collaborate. Huge number of interconnected devices and enormous data open new opportunities to create services capable of bringing tangible benefits to the economy, environment, society, and individual citizens. Owing to the massiveness of IoT and inadequate data security, the impact of security breaches may turn out to be humongous leading to severe impacts. Traditional security approaches against the most prominent attacks are insufficient. Therefore, enabling the IoT devices to learn and adapt to various threats dynamically and addressing them proactively is the need of the hour. In this regard, machine learning (ML) techniques are employed that provide both intelligence and reconfigurability to various IoT devices.

Industrial processes are integrated with full-scale process automation with the emergence of new digital industrial technology, thereby reshaping the future of industrial revolution. In today's era, Industrial IoT has severe impact in changing the industrial digital ecosystem by completely redefining the way machines, enterprises, and stakeholders connect and interact with each other. Connected and smart factories that rely on the machineries transmitting real-time data enable industrial data analytics for improving productivity and operational efficiency. Moreover, IoT finds a huge range of applications in the healthcare sector. Smart healthcare can reduce risk, cost, and time of IoT deployments in healthcare, thereby enhancing the patient care and efficiency. IoT-based healthcare systems can be used to realize personalized access to effective clinical decision-making and vital medical data for every individual with uninterrupted and self-paced sensing. However, this IoT-based healthcare system faces innumerable challenges with respect to accuracy and cost of medical sensors, huge volume of generated data, assorted wearable devices, nonstandard IoT system architectures, and interoperability issues. These challenges and requirements bring forth a huge range of opportunities to examine and discover new intelligent medical systems, algorithms, applications, and new conceptual theories in IoT-based healthcare domain. Another important issue linked to IoT that the researchers/scientists need to focus on is the privacy and security requirements of the sensor-generated data from misuse, theft, or unfortunate losses. To this end, blockchain has received enormous attention both from industry and from academia. Therefore, this book also intends to embrace the role of blockchain in IoT security.

Wider selection of topics for this book is leading towards thorough understanding of reader for IoT basics, security and privacy issues associated with IoT, IoT adoptions in industry and healthcare sectors, integration of IoT with big data and cloud, and IoT security through blockchain and 5G Internet of Things. This book investigates the security and privacy issues associated with various IoT systems along with exploring various ML-based IoT security solutions.

Editors

Dr. Sudhir Kumar Sharma, PhD, is currently a Professor in the Department of Computer Science, Institute of Information Technology and Management affiliated to GGSIPU, New Delhi, India. He has extensive experience for more than 19 years in the field of Computer Science and Engineering. He obtained his PhD degree in Information Technology from USICT, GGSIPU, New Delhi, India. Dr. Sharma obtained his MTech degree in Computer Science and Engineering in 1999 from the Guru Jambheshwar University, Hisar, India, and MSc degree in Physics from the University of Roorkee (now IIT Roorkee), Roorkee, in 1997. His research interests include machine learning, data mining, and security. He has published a number of research papers in various prestigious international journals and international conferences. He is a life member of CSI and IETE. Dr. Sharma is an Associate Editor of the *International Journal of End-User Computing and Development* (*IJEUCD*), IGI Global, the United States. He is a convener of ICETIT-2019.

Bharat Bhushan, MTech, is an Assistant Professor of Computer Science and Engineering (CSE) at HMRITM, GGSIP University, New Delhi, India. Currently, he is also serving as a Managing Director, Gatevilla, Allahabad, Uttar Pradesh, India, and pursuing a PhD degree specializing in wireless sensor Networks at the Birla Institute of Technology, Mesra, India. He received his undergraduate (BTech) degree in CSE with distinction in 2012 and received his master's (MTech) degree in Information Security with distinction in 2015 from Birla Institute of Technology, Mesra, India. He earned numerous international certifications such as Cisco-certified Network Associate, Cisco-certified Network Professional Trained, Cisco-certified Entry Networking Technician, Microsoft-certified Technology Specialist, Microsoft-certified IT Professional, and Red Hat-certified Engineer. He has published more than 74 papers in different national and international conferences, and journals, including wireless personal communications (Springer) and wireless networks (Springer). He was selected as a Reviewer/Editorial Board Member for several reputed international journals. He also served as a Speaker, Session Chair, or Co-chair at various national and international conferences. In the past, he worked as a Network Engineer for HCL Infosystems Ltd. He has qualified for four GATE examinations (GATE 2013, 2014, 2015, and 2016) and gained several years of experience in imparting quality education to students. He is a member of IEEE.

Professor Dr. Narayan C. Debnath, PhD, DSc, is currently the Founding Dean of the School of Computing and Information Technology at Eastern International University, Vietnam. He is also serving as the Head of the Department of Software Engineering at Eastern International University, Vietnam. Dr. Debnath has been the Director of the *International Society for Computers and their Applications* (*ISCA*) since 2014. Formerly, Dr. Debnath served as a Full Professor of Computer Science at Winona State University, Minnesota, the United States, for 28 years (1989–2017).

He was elected as the Chairperson of the Computer Science Department at Winona State University for three consecutive terms and assumed the role of the Chairperson of the Computer Science Department at Winona State University for seven years (2010–2017). Dr. Debnath earned a Doctor of Science (DSc) degree in Computer Science and also a Doctor of Philosophy (PhD) degree in Physics. In the past, he served as the elected President for two separate terms, Vice President, and Conference Coordinator of the *International Society for Computers and their Applications*, and has been a member of the ISCA Board of Directors since 2001. Before being elected as the Chairperson of the Department of Computer Science in 2010 at Winona State University, he served as the Acting Chairman of the Department of Computer Science. Dr. Debnath received numerous honors and awards while serving as a Professor of Computer Science during the period 1989–2017, including the Best Paper Award in the field of networking at the *2008 IEEE International Symposium on Computers and Communications*. During 1986–1989, Dr. Debnath served as an Assistant Professor of the Department of Mathematics and Computer Systems at the University of Wisconsin-River Falls, the United States, where he was nominated for the National Science Foundation (NSF) Presidential Young Investigator Award in 1989. Dr. Debnath is an Author or Co-author of more than 425 publications in numerous refereed journals and conference proceedings in computer Science, information science, information technology, system sciences, mathematics, and electrical engineering. Professor Debnath has made numerous teaching and research presentations at various national and international conferences, industries, and teaching and research institutions in Africa, Asia, Australia, Europe, North America, and South America. He has offered courses and workshops on software engineering and software testing at universities in Asia, Africa, Middle East, and South America. Dr. Debnath has been a visiting professor at universities in Argentina, China, India, Sudan, and Taiwan. He has been maintaining an active research and professional collaborations with many universities, faculty, scholars, professionals, and practitioners across the globe. Dr. Debnath has been an active member of the ACM, IEEE Computer Society, Arab Computer Society, and a senior member of the ISCA.

List of Contributors

Manisha Agarwal
Department of Computer Science
Banasthali Vidyapith
Vanasthali, Rajasthan, India

Shefali Arora
Computer Engineering
Netaji Subhas Institute of
 Technology
Delhi, India

Sambit Bakshi
National Institute of Technology
Rourkela, Odisha, India

M.P.S. Bhatia
Netaji Subhas Institute of
 Technology
Delhi, India

P. Bhuvaneswari
School of Computer and Information
 Sciences
University of Hyderabad
Hyderabad, Telangana, India

Koyela Chakrabarti
Department of Computer Science and
 Engineering
West Bengal University of
 Technology
Kolkata, West Bengal, India

Suvamoy Changder
Department of Computer Science and
 Engineering
NIT Durgapur
Durgapur, West Bengal, India

Debbrota Paul Chowdhury
National Institute of Technology
Warangal, Telangana, India

I.J. de Zwarte
Blackbox Measurements
Netherlands

Narayan C. Debnath
School of Computing and Information
 Technology
EIU, Thu Dau Mot, Vietnam

Bhoomi Gupta
Department of Information Technology
Maharaja Agrasen Institute of
 Technology
Delhi, India

Kalpana Gupta
Centre for Development of
 Advanced Computing (C-DAC)
Noida, Uttar Pradesh, India

Sachin Gupta
Department of Computer Science
 Engineering
MVN University
Aurangabad, Haryana, India

J.S. Gusc
Faculty of Economics and Business
 Accounting
University of Groningen
Groningen, Netherlands

U. Hariharan
Department of Information Technology
Galgotias College of Engineering
 and Technology
U.P, India

S. Jarka
Warsaw University of Life
 Sciences
Warsaw, Poland

Rahul Johari
SWINGER (Security, Wireless
   IoT Network Group of
   Engineering and Research) Lab,
   USICT
GGSIP University
Dwarka, Delhi, India

Gurjot Kaur
Netaji Subhas Institute of
   Technology
Delhi, India

Ishveen Kaur
SWINGER (Security, Wireless IoT
   Network Group of Engineering and
   Research) Lab, USICT
GGSIP University
Dwarka, Delhi, India

Ila Kaushik
Department of Information Technology
Krishna Institute of Engineering &
   Technology
Ghaziabad, Uttar Pradesh, India

Naghma Khatoon
Faculty of Computing & Information
   Technology
Usha Martin University
Ranchi, Jharkhand, India

Gautam Kumar
Department of Computer Science
NIT Rourkela
Rourkela, Odisha, India

Santosh Kumar
Computer Science and Information
   Technology
ITER-SOA
Bhubaneswar, Odisha, India

Rakesh Kumar Singh
Department of Information Technology
IGDTUW
Delhi, India

Manorama Mohapatro
Department of Computer Science and
   Engineering
BIT Mesra
Ranchi, Jharkhand, India

Shrid Pant
CAITFS, Department of Information
   Technology
Netaji Subhas University of Technology
New Delhi, India

A. Parlinska
Warsaw University of Life
   Sciences – SGGW
Warsaw, Poland

Nilotpal Pathak
Department of Information Technology
Galgotias College of Engineering and
   Technology
U.P, India

K. Rajkumar
Department of Information Technology
Galgotias College of Engineering and
   Technology
U.P, India

Ratnakirti Roy
Department of Computer Applications
BCREC
Durgapur, West Bengal, India

Pankaj Kumar Sa
National Institute of Technology
Rourkela, Odisha, India

Abhishek Sharma
Department of Computer Science and
 Engineering
HMR Institute of Technology &
 Management
Delhi, India

Deepak Kumar Sharma
CAITFS, Department of Information
 Technology
Netaji Subhas University of
 Technology
New Delhi, India

Nikhil Sharma
Department of Computer Science and
 Engineering
HMR Institute of Technology &
 Management
Delhi, India

Mehul Sharma
CAITFS, Department of Information
 Technology
Netaji Subhas University of
 Technology
New Delhi, India

Itu Snigdh
Birla Institute of Technology Mesra
Ranchi, Jharkhand, India

E. Stawicka
Warsaw University of Life
 Sciences – SGGW
Warsaw, Poland

Nagender Kumar Suryadevara
School of Computer and Information
 Sciences
University of Hyderabad
Hyderabad, Telangana, India

Edyta Karolina Szczepaniuk
Military University of Aviation
Dęblin, Poland

Hubert Szczepaniuk
Warsaw University of Life Sciences -
 SGGW (WULS-SGGW)
Warsaw, Poland

Reena Tripathi
SWINGER (Security, Wireless
 IoT Network Group of
 Engineering and Research) Lab,
 USICT
GGSIP University
Dwarka, Delhi, India

Veenu
Division of Computer Engineering
Netaji Subhas Institute of
 Technology
Delhi, India

Vidushi
Department Computer Science and
 Engineering
Banasthali Vidyapith
Vanasthali, Rajasthan, India

1 The Convergence of IoT with Big Data and Cloud Computing

K. Rajkumar, Nilotpal Pathak, and U. Hariharan
Galgotias College of Engineering and Technology

CONTENTS

1.1 INTRODUCTION

The digital transformation has made technologies such as big data, the Internet of Things (IoT), and cloud computing to be the top most agenda in most businesses. Formally, each of the technologies was sophisticated, and successful businesses utilizing these technologies were accomplished. However, in modern society, each of the three technologies has its necessity. Even though big data, IoT, and cloud computing evolved independently, they now become more intertwined. Due to large modifications and developments, the three technologies have shown a persistent convergence, which has helped to make techniques highly capable [1,2]. Big data, cloud computing, and IoT are currently trending aspects in most organizations worldwide. During the time of transition in the IT industry, where the data is growing at an unconstrained and exponential rate. As mentioned above, most organizations and users become more data-centric. It has been a standard belief by academicians, researchers, and, therefore, the businessmen that the planet possesses the age of massive data. This chapter provides a detailed review of big data analytics in the cloud platform; cloud-based IoT and its applications; big data analytics and its applications in IoT; challenges of internet- and cloud-based IoT applications; big data issues in gathering, governance, General Data Protection Regulation (GDPR), security, and privacy; machine and deep learning techniques in IoT and cloud; and other issues related to the integration of IoT, cloud computing, and big data. The three technologies are combined to solve the real-time problems.

1.1.1 BIG DATA

Big data is employed to summarize information that can't be in a position to prepare the preexisting knowledge, how, techniques and theories. Many users focus on extensive details about any sort of its feature, which challenges the constraints of a system's electrical capacity in addition to the requirements of a company. Additionally, big data has the capacity to make use of previous details for providing and enhancing the information (i.e., data) in the long term. In big data analytics, the particular information gets to be corrected and also it should be exact. Hence, big data analytics depends on succeeding predictions. Significant details and technologies try to locate, consistently modifying information to build associations in between producing information. Companies that realize the interconnectedness can enjoy the info within serious details, hence unlocking the actual worth of this technology [1,3,4]. Beneficial aspects of big data analytics provide extensive details for many companies, but on the flip side, it poses problems to an entity as well. In order to draw out the actual worth of it, important details must be prepared and examined promptly. Additionally, the outcomes of the evaluation are obtained in a fashion; they can affect company choices. Besides the above, extensive details may additionally be considered the flood of digital details coming from different digital sources of energy, including social networks, email, videos, Internet, mobile phones, numerical modeling, scanners, digits, and sensors. Extensive details are now being typically found within the type of texts, geometric, pictures, video clips, or maybe a mix of every one of the types. The information can be indirectly or directly associated with geospatial information. The information within the area of critical information is recognized by value, veracity, variety, velocity, and volume [5,6].

It has shifted the technique of information from the conventional fixed method to a far more speeding-up technique as a result of the evolution of man's understanding and technologies' comprehension of the information. Ever since the volume of information is developing explosively and also stretches past a person's electrical capacity of dealing with big sets of information, adapting the necessary information engineering is unavoidable. Volume is probably the most typical explanation of big data information. Velocity– another attribute of serious details– describes the quick model as well as the transmission of details throughout the web as exemplified by information compilation out of social networking sites, a wide array of receptors, including both global and atomic information. Additionally, it involves the transmission of information from the receptors to supercomputer and act as a choice maker. However, it suggests the various forms of information and also the product as well as buildings whereby information is attained. The integrity of serious details concentrates on the range of quality, reliability, and trustworthiness of the information, which is essentially important, and provides a specific focus on the study and choice assistance programs that enhance the day of ours, companies, and work [5,7].

From the viewpoint of company, big data era and evolution were envisioned when the reputable advantages such as originality, contests, and efficiency are based on the MC Kinsey article. This particular point of view is actually grounded in the power of big data to derive the development and company revenues of brand-new possibilities. Significant information is expected to enhance the preexisting companies to approximately60% as well as to promote vast amounts of bucks to succeeding companies within the subsequent ten years. Huge detail users' possibilities are excellent, and there are switches within the digital planet area regarding how to follow, feel, and deliver the results. Recent investigations on the processing of big data are centered on the division and stream-based processing. Even though cloud computing emerged significantly previously compared to significant details, it's a brand-new computing paradigm that concentrates on supplying computation variables, including suppleness, pooled online resources, need, and self-service on gain access to [1,8].

1.1.2 INTERNET OF THINGS (IoTs)

Web of items is the networking of actual physical items that allow these actual physical items to gather the information and to swap it. This particular networking arrives in concert via the hookup of devices as well as receptors. Inside IoT, a cloud is a vital part, which means which interconnection involving the systems is essential for a premium price and sufficient function. This is simply because the web inside IoT won't function whenever a cloud is actually absent. Web of items is easily the most varied; the majority of considerable, and the substantial majority of, know-how is a result of engineering fashion that evolves [9,10]. The sophisticated receptors and their web host products, including movable cell phones and wellness monitors, are often attached inside a cyber-physical phone system to determine, sometimes, an area of people, action of cars, vibration of the printer, heat, precipitation, moisture as well as chemical-based switches within the environment. Figure 1.1 [7] shows the

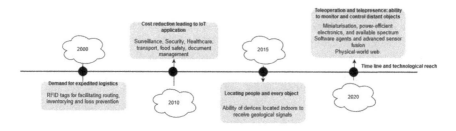

FIGURE 1.1   Technology development in the IoT.

growth of IoT. The web of items captures and makes brand-new domains together with creating information routes all over the world with geographical footprints coming from the interconnected movable products, private PCs, receptors, and also digital cameras. Several receptors on it create big data, which has abundant spatio-temporal information. Big detail systems along with IoT have a wide range of uses. These apps incorporate much better item-type managing, better and also a regular criminal exploration, and improve farming efficiency. In the present planet, the entire company, as well as the functionality of businesses, counts on the net of the issues.

1.1.3   CLOUD COMPUTING

Cloud computing is the technological innovation that will be modified if the demand develops to shop, progress, and evaluate huge details. The cloud may be described as the large-scale division of computing paradigm, offering computing, network, online storage, and other resources. These solutions are supplied to the applications as a degree of quality every time they're needed. Cloud is attractive to owners as a result of its decreased cost, superfast suppleness, accessibility, and effective control [3]. Big data and data science analytics are available straight into the photo when businesses wish to increase success and worth out of great details. Web of items stands out as the niche comprising set analysis, preparation, and cleansing. It is the method that is utilized to remove information as well as insights about information. It will give rise to the idea of serious detail analytics, and it is the process of gathering, organizing, and evaluating large sets of data to build information that is helpful [1,3].

1.1.4   CLOUD COMPUTING

With all the models having a huge quantity of information, cloud computing plays a pivotal role in maintaining storage space and also in managing that particular infor-mation. It is not just around the development of serious details, but additionally, around the development of information analytics open source system like Hadoop [11,12]. Using Hadoop, brand-new possibilities are available in cloud computing. Thus, the program suppliers such as Amazon Web Services (AWS), Microsoft, and Google are selling their very own significant details about devices in a cost-efficient fashion that will be scalable for various kinds of organizations. They particularly created a brand-new program design that is termed as platform as a service *(PaaS)*. PaaS is a scalable and faster method to incorporate data in structured, unstructured, and semi-structured forms; analyze them, change them; and imagine them inside the period, which is illustrated in Figure 1.2.

1.1.4.1 The Benefits of Combining the Big Data and Cloud Computing

- The cloud computing atmosphere normally has a number of PC user terminals and assisting suppliers. By the compilation terminals, the person records the information with the large information equipment. On the flip side, without the program provider, it preserves, shops, and processes the fundamental information. Thus, cloud computing supplies huge information infrastructure. The infrastructure needs to provide on-demand offerings and energy to confirm uninterrupted program;
- Since the cloud planet is scalable, it can provide suitable details about managing remedy regardless of the amount of all of the information. When required, the cloud computing program provider may also provide protection policies as per the person's requirement;
- Identity managing and entry management are two main problems while coping with confidential business information. Cloud computing is able to meet up with the necessary protection utilizing a super-easy software program–user interface by abstracting inner information on the info. Furthermore, this particular promise provides full confidentiality of consumer information and just offers permission to access the authorized customers;

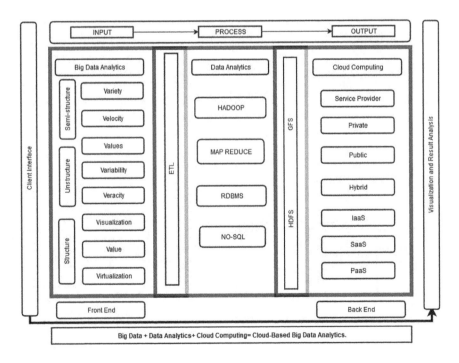

FIGURE 1.2   Service model for data analytics.

- Big details for information processing could be placed worldwide, but keeping these kinds of massive servers in various places is expensive in terms of business. As cloud computing is able to maintain information via geographically dispersed and also virtual servers, it considerably decreases the price of serious information processing;
- Cloud computing utilizes a high-level software program and programs that don't rely on the effectiveness of the person's products. In addition, it depends upon the system servers and their energy [3,4]. On the other hand, cloud computing utilizes private sources for maintaining serious information, which will probably be determined by the person's requirement. Thus, large details of cloud computing program are helpful;
- Cloud computing allows high-speed detail flow to the system. Because of this, it leads to quicker and major detail processing.

1.1.5 RELATIONSHIP BETWEEN INTERNET OF THINGS AND BIG DATA

IoT provides a chance to improve functioning in a number of sectors to allow an interaction between people and machines (M2H), and also between machines and equipment (M2M). It shows a sufficient range of development until right now. In the majority of the instances, sensor-generated details are given on the large information process for ultimate accounts and evaluation that are resulted from it. Thus, the interrelation between the IoT and big data is shown in Figure 1.3.

1.1.5.1 The Benefits of Combining the Big Data and Cloud Computing

- Data storage;
- Integration;
- Analytics;
- IT business provides numerous possibilities when compared to the obstacles. IoT is projecting a potential future of 19 dollars trillion market segments for the web market within the coming ten years, which opens more advancements and a lot of studies within each big detail's domain and IoT.

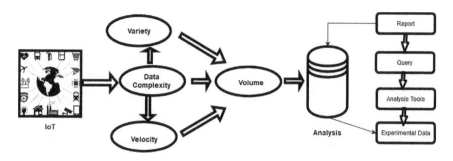

FIGURE 1.3 Relationship between IoT and big data.

1.1.6 Relationship between Internet of Items and Cloud Computing

The web of items has developed a brand-new idea for the Internet community. This allows interaction between a number of items, including

- Smart devices;
- Mobile gadgets;
- Others and sensors.

IoT offers a good interaction involving the most aspects of its structure. The components of IoT are as follows [10,13]:

- Objects;
- Gates;
- Network infrastructure;
- Cloud infrastructure.

Several advantages of utilizing IoT and cloud computing together are as follows:

- In a cloud infrastructure, you are able to deploy programs to approach as well as to evaluate information rapidly, and generate choices quickly [10];
- It is believed that nearly 4.4 trillion GB information will be produced by the entire year 2020.There is a little doubt which it is going to put enormous stress on its infrastructure. Thus, there is a requirement to decrease this particular large strain as well as to provide an answer to convey the information. Cloud computing, however, gives sufficient functionality and scalability to retailer, and provides an enormous amount of information;
- IoT and cloud computing feature a complementary connection. While IoT yields huge quantities of information, many cloud suppliers enable information transfer through the World Wide Web (www) that helps to get around the information [14];
- Cloud computing helps you to collaborate in IoT advancement. Utilizing cloud wedge, IoT designers are able to keep the information remotely, and then entering the information is very easy;
- Cloud computing helps you to progress, keep track of, and provide the analytics of its products;
- IoT products that make use of typical Application Program Interface (API) along with back-end infrastructure are able to get essential protection revisions immediately through cloud; at the moment, any kind of protection breach occurs within the infrastructure. This particular mixed attribute of cloud and IoT computing is a crucial parameter in consumer protection and secrecy.

1.2 THE CONVERGENCE OF THE INTERNET OF THINGS, CLOUD AS WELL AS BIG DATA COMPUTING

As a result of the above-mentioned reasons, we are able to get the interdependency between the three mutually highly sought-after solutions. Cloud computing plays a role in providing commonplace for big data and IoT to work, and data as a concept is definitely the analytic wedge of all of the information, which is shown in Figure 1.4.

Based on International Data Corporation (IDC), over 90% of IoT information will likely be hosted on the cloud wedge within the coming five years, due to the following reasons:

1. An overwhelming quantity of IoT information development will nourish the fundamental information methods;
2. The intricacy of information blending inside IoT is less, which is considered as one of the requirements for its optimization [11,15]. In case of the IoT programs as well as information work as part of silos, we won't grab the complete likelihood of it. Thus, in order to improve insights and generate choices, blending info (data) coming from different energy sources is definitely the most effective method.

1.2.1 THE CONVERGENCE OF BIG DATA, CLOUD COMPUTING, AND DATA SCIENCE

Big data analytics allow an online business to create the proper options following the examination of useful info. Cloud computing provides interesting large details along with the likely ways enlightened in various planet issues in the domains such as industry, business, astronomy, and social science. The obstruction in the connection between serious details, cloud, and with information science is shown in Figure 1.5.

As a consequence of significant measurement, higher velocity, and then vivid type, large details take a wedge, which can keep controlling and processing it [11].

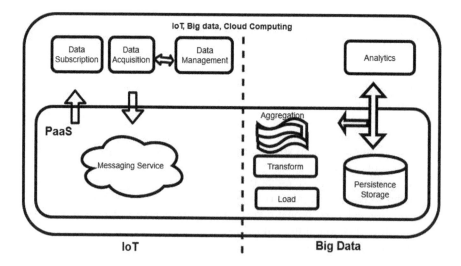

FIGURE 1.4 Convergence of IoT, big data, and cloud computing.

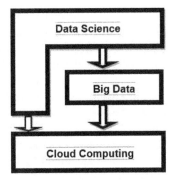

FIGURE 1.5   Relationship between big data, cloud computing, and the IoT.

Cloud has the wedge needed for storage space, processing, and controlling serious details in a more effective manner. This means which great details reside in the cloud. It, however, loves crude engine oil since it must be filtered and enhanced to unlock the worth of it, and it must be much more functional. In order to increase the worth as well as to derive insights obtained from great details, analytics ought to be carried out on the information. Web on large details analytics and items is involved along with this technique. Furthermore, the materials along with algorithms of info science in addition to info and analytics live in the cloud. The detailed analytics are actually carried out on the fundamental information residing within the cloud in which cloud functions as the wedge [16,6,17].

When it comes to dealing with some company situation, the small business needs to make an effort to reply to the doubting of what, where, and how cloud; large details; and the web of items may be worn interdependently to get the greatest worth for your company. Understanding how you can use the three concepts can help a company crop huge quantities of insights, worth, and productivity. Cloud engineering answers the doubting of wherever running a business. It provides answers on where you can hold information along with where you can provide the way to thing to do information. This is built feasible through the infrastructure for processing as well as for storage space of big data [18,19]. Big data, in contrast, answers the doubting of exactly how. Large information is liable for answering the doubting of how you can manage a tremendous quantity of information, and how you can keep huge quantities of information and the way to thing to do the overwhelming quantities of information. Additionally, it provides guidance on how you can do something via cleaning up as well as examining considerable details [20]. The doubting of what is additionally clarified by information science. This allows a predictive analytics of what'll come about as well as a prescriptive analytics of what needs to come about.

As stated previously, big details and cloud computing are actually unique areas that developed individually with awhile now. They, however, started to be interdependent with companies depending on all of them because of their earnings as well as being successful. Even though the principles of great details, as well as web of issues, were around for a very long time, the true program of theirs began fairly recently. The cloud computing discipline likewise traces back again to the 1960s intergalactic laptop networking, but has, however, undergone several phases just before it grew to become a mainstream business need. Nowadays, the systems are actually intertwined [21]. The need for serious details can't be actualized without having both IoT and cloud OS. Web of items allows the number of serious details to rise. Furthermore, the adoption of serious details compels an action towards cloud technologies. Exploration by IDC stipulates this within the consequent five years; a great deal of over 90% of all Internet for merchandise info will probably be hosted by this system provider.

Several system providers use, for instance, cloud computing, which is a term used to describe the diminished intricacy of supporting the web of items, hence leading to information blending. This means which any kind of group is looking for in order to transform the web of the item's information and also uses its prospective necessity within the very first spot that completely embraces cloud-based methods. Projections within the IT business are actually rising so that the amount of its products is actually expected to grow to 20 billion by 2020, while the important information sector is actually anticipated to get a really worth of US$66. 8 billion. The two sectors have been classified since the fast-growing sectors within the IT market are the most crucial sectors for the technological feature [6,22]. The convergence of the three technologies is actually creating huge alterations towards the dependency on interconnected devices as well as on the information they produce. The three technologies have shifted the companies originating from a product-oriented view to an information-based effect orientation. The interconnections between these technologies are shown in Figure 1.6 [23].

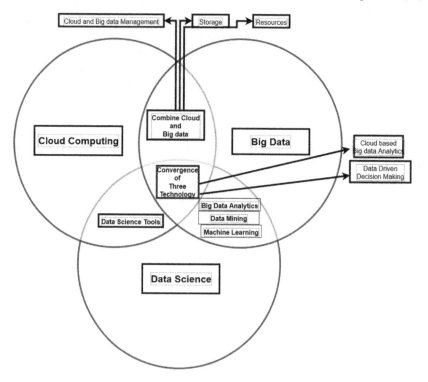

FIGURE 1.6   The convergence of big data, cloud, and data science.

1.2.2 CONVERGENCE OF BIG DATA AND CLOUD COMPUTING

The areas of great detail and cloud computing go hand in hand. Hence, they have got a complimentary connection. Large details and cloud assistance will minimize the pride of ownership and also are worth out of the business. Large details are actually the systems that record the importance of information using a serious weighing machine, which is affordable. Nowadays, cloud computing technology guarantees rapid analytics with a much-reduced price as compared to during the former days or weeks. Thus, in every business, no matter the sizes, cloud computing can derive a lot more insights coming from information than ever before seasons. This, however, acquires the businesses to get as well as put a lot more details, hence producing much more requirement for processing strength as well as steering a virtuous group. With the help of a system, the cloud supplies almost everything by eliminating intricacy as well as difficulties that are involved in creating a scalable, flexible, self-service program [10,2,24]. Precisely, the same characteristic is applicable to substantial details processing because it hides the intricacy of large-scale sent-out processing using the conclusion PC user in a way comparable to a cloud. Thus, the primary reason behind the mass adoption of the cloud and big details would be the simplification supplied by the two systems.

The two areas of technologies are actually complementary; hence, they develop a dialectical connection. Cloud computing is actually among the fashion to come down with technological advancements, whereas serious details are actually an unavoidable trend of fast improvement associated with a contemporary info world. Present strategies, as well as cloud computing systems, are actually required to resolve difficulties associated with huge details. Furthermore, the cutting edge of serious details systems facilitates the uses of the web of items and cloud computing of issue systems. This is actualized by installing the land surface for the systems, thereby marketing and allowing them to be used within the in-depth methods. Among easier terminologies, the web of items may be considered the king; big details may be called as the queen, and also, cloud may be considered as the palace [20,25,15].

Cloud computing via sending out storage space engineering enables the real managing of big details. Cloud parallel computing capability is able to better the effectiveness of obtaining and examining large details. The information within the issue must be proactive details; hence, the compilations and the evaluation of information are actually useless without the proper steps. Because of the freedom of cloud computing, dissemination, analysis, and the collection of outcomes plus steps are much simpler. Shifting information to the cloud is actually sensible as serious details call for a great deal of room to improve its effectiveness. Nearly all businesses depend on the cloud for their larger details analytics. The gains of implanting large details technologies via cloud computing are actually less expensive in deep hardware, processing, and capability to experiment [25,15]. Companies have a chance to select the ideal model from various types of cloud computing based on their demand and advantages provided by the program clothes aiders. Many business organizations choose designs that are helpful within price cost savings, whereas others are actually worried about information protection and loss in command.

1.2.3 CONVERGENCE OF BIG DATA, DATA SCIENCE, AND CLOUD COMPUTING

Data science is definitely the secret sauce of a business with regard to using large details together with the goal of increasing earnings as well as worth [26]. Use of printer learning as well as information mining resources on great details is a consequence of serious details analytics. This may be conveyed as follows.

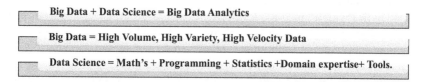

Big Data + Data Science = Big Data Analytics

Big Data = High Volume, High Variety, High Velocity Data

Data Science = Math's + Programming + Statistics +Domain expertise+ Tools.

Regardless of the dimensions of a business, it's confronted with a big struggle of examining information and storage space of information. The number of serious details produced by an enterprise will keep on speeding up as time passes. Cloud is packaged straight into the photograph on the best goals of a business with regard to saving large details securely and economically. Consequently, groups have implemented the pattern of getting competent details, analysts, information engineers, and also information researchers. These industry experts must have capabilities, such as analysis and stats, in addition to programming as they are more likely to concentrate on different ways by which the business organizations go shopping info. An information researcher analyzes the various kinds of information which are actually kept in the cloud. Information researchers are essential for businesses because the increased serious information-making businesses shop big sets of information within the Internet wedge [1]. Cloud is key because the storage space has today been built less costly, not to mention you'll find amenable supply to us as well as equipment readily available for information researchers. Cloud-based analytics can be expressed as follows.

Cloud + Big Data + Data Science = Cloud-Based Big Data Analytics.

The intersection of the three system's prospects provides cloud-based major details analytics and information-pushed generating choices. The information-pushed generating choice may be the procedure of generating choices based on an evaluation of information quite compared to instinct. With the help of an information-pushed determination procedure, the small business can derive higher quantities of evaluation as well as insights from the company. This can additionally result in earnings and decreased expenses, and is a lot more effective.

1.2.4 BENEFITS OF CONVERGENCE OF BIG DATA, INTERNET OF THINGS, AND CLOUD COMPUTING

The convergence of the three systems implies lots to the market as they've assisted in changing the IT industry in many ways. First of all, by means of this particular convergence, the return on buy for your business organizations is actually deemed to improve. By means of IoT, large details analytics, and cloud computing, commercial enterprises have converted within the manner they produce printers. Information analytics that are backed through the development of computing strength provides excellent guidance to companies as they have been empowered by it to draw out the optimum advantage in the information, hence obtaining a very good insight. As businesses make an effort to generate much more importance out of the data of theirs, they produce an information lake as part of the brand-new currency of theirs [27,28].

On the flip side, information storage space continues to be challenging for many businesses. It has made people to change in the cloud as it's a much less complicated framework, and it is additionally linked to lesser expenses [16,11]. The return on the buy (ROI) of companies will likely be increased in case a business has additional details on providing, on the one hand, a secure storage space wedge and, on the other hand, the ideal evaluation of this kind of details. The price of deploying the systems is actually expected to lessen with a period, which often decreases the price of computation for your business organizations. Reduced expenses imply additional money for an increase as well as internet business buy rise in the shareholder values.

Next, the convergence of the three systems is actually advantageous within generating the healthcare business wiser. The blend of this quality of great details and cloud computing can easily reshape the coming model of the e-health phone system. Within the well-developed segment, large details can result in the transformation out of a hypothesis-driven study of information-driven evaluation. The overwhelming level of individual information that is collected from the person captures as well as receptors coming from various health-related solutions takes a wedge for storage space. As a consequence of this demand for just a wedge for storage space, cloud computing plays a vital role. Furthermore, the evaluation of the information has empowered a wiser healthcare process with personalized diagnostics as well as help [10]. Because of the controlling as well as the evaluation of non-trivial and trivial junctions in between various sensor indicators as well as current huge details, brand-new means are attained that assist with in supplying remote diagnostics, a clear knowledge of sicknesses, and also the improvement of revolutionary ways for therapeutics [29].

Furthermore, while using the three systems that are intertwined, self-service analytics are going to be increasing. As technological growth expands, people are going to be much less dependent on the IT division. It's simply because virtually all of the features that were formally managed through the IT division are actually in a position to be managed by way of information integration and hands-free operation. This means which info equipment that is huge will more and more be straightforward and self-sufficient to perform the basic features. It is going to be usual to get analytics as being a program.

Additionally, via the convergence of three systems, the manufacturing web of items (industrial IoT (IIoT)) will probably be gained. The manufacturing web of items is in the cutting edge of using serious details, the web of items, and cloud computing. As a result of this particular explanation, it's presently getting cultivating focus out of both the business and the academicians. Manufacturing web of items consists of various products that are interconnected, having a perspective of keeping track of, collecting, swapping, examining, and immediately acting on info to moderate the manufacturing product's conduct along with the manufacturing ecosystem [23].

With this perspective, the convergence of big data, cloud computing, and IoT within IIoT is actually a crucial part.

Additionally, edge computing is going to be generally used via the convergence of three systems. It's turned out to be a need to act on the real-time details. As a consequence of this deployment of three systems, the benefits of the real-time actions increase. Real-time on-demand activity continues to be empowered via great details and IoT. The advantages of computing are now well known in this particular perspective.

1.3 REAL-TIME PROBLEMS TO BE SOLVED BY THE COMBINATION OF THESE THREE TECHNOLOGIES

- **Delivering Value to the Customer:** Determining the issue declaration is specifically proportional to the good results of an IoT setup. But this is what many IoT program suppliers miss away. They need to comprehend exactly how the remedies can influence the effectiveness, client satisfaction, and efficiency within the long term. This particular complete cycle must have excellent retrospection, not to mention that presently there can't be a larger setup obstacle than this particular gap of comprehending the consumer trouble declaration. Thus, IoT specialists need to figure out the primary key functionality signs to calculate through an IoT remedy.
- **Hardware Compatibility Issues:** Information recording mainly arises through different receptors, like Programable Logic Controllers (PLCs), that are linked to IoT gateways to gather and transmit information on the cloud. Businesses have to meticulously recognize the gear, hardware, and current history devices determined by the objectives of their and company results. When you find history devices that don't possess the reported PLCs as well as the receptors required, the IoT setup problem becomes additionally crucial. Putting in outside receptors on the history devices is fast labor in existence, although it won't be a complete evidence, which makes it an extremely demanding job. Thus, determining the actual physical equipment as well as knowing the connected compatibility problems before the IoT setup is extremely suggested.
- **Data Connectivity Issues:** This is perhaps probably the most disregarded obstacle since information connectivity has immensely enhanced. Nevertheless, at this time, there continues to exist in several places that information connectivity is an IoT setup task. It calls for exactly how IoT products converge to the cloud and the gateway as well as what information structure they will produce. Nearly all IoT gateways on the market are agreeable with Wi-Fi/LAN and GPRS, but history products rely on PLCs, Remote Terminal Units (RTUs), and telemetry methods that come up with information. Thus, the necessity is designed for a good advantage level, which translates transportation as well as information structure protocols to transmit information on the IoT wedge [26]. Determining the appropriate mixture of the protocols before proceeding with having an IoT setup can help within living quite a distance [30].
- **Incorrect Data Capture Difficulties:** We need to think that the whole setup didn't confront the method, and some difficulties are running, but presently

there will be the problem of incorrect detail record sneaking upward. Due to a few untoward events or maybe the failure of the software program to deal with particular anomalies in a deep operate period, incorrect details become captured. This brings about incorrect analytics that could not assist within snapping a much better choice. This particular IoT setup task could be a significant choice influence towards the businesses in addition to the buyers [31].

• **Analytics Challenges:** The actual worth of an IoT option would be discovered via actionable insights produced from the gathered-up IoT information. This calls for a high-performance analytics wedge that is able to deal with the generous quantity of information to become put into the answer with a later on issue. Information analytics associates have to continue to keep this particular view while devising the youth setup structure to entail information processing, cleaning, and representation. As a result, making sufficient room for extensibility to incorporate a predictive or real-time analytics to an IoT remedy is able to assist in resolving this particular crucial IoT setup task.

• **Data Security Issues:** With occurrences of many ransomware strikes lately, enterprises and clients are concerned about information protection. There is additionally a possibility of business espionage to get an intellectual home. Thus, IoT program suppliers have to make certain which the data of theirs is going to be healthy. These information protection problems may be solved by installing an extensive governance function, which supplies protected entry to vulnerable data and accounts. This particular stage of preparation, which describes different details associated with protection policies, is vital for effective IoT setup [18,32,33].

Convergence of these technologies gives the way to fix the issue. Every sector has issues that individuals do not understand the remedies too. Individuals often need to invest a very long period finding out exactly what the option would be, or maybe they need to squander your time and effort awaiting somebody different, which does understand the solution on the issue. The Internet and Big Data of Everything are creating goods as well as solutions that will instantly present you with the solution to the issue of the development of IoT. For instance, in case a computer inside and outside the factory malfunctions, you will instantly get directions on exactly how to fix it.

1.4 TECHNOLOGY CHALLENGES AS A RESULT OF THE CONVERGENCES

Even though the convergence of big details, the web of items, and cloud computing has found to become the option for your coming model, different specialized difficulties emerge, which are disadvantageous to the visitors. You'll find several problems to come down with serious details analytics, online, and the cloud of items that, when disregarded, might be of considerable injury to the company. The growing overflow of information within the huge information era has brought approximately overwhelming obstacles. To solve the fundamental information troubles, we depend on contemporary significance and cloud computing systems [7]. Several issues to come down with cloud-based major details analytics are actually displayed in Figure 1.7.

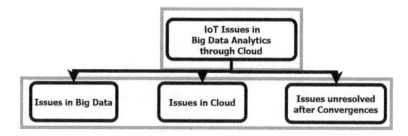

FIGURE 1.7   Categorization of issues.

1.4.1   ISSUES IN BIG DATA AND THE INTERNET OF THINGS

Big data is not only associated with opportunities, but also poses challenges to the users. This implies that a suitable means should be applied to overcome the challenges and gain valuable insights from the big data. One of the issues related to big data is heterogeneity. Due to the production of a variety of data from multiple sources such as sensors, smart devices, and social media, the art of storage, processing, and management of such data has become complicated. This is because the data is produced in raw, structured, semi-structured, and unstructured formats [10,27,34]. Some levels of heterogeneity exist in type, structure, semantics, organization, granularity, and accessibility of the datasets. As a result of variety and heterogeneity, data representation is a major challenge.

Another issue that arises from interaction of the IoT with big data is the storage issue. The rapid growth of data has restricted the capacity of the existing storage technologies that help in data storage and management. In the past few years, the structure of Relational Data Base Management System (RDBMS) was relied on the traditional storage system. However, each storage system has its limitations when it comes to data storage, which is a disadvantage in the storage and management of big data [15]. There is a need to develop a storage architecture in order to manage and store a large dataset. Storage architecture is suitable as it is highly efficient when availability and reliability are required in storing and managing large datasets.

Additionally, big data also leads to a data transportation issue. In this challenge, it has been estimated that the time taken to transmit the data from a collection or storage point to a processing point is lesser than the time taken to process the data. Currently, disk technology is limited to 4 terabytes per disk. This implies that 1 Exabyte would need 25,000 disks. An Exabyte of data can be processed in a single computer system; however, it would not be able to attach the required number of disks directly. Access to the data is a challenge, and it would overwhelm the existing communication networks. For instance, 1 gigabyte per second has an effective, sustained transfer rate of 80%, and the sustainable bandwidth is about 100 megabytes. Transferring of an Exabyte will thus take about 2800 hours if we assume that a sustainable transfer is maintained [35].

The other issue that arises in big data is the data management issue. Big data and big data management work together to achieve business and technology. As mentioned earlier, the great information is about the high-volume, high-velocity data that comes from numerous sources and is multi structured. Data management entails the collection and storage of data together with its collection and storage.

This discipline covers a couple of data fields such as warehousing, data integration, content management, event processing, and database administration [11]. Old data and new big data vary with respect to content, structure, and intended use as each of these categories has more variations within them. The most difficult problem to address in big data is the aspect of management. Management of the data issues such as access, metadata, utilization, updating, governance, and reference has proven to be a significant challenge, and there is yet no perfect solution for big data management. Another technological challenge arising from big data is the processing and data security issues. New analytics algorithms and extensive parallel processing are necessary for the effective processing of the exabyte of data to ensure timely and actionable information. While processing a query in big data, speed is a significant demand. This, however, may cost much time as the whole database cannot be traversed in a short period. In such a case, the index will be the optimal and desirable solution when there is a combination of an appropriate index for big data and up-to-date preprocessing technology. Due to the challenge of limited capacity to maintain and analyze massive datasets, service providers and business owners rely on professionals and tools to analyze the data [25,36]. Such reliance increases the potential safety risks of the data. The analysis of big data should only be delivered to a third party on the condition that there are protective measures aimed at protecting the sensitive data.

The other issues arising from big data are scalability and response time. The management of large and rapidly increasing volumes of data is a challenging issue. Scalability is expressed in three aspects, namely, data volume, hardware size, and concurrency. The analytical system of big data must support the present and future datasets. The larger the data, the more the time required to analyze the data.

1.4.2 ISSUES IN CLOUD

Since users are still skeptical about the authenticity of the cloud, the current adoption of cloud computing is linked with many challenges. According to a survey by IDC in 2008, the major problems that prevent cloud efficiency are secure, costing and charging models, service-level agreements, and cloud interoperability. The issue of security has limited cloud computing, which is acceptable. Many security concerns are linked to cloud computing [23]. These issues include security issues as a result of cloud providers and security issues faced by the customers. Additionally, the cost of transferring an organization's data between different stakeholders through the cloud is estimated to be high [12,37]. This is a challenge, especially for consumers who use a hybrid cloud deployment model where the organization's data is distributed among a variety of clouds [38].

1.5 TECHNOLOGY CHALLENGES AS A RESULT OF THE CONVERGENCES

Although many customers have significantly recognized cloud computing, the study on serious details within the cloud has stayed within its first stages. Generally, there stay a number of unsolved concerns that have carried on to come via the use of the systems. The convergence of the three systems has resolved the majority of the issues, but presently there continues to be a percentage of the problems unsolved following the convergence [4,6].

Several problems include heterogeneity that takes around problems, including seller lock-in technologies along with licensing issues that are related. Information integrity is actually an important part of information protection, which stipulates that information must simply be altered by authorized people in order to stop misuse; information quality, which continues to be a problem because of the origins of information coming from several energy sources that can't be verified; privacy, because the individuals' info is subject to scrutiny, hence offering rise to worries on loss, stealing, and profiling of control [39,40]; governance in which policies, concepts, and also frameworks are involved. All the aforementioned problems hit a balance between danger and value, because authorized problems produce large obstacles. Exactly, various nations have laws that are different and laws to help you attain protection and privacy [15]. Nevertheless, the laws don't provide sufficient shelter for individuals' information.

1.5.1   CHALLENGES OF ADOPTING INDUSTRIAL IoT

- **Interoperability:** As per the IoT Nexus survey, 77% of IoT experts considered interoperability as the most significant obstacle within the industrial Internet. The production earth is flooded with devices and protocols that are still to become interconnected, and the majority of them are frequently not interoperable [20,14]. And so, linking the history manufacturing methods and guaranteeing interoperability in between them is a task.
- **Security:** As production procedures are starting to be wiser (with the usage of Supervisory Control and Data Acquisition (SCADA) systems), the generation procedures have become much more technology-driven, of the terminology of wireless M2M solutions. The majority of the attached models write about info straight to the cloud and therefore are subjected to protection risks as well as strikes [23]. Quite only, 'any thing' or maybe "device" or maybe "asset" that is managed through the system or maybe the web is susceptible to hacks as well as strikes [40–44].
- **Data Transfer and Analysis:** In a computing atmosphere, receptors or advantageous equipment creates and gathers a massive quantity of information. However, they don't possess the computing strength and storage space assets to do superior analytics in addition to machine learning activities. Furthermore, it's hard for important four detail chunks to thing to make the information as it requires a lot more time and energy to react [1]. And so, to transmit vulnerable details to the web for carrying out a vital evaluation typically turns into a task [45,46,37].
- **IT and OT Convergence:** In business 4.0, the integration of serotonin (information technology) with the OT (operational know-how) is tough to attain since there is an enormous engineering gap in between them.

It's very older with well-defined policies, but OT is a forthcoming pattern, plus it hasn't usually been networked engineering. And so, frequently, it becomes hard to arrange to produce operations with IT systems [34].

1.6   CONCLUSION & FUTURE WORK

The aspects of big data, IoT, and cloud computing have emerged in the past few years, hence providing more data and opportunities that have enabled research and decision support applications in almost all areas of organizations. It's the possibility to revolutionize numerous areas of the world of ours. Extensive details refer to the solutions as well as the methods that involve huge, fast, and heterogeneous alterations for standard solutions, abilities, as well as infrastructure with the target of making sure effectiveness. Cloud computing, on its part, provides dynamically scalable and virtualized resources over the Internet. A convergence of the three technologies has created new opportunities in the health industry, self-service analytics, edge computing, and higher return on investment. However, the convergence of the technologies has resulted in challenges in matters of security issues, governance, privacy, quality of data, heterogeneity, data integrity, and infrastructure that need to be addressed before its potential can fully be realized.

Technological advancements were shaping the convergence of cloud computing, IoT, and big data. They require futuristic applications to the next level. With the connected and digital platform, businesses obtain a lot more customer information than in the past. Traditionally important detailed strategies are restricted towards the user's understanding of a person, therefore analyzing historical information as well as giving insights within the behavior. Even insufficient proficient energy is more restricting the product quality of all the insights. Developments to come down to human-made intelligence recently are allowing designers to uncover the concealed connection between pieces of information, thereby considerably facilitating information and analytics procedures; making use of little details feedback; and also handling important details and general performance problems.

The work of end-user industries, including electronics and semiconductors, power and energy, drugs, auto, weight machines, and metals production, as well as beverages and food, is reduced by this platform. The extensive uses of big data, IIoT, and robotics are fueling the growth of the AI (artificial intelligence) infrastructure market for deep learning technology [47,48]. The quantity of information facility suppliers as well as cloud businesses is apt to improve owing to the high effectiveness and economies of scope provided by cloud computing. Cloud program suppliers provide solutions to many clients through one common shared infrastructure (i.e., gear for functions, data storage, networking, and then hardware), and also by assisting businesses, you can conserve their IT infrastructure cost [49]. Biometrics are among the primary key solutions for increasing traction within the automobile business. Auto biometric classes include things like integrated (biosensors in seats and wheels), brought-in (smartphones, sensible watches, eyeglasses, along with health and fitness bands), cloud-driven info, and for analytics [50].

REFERENCES

1. Cuzzocrea, A. G. (2013). "Managing Data and Processes in Cloud-Enabled Large-Scale Sensor Networks: State-Of-The-Art and Future Research Directions." In *2013 13th IEEE/ACM International Symposium on Cluster, Cloud and Grid Computing*, pp. 583–588. IEEE, Delft, The Netherlands.
2. Alansari, Z., Anuar, N. B., Kamsin, A., Soomro, S., Belgaum, M. R., Miraz, M. H., and Alshaer, J. (2018, August). "Challenges of Internet of Things and Big Data Integration." In *International Conference on International Conference on Emerging Technologies in Computing 2018 (iCETiC'18)*. Springer-Verlag, London Metropolitan University, London, UK.
3. Fox, G. C., Kamburugamuve, S., and Hartman, R. D. (2012). "Architecture and Measured Characteristics of a Cloud-Based Internet of Things." In *International Conference on IEEE Collaboration Technologies and Systems (CTS)*, pp. 6–12. IEEE, Denver, CO.
4. The 3rd Generation Partnership Project (3GPP), TS 23.888B System Improvements for Machine Type Communications (MTC), Version 11.0.0, 2012.
5. Ochian, A., Suciu, G., Frate, O., and Suciu, V. (2014). "Big Data Search for Environmental Telemetry." In *IEEE International Black Sea Conference on Communications and Networking (Black Sea Com)*, pp. 182–184. IEEE, Odessa, Ukraine.
6. Aghabozorgi, S. A. (2015). "Time-Series Clustering – A Decade Review." *Information Systems*, 53 (C), 16–38.
7. Cheatham, M. (2015). "Privacy in the Age of Big Data." In *The 2015 International Conference on Collaboration Technologies and Systems (CTS)*, pp. 334–335. Atlanta, GA.
8. Rivera, D., Cruz-Piris, L., Lopez-Civera, G., de la Hoz, E., and Marsa- Maestre, I. (2015, October 2015). "Applying a Unified Access Control for IoT-Based Intelligent Agent Systems," In *2015 IEEE 8th International Conference on Service-Oriented Computing and Applications (SOCA)*, pp. 247–251. IEEE, Rome, Italy.
9. Vermesan, O., Friess, P., Guillemin, P., Serrano, M., Bouraoui, M., Freire, L., Kallstenius, T., Lam, K., Eisenhauer, M., Moessner, K., and Spirito, M. (2016). "IoT Digital Value Chain Connecting Research, Innovation, and Deployment." *Digitising the Industry Internet of Things Connecting the Physical, Digital and Virtual Worlds*, 49, 15–129.
10. Wang, L., and Alexander, C. A. (2016, August 11). "Big Data Analytics and Cloud Computing in the Internet of Things." Published by the American Institute of Science. This Open Access Article is under the CC BY license. http://creativecommons.org/licenses/by/4.0/.
11. Dior, A. A., and Chilamkurti, N. (2018). "Distributed Attack Detection Scheme Using Deep Learning Approach for the Internet of Things." *Future Generation Computer Systems*, 82, 761–768.
12. Abeshu, A., and Chilamkurti, N. (2018, February). "Deep Learning: The Frontier for Distributed Attack Detection in Fog-to-Things Computing." *IEEE Communication Magazine*, 56, 169–175.
13. Aggarwal, S., Gulati, R., and Bhushan, B. (2019). "Monitoring of Input and Output Water Quality in Treatment of Urban Waste Water Using IOT and Artificial Neural Network." In *2nd International Conference on Intelligent Computing, Instrumentation and Control Technologies (ICICICT)*. doi: 10.1109/icicict46008.2019.8993244.
14. Zhu, J., Chan, D. S., Prabhu, M. S., Natarajan, P., Hao, H., and Bonomi, P. (2013). "Improving Website Performance Using Edge Servers in Fog Computing Architecture." In *7th IEEE International Symposium on Service-Oriented System Engineering (SOSE)*, pp. 320–323. Redwood City, CA.

15. Alansari, Z., Soomro, S., Belgaum, M. R., & Shamshirband, S. (2018). "The Rise of the Internet of Things (IoT) in Big Healthcare Data: Review and Open Research Issues." In *Progress in Advanced Computing and Intelligent Engineering*, pp. 675–685. Springer, Singapore.

16. Abouzeid, A. K.-P. (2009). "HadoopDB: An Architectural Hybrid of MapReduce and DBMS Technologies for Analytical Workloads." *Proceedings of the VLDB Endowment*, 2 (1), 922–933.

17. Luan, T. H., Gao, L., Li, Z., Xiang, Y., Wei, G., Sun, L. (2015). "Fog Computing: Focusing on Mobile Users at the Edge." arXiv: 1502.01815.

18. Bagade, P., Banerjee, A., Milazzo, J., and Gupta, S. K. S. (2013). "Protect Your BSN: No Handshakes, Just Namaste!" In *Proceedings of 2013 IEEE International Conference on Body Sensor Networks (BSN)*, pp. 1–6. IEEE, Cambridge, MA.

19. Arora, A., Kaur, A., Bhushan, B., and Saini, H. (2019). "Security Concerns and Future Trends of Internet of Things." In *2nd International Conference on Intelligent Computing, Instrumentation and Control Technologies (ICICICT)*. doi: 10.1109/icicict46008.2019.8993222.

20. Chen, C. P.-Y. (2014). "Data-Intensive Applications, Challenges, Techniques and Technologies: A Survey on Big Data." *Information Sciences,* 275, 314–347.

21. Balakrishna, C. (2012). "Enabling Technologies for Smart City Services and Applications." In *6th International Conference on Next Generation Mobile Applications, Services and Technologies (NGMAST2012)*, pp. 223–227. IEEE, Paris, France.

22. Sharma, M., Tandon, A., Narayan, S., and Bhushan, B. (2017). "Classification and Analysis of Security Attacks in WSNs and IEEE 802.15.4 Standards: A Survey." In *2017 3rd International Conference on Advances in Computing, Communication & Automation (ICACCA)* (Fall). doi: 10.1109/icaccaf.2017.8344727.

23. Riggins, F. J., and Wamba, S. F. (2015, January). "Research Directions on the Adoption, Usage, and Impact of the Internet of Things Through the Use of Big Data Analytics." In *2015 48th Hawaii International Conference on System Sciences (HICSS)*, pp. 1531–1540. IEEE, Kauai, HI.

24. Kahn, E. (2014). "Natural Language Processing, Big Data, Bioinformatics, and Biology." *International Journal of Biology and Biomedical Engineering*, 8, 107–117.

25. Alansari, Z., Anuar, N. B., Kamsin, A., Soomro, S., and Belgaum, M. R. (2017). "Evaluation of IoT-Based Computational Intelligence Tools for DNA Sequence Analysis in Bioinformatics." In *International Conference on Advanced Computing and Intelligent Engineering 2017.* Springer, India.

26. Saarika, P., Sandhya, K., and Sudha, T. (2017). "Smart Transportation System Using IoT." In *Proceedings of the 2017 IEEE International Conference on Smart Technologies for Smart Nation (SmartTechCon)*, 17–19 August 2017, pp. 1104–1107. Bangalore, India.

27. CISCO: Visual Networking Index Global Mobile Data Traffic Forecast 2014–2019, CISCO Whitepaper, 2015.

28. Shah, T., Rabhi, F., and Ray, P. (2015). "Investigating an Ontology-Based Approach for Big Data Analysis of Inter-Dependent Medical and Oral Health Conditions." *Cluster Computing,* 18, 351–367.

29. Jara, A. J., Genoud, D., and Bocchi, Y. (2014). "Sensors Data Fusion for Smart Cities with KNIME – A Real Experience in the Smart Santander Testbed." In *2014 IEEE World Forum on Internet of Things (WF-IoT)*, pp. 173–174. IEEE, Seoul, South Korea.

30. Soni, S., and Bhushan, B. (2019). "A Comprehensive Survey on Blockchain: Working, Security Analysis, Privacy Threats and Potential Applications." In *2nd International Conference on Intelligent Computing, Instrumentation and Control Technologies (ICICICT)*. doi: 10.1109/icicict46008.2019.8993210.

31. Sinha, P., Rai, A. K., and Bhushan, B. (2019). "Information Security Threats and Attacks with Conceivable Counteraction." In *2nd International Conference on Intelligent Computing, Instrumentation and Control Technologies (ICICICT)*. doi: 10.1109/icicict46008.2019.8993384.

32. Dsouza, C., Ahn, G.-J., and Taguinod, M., (2014). "Policy-Driven Security Management for Fog Computing: Preliminary Framework and a Case Study." In *15th IEEE International Conference on Information Reuse and Integration (IRI)*, pp. 16–23. IEEE, Redwood City, CA.

33. Stojmenovic, I., and Sheng, W. (2014). "The Fog Computing Paradigm: Scenarios and Security Issues." In *Federated Conference on Computer Science and Information Systems (FedCSIS)*, pp. 1–8. IEEE, Warsaw, Poland.

34. Tiwari, R., Sharma, N., Kaushik, I., Tiwari, A., and Bhushan, B. (2019). "Evolution of IoT & Data Analytics Using Deep Learning." In *International Conference on Computing, Communication, and Intelligent Systems (ICCCIS)*. doi: 10.1109/icccis48478.2019.8974481.

35. L'heureux, A., Grolinger, K., Elyamany, H. F., and Capretz, M. A. M. (2017, April). "Machine Learning with Big Data: Challenges and Approaches." *IEEE Access,* 5, 7776–7797.

36. Nishio, T., Shinkuma, R., Takahashi, T., and Mandayam, N. B. (2013). "Service-Oriented Heterogeneous Resource Sharing for Optimizing Service Latency in the Mobile Cloud." In *1st ACM International Workshop on Mobile Cloud Computing & Networking*, pp. 19–26.

37. Celesti, A., Galletta, A., Carnevale, L., Fazio, M., Lay-Ekuakille, A., and Villari, M. (2018). "An IoT Cloud System for Traffic Monitoring and Vehicular Accidents Prevention Based on Mobile Sensor Data Processing." *IEEE Sensors Journal,* 18, 4795–4802.

38. Ottenwälder, B., Koldehofe, B., Rothermel, K., and Ramachandran, U. (2013). "MigCEP: Operator Migration for Mobility Driven Distributed Complex Event Processing." In *7th ACM International Conference on Distributed Event-Based Systems*, pp. 183–194. ACM, New York, NY.

39. Sang, K. S., Zhou, B., Yang, P., and Yang, Z. (2017). "Study of Group Route Optimization for IoT Enabled Urban Transportation Network." In *Proceedings of the 2017 IEEE International Conference on Internet of Things (iThings) and IEEE Green Computing and Communications (GreenCom) and IEEE Cyber, Physical and Social Computing (CPSCom) and IEEE Smart Data (SmartData)*, 21–23 June 2017, pp. 888–893. Exeter, UK.

40. Banerjee, A., Gupta, S., and Venkatasubramanian, K. K. (2013). "PEES: Physiology-Based End-to-End Security for mHealth." In *The Wireless Health Academic/Industry Conference, Baltimore, MD*, pp. 1–8.

41. Stojmenovic, I., (2014). "Fog Computing: A Cloud to the Ground Support for Smart Things and Machine-to-Machine Networks." In *Telecommunication Networks and Applications Conference (ATNAC)*, pp. 117–22. IEEE, Southbank, Australia.

42. Saad, W., Abbes, H., Jemni, M., and Cerin, C. (2014). "Designing and Implementing a Cloud-Hosted SaaS for Data Movement and Sharing with Slap OS." *International Journal of Big Data Intelligence,* 1 (2), 18–35.

43. Bagade P., Banerjee, A., Gupta, S. K. S. (2015) Evidence-Based Development Approach for Safe, Sustainable and Secure Mobile Medical App. In: Mukhopadhyay S. (eds) Wearable Electronics Sensors. Smart Sensors, Measurement and Instrumentation, vol 15. Springer, Cham.

44. Malik, A., Gautam, S., Abidin, S., &Bhushan, B. (2019). "Blockchain Technology-Future of IoT: Including Structure, Limitations and Various Possible Attacks." In *2nd International Conference on Intelligent Computing, Instrumentation and Control Technologies (ICICICT)*. doi: 10.1109/icicict46008.2019.8993144.

45. Jindal, M., Gupta, J., and Bhushan, B. (2019). "Machine Learning Methods for IoT and Their Future Applications." In *International Conference on Computing, Communication, and Intelligent Systems (ICCCIS)*. doi: 10.1109/icccis48478.2019.8974551.

46. Goel, A. K., Rose, A., Gaur, J., and Bhushan, B. (2019). "Attacks, Countermeasures and Security Paradigms in IoT." In *2nd International Conference on Intelligent Computing, Instrumentation and Control Technologies (ICICICT)*. doi: 10.1109/icicict46008.2019.8993338.

47. Bansod, G., Raval, N., and Pisharoty, N. (2015, January). "Implementation of a New Lightweight Encryption Design for Embedded Security." *IEEE Transactions on Information Forensics and Security,* 10, 142–151.

48. Zhang, D., Han, X., and Deng, C. (2018, September). "Review on the Research and Practice of Deep Learning and Reinforcement Learning in Smart Grids." *CSEE Journal of Power and Energy Systems,* 4, 362–370.

49. Nishiguchi, Y., Yano, A., Ohtani, T., Matsukura, R., and Kakuta, J. (2018, February). "IoT Fault Management Platform with Device Virtualization." In *2018 IEEE 4th World Forum on the Internet of Things (WF-IoT)*, pp. 257–262. IEEE, Singapore.

50. Doshi, R., Apthorpe, N., and Feamster, N. (2018, May). "Machine learning DDoS Detection for the Consumer Internet of Things Devices." In *2018 IEEE Security and Privacy Workshops (SPW)*, pp. 29–35. IEEE, San Francisco, CA.

2 Cybersecurity Management within the Internet of Things

Hubert Szczepaniuk
Warsaw University of Life Sciences

Edyta Karolina Szczepaniuk
Military University of Aviation

CONTENTS

2.1 INTRODUCTION

The Internet of Things (IoT) revolutionizes various areas of human life. New-generation intelligent systems equipped with sensors are used, among others, in logistics, automotive industry, construction, healthcare, agriculture, education, industry, and households. Undoubtedly, dynamic development of IoT technology brings about significant benefits and increase in effectiveness; however, simultaneously, it becomes a source of a new kind of threats (see, e.g., [1,2]). This is confirmed by the "Cyber-Telecom Crime Report 2019" developed by the European Cybercrime Centre and Trend Micro Research. Within the report, among others, an analysis of telecommunication crimes aimed at the IoT has been developed [3]. It is to be noted that the IoT ecosystems develop faster than the regulations and norms within the scope of security. A wide spectrum of IoT effects, its complex architecture, and lack of standardized security norms justify an urgent necessity to perform research on cybersecurity of the IoT.

The subject of research is the IoT within the context of cybersecurity management. IoT constitutes a complex megasystem comprising multiple subsystems. The functional and organizational complexity of IoT in the aspect of cybersecurity management constitutes an interdisciplinary research subject. Theoretical basis of the discussed issue originates in multiple academic fields, e.g., in IT, management and quality sciences, security sciences, and legal sciences. The main goal of this chapter is to develop a concept of a model of IoT cybersecurity management and to indicate the recommended solutions facilitating an increase in the level of IoT security. Achieving the adopted research goal requires realization of the following specific objectives:

- To characterize IoT architecture in the aspect of cybersecurity;
- To define IoT cybersecurity within the systemic approach;
- To identify IoT cybersecurity threats.

The presented research issue is of complex and interdisciplinary specificity; therefore, a research process model that is proper for systemic analysis has been adopted (see, e.g., [4]). Within the aspect of research techniques and methods, analysis and synthesis of the source literature, theoretical modeling, systemic analysis, and mathematical modelling were referred to. This chapter is comprised of five subsections. In Section 2.2, theoretical basis of IoT cybersecurity management was defined. In Section 2.2.1, IoT architecture within the aspect of cybersecurity was analyzed. Moreover, an expansion of the IoT-layer architecture was developed toward multicomponent architecture, which is significant with respect to cybersecurity analysis. In Section 2.2.2, IoT cybersecurity systemic approach was realized, based on the Clemens formal mathematical model (see, e.g., [5,6]), and sets of security requirements for specific layers of the IoT architecture were defined. In Section 2.3, the authors have distinguished threats to IoT based on previous research results. While defining threats, three main levels resulting from the classic theory of security were taken into consideration: the social level, the information level, and the technical level. Finally, the defined threats were detailed into specific IoT architecture layers. In Section 2.4, an IoT Cybersecurity Management Model was developed. Referring to systemic analysis and based on a mathematical model of security management, the IoT Cybersecurity Management Model was considered. For the purpose of presenting the model, a theoretical modeling method was used. The proposed model includes cybersecurity threats, security requirements, security measures, multilateral cooperation of the institutions, and enterprises involved in security management of the IoT, improvement of IoT cybersecurity, and rules of safe exploitation of hardware and software for the end users. The proposed IoT Cybersecurity Management Model may constitute a general reference model for the purpose of designing and improving specific components of the security architecture of IoT. This chapter ends with a summary where synthetic conclusions from the conducted research can be found.

2.2 THEORETICAL BASIS OF IOT CYBERSECURITY MANAGEMENT

2.2.1 IoT Architecture in the Aspect of Cybersecurity

Similarly in case of definitional analysis, there are various concepts of IoT architecture in the source literature. Disregarding the research approach, three key elements that constitute an IoT system can be recognized:

- End devices equipped with sensors enabling the recognition of signals originating from the surrounding environment and recording them;
- Subsystems enabling reception and processing of data to end devices;
- Computer network infrastructure enabling the communication between specific IoT objects.

As can be seen from the above, IoT systems are embedded in computer network structures. In IT, the architecture of computer networks is described according to a template ISO-OSI (Open System Interconnection) model, which contains seven layers cooperating with each other in a strictly defined manner. Therefore, also attempts to represent IoT architecture based on a layer model are a natural step. In the source literature, various approaches within this scope can be found. In the beginning stage of IoT development, a three-layer IoT architecture model has been developed (see, e.g., [7–11]). One may assume that this model results from the indicated key elements constituting IoT systems and covers the layers of perception, network, and application. The perception layer is responsible for the recognition of signals from the surrounding environment and their recording. End devices constituting IoT architecture are equipped in sensors that can register data from the environment in which they operate (e.g., barcode scanners, temperature, humidity, light, laser, RFID (radio frequency identification) sensors, GPS, and cameras). Architecture of this layer is organized according to functional and beyond functional requirements of the application layer. Modern requirements of IoT generate the requirement for new-generation sensors that not only register binary data but also can realize the initial signal selection. The purpose of this sensor is to lower the cost of transmission and to enhance later processing of data in the application layer. The application layer implements software and services enabling an integral exchange of data between specific IoT network nodes and their processing and storing. In this layer, various IT technologies enabling realization of the following functionalities, among others, are utilized: databases, data warehouses, artificial intelligence algorithms, machine learning, and cloud computing. The network layer provides communication between the perception and the application layer, and it is responsible for sending information gathered by the end devices. In this layer, classic IT solutions within the aspect of computer networks are utilized efficiently. Various wire and wireless transmission media, TCP/IP and UDP/IP protocols, global system for mobile communications (GSM) transmission technologies, 802.11, Bluetooth, 6LoWPAN, and ZigBee standards are widely used. Dynamic development of IoT becomes a necessity to develop a three-layer architecture. Due to this, researchers have expanded the current model with an additional supporting layer [8,12]. The supporting layer is responsible for security of IoT architecture via mechanism of users and information verification, and it sends data to the network layer [8]. The next stage of IoT architecture development is a five-layer model [13,14]. It is understood that the five-layer architecture realizes IoT requirements better. The new architecture is comprised of three layers like in previous models (perception, transport, and application layers) and two new layers (processing and business layers). The business layer is responsible for global IoT management, including realization of system business logic and providing user privacy. The business layer focuses on the added value which is to be provided by implementation and utilization of an IoT system. On the other hand, the processing layer (also referred to in the literature as middleware layer) is responsible for processing data received from the perception layer, often in the context of heterogeneous Big Data sets. In this case, various database processing methods in a computing cloud and Big Data analysis technologies are utilized. Next, processed data is delivered to the application layer, which in the five-layer model is responsible for providing services to users.

It is worth indicating that in the source literature, other IoT architecture models can be found. One of them is based on the structure of human brain and nervous system [15]. However, it will not be discussed within security categories in this chapter. The depicted multilayer IoT architectures consider general system operating logic. IoT is of complex architecture covering a wide set of cooperating IT technologies. For the purpose of analysis of security-related aspects, it is necessary to provide further details of the three-layer IoT architecture. Mohamad Noor and Hassan have detailed the three-layer IoT architecture in the following way [16]:

- In the application layer, the application support layer and IoT application are assigned;
- In the network layer, the access network, core network, and local area network are assigned;
- In the perception layer, the perception node and perception network are assigned.

Thanks to details of the three-layer model, the indicated authors were able to image a typical IoT security architecture. In this chapter, the authors propose a different approach based on systemic decomposition of the five-layer IoT model in the aspect of multicomponent architecture. The basic assumption is the classic definition of a computer system, according to which it constitutes a cooperating system of two key elements: hardware and software. Therefore, let us discuss IoT architecture within the context of a computer system. While designing computer systems of the complex architecture, the functional and nonfunctional requirements play a key role. Their proper definition increases the guarantee of realization of business logic and maximizes the project-added value. In the proposed multicomponent architecture, the task of defining requirements is realized by the business layer, which determines the organization of further lower layers. In each layer of IoT architecture, its decomposition into cooperating and interrelated hardware and software component subsystems is possible. Sets of components of the proposed model apply functionality of the classic five-layer IoT architecture. The decomposition is depicted in Figure 2.1.

The multicomponent model depicted in Figure 2.1 constitutes a template of architectural layout. Specific IoT exemplifications will cover an individual set of components depending on the requirements posed by management system and logic in the business layer; however, the application layer often operates in the client–server architecture. For the implementation of such architecture, server software realizing key IoT system logic and client software enabling access to the provided services are required. Both server software and client software require hardware on which they run. Therefore, among hardware components in the application layer, there are

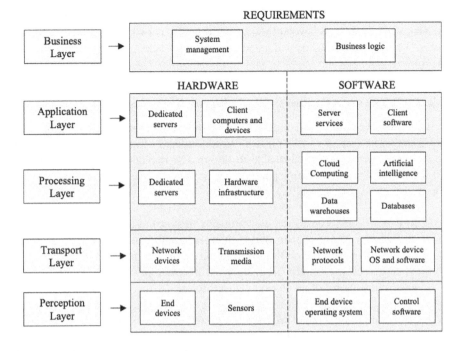

FIGURE 2.1 Decomposition of IoT-layer architecture toward the multicomponent model. (Own work.)

dedicated servers, client computers, and devices that are able to run client applications. The processing layer requires dedicated hardware servers and additional infrastructure on which database management systems (DBMS), data warehouses along with data extraction software, artificial intelligence, and machine learning tools run, as well as cloud computing services. The transport layer within the aspect of hardware covers the components comprising network devices and transmission media. Network devices may create a typical computer network architecture. In IoT ecosystems, devices of the ISO/OSI data link model layer may be used (bridge, hub, and switch) and devices of the network layer (router) act as a third-layer switch. Software controlling the indicated devices and their operating systems implement a network communication based on the communication protocols, such as TCP, UPD, and IP. Hardware-wise, the perception layer includes the components comprising sensors and end devices. Usually, the end devices and sensors are equipped with dedicated controlling software or built-in operating systems. The significance of software components in the perception layer will seriously increase in the future. The reasons for these state of affairs are found in the need for intelligent sensors that can realize the initial selection of registered signals, rejection of invalid data, and their initial processing. All these options can be realized by the built-in operating systems of end devices or the controlling software. The model depicted in Figure 2.1 will constitute a basis for further considerations. Decomposition of IoT systems into cooperating hardware and software components is critical due to security analysis regarding vulnerability and attack vectors. From Figure 2.1, one may state that IoT systems strongly exceed beyond the classic computer network theory. The application of security mechanisms dedicated for computer networks may turn out to be insufficient.

2.2.2 IoT CYBERSECURITY IN SYSTEMIC APPROACH

In the source and practical solutions literature, there are multiple definitions of security functioning (see, e.g., [17,18]). Due to the method of organizing and rules of operating of the IoT, it is reasonable to design and implement solutions enabling IoT security within the systemic perspective. In the systemic analysis, there are two approaches to security prevailing [19]:

- System security considered as a property (feature) of an examined object, characterized by resistance toward emergence of threats; however, the focus concentrates on vulnerabilities toward the occurrence of security incidents;
- System security defined in categories of possibility to protect internal values (resources) of an object against threats.

In the above definition, system security is analyzed within the context of system attributes that determine its reliability and efficiency of operating in case of occurrence of a threat. Evaluation of security of an examined system should then be realized in relation to probable threats. Regarding information gathered, processed, and sent in ICT systems, multiple models formally describing the system security can be assigned [20]:

- Access control models, e.g., the Bell–LaPadula model (see, e.g., [21,22]), Biba model (see, e.g., [6,23]);
- Models of theoretical restrictions of information protection systems, e.g., the Graham-Denning model (see, e.g., [24]), Harrison–Ruzzo–Ullman model (see, e.g., [25]);
- Information protection in database models, e.g., the SeaView model (see, e.g., [26]), Wood model (see, e.g., [27]).

One of the general models describing the security is the Clements model, in which a security system is described with the following set [5,6]:

$$S = \{O, Z, B, R, P\} \tag{2.1}$$

Key:

O – set of objects which are subject of threat;
Z – set of threats;
B – set of security (protection) measures;
$R \subseteq Z \times O$ – set of penetration paths;
$P \subseteq Z \times B \times O$ – set of penetration paths protected against attack.

According to the presented model (2.1), within the system there is a set of objects $O = \{o_1, o_2, o_3, ..., o_n\}$, which can be affected by threats $Z = \{z_1, z_2, z_3, ..., z_n\}$. Then, set of points vulnerable for an attack (system penetration paths) can be expressed with a Cartesian product $R \subseteq Z \times O$. Within a system, there is also a set of security (protection) measures $B = \{b_1, b_2, b_3, ..., b_n\}$. Thus, a set of system penetration paths that are protected with at least a single security measure can be characterized with a relation $P \subseteq Z \times B \times O$ (see, e.g., [6]). An element of a system can be affected by multiple threats, and a single threat can affect multiple objects. Identification of threats can significantly minimize risk of their occurrence because it enables the utilization of adequate security measures. The system is coherent (fully secured) if for every possible system penetration path, there is a protection measure. In practice, fully secured systems are rarely encountered. Nevertheless, coherence of a security system can facilitate decision-making within the aspect of risk management and by adopting acceptable security level. Realization of these assumptions requires systemic approach due to the fact that both system environment elements and cause–effect relations occurring between system elements are volatile in time. Referring the above deliberations to the IoT, one can assume that IoT cybersecurity indicates the ability to provide security of realization of services of proper quality and ability to protect information legally protected against unauthorized access and undesirable interference. A security incident regarding the elements of IoT can be of various scales of strength and of diversified effects of destructive influence (see, e.g., [28]). Threats regarding IoT can be related to unauthorized access to personal data, which may lower the level of quality of IoT services by disruptions in the process of their providing or disabling service provision entirely. From the point of view of data protection and security of realized services requirements, IoT elements, like other network systems, should fulfill the attributes of information security (Figure 2.2).

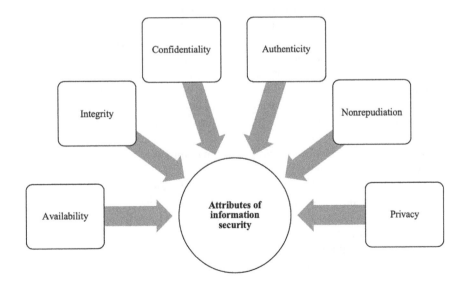

FIGURE 2.2   Attributes of information security. (Own work.)

From Figure 2.2, it is noted that in order to provide secure functioning of the IoT, the following conditions have to be fulfilled jointly:

- **Availability**: Providing availability of data and services at any moment on demand of authorized persons (or systems).
- **Integrity**: Ensuring validity of data, i.e., preventing introducing unauthorized changes.
- **Confidentiality**: Ensuring that data and services are available only for authorized persons.
- **Authenticity**: Ensuring that user or resource identity is compliant with the declared one [29].
- **Incontestability**: Ensuring the possibility of trusted verification of subjects involved in data exchange.
- **Privacy**: Providing a unit with a right to decide about the range of accessibility of data it is concerned with (see, e.g., [30]). Ensuring that a service does not perceive client's data automatically [31].

Due to the above, IoT cybersecurity is related to the ability to protect data and realization of services while simultaneously maintaining the attributes of information security. It is necessary to underline that the development and popularizing of IoT and processing of large amounts of data require adequate mechanisms for ensuring security. A critical requirement of IoT is the connection of the devices that enable the realization of specific tasks, e.g., identification, communication, and processing of information. IoT receives, processes, and sends information via networks from the end nodes of IoT (e.g., RFID, sensors, gate, intelligent devices). The IoT should provide protection both for users and for their data, applications, and networks. Therefore, for designing functional framework of IoT, the following must be taken into account: technical factors (e.g., identification techniques, communication methods, network technologies), protection and security (e.g., information confidentiality, transmission security, privacy protection), and business issues (e.g., business models or processes) [31]. Taking into account IoT architecture and the characterized elements of system security, the following requirements for the IoT cybersecurity are proposed (Table 2.1).

Summarizing, cybersecurity of the IoT should be perceived within the context of data protection and providing proper quality of delivered services. It is proposed to define cybersecurity of the IoT as a state and a process in which [32]:

- Security of information is achieved and maintained on the assumed level of confidentiality, integrity, and availability;
- Security of provided services is achieved and maintained on the assumed level of reliability, availability, and integrity of services;
- Authenticity and accountability of subjects consider authorization of users using specific information or services;
- Elements constituting IoT are characterized by the ability to protect against current and future disruptions (threats) regarding functioning or loss of specific values – system is resistant toward threats (internal, external, accidental, targeted);

TABLE 2.1
Security Requirements in IoT

Layer	Security Requirement in IoT
Business layer	• Security policy • Security education • Information security management
Application layer	• Privacy protection • Authentication/key agreement
Processing Layer	• Secure cloud computing • Authentication/authorization
Transport layer	• Communication security • Real-time monitoring and security tools (firewall, IDS/IPS, SIEM) • Transport encryption (cryptographic protocols, certificates, identity verification) • Identity authentication
Perception layer	• Protecting sensor data • Lightweight encryption • Access control • Physical security of IoT devices • Authentication/authorization • Key agreement

Own work.

- Information and service users are aware of threats and are not vulnerable regarding them;
- Perpetrators of security incidents (also internal attackers) have restricted possibilities of using cyberspace for generating threats by utilizing security system vulnerabilities.

2.3 THREATS FOR IOT SECURITY

For the purpose of proper identification of threats regarding the IoT, it is reasonable to develop theoretical basis originating in classic theory of security sciences. The two key issues requiring translation into new types of issues related to IoT security are as follows:

- Levels of cyberspace security;
- Types of threats.

From the point of view of security sciences, it is possible to assign the following levels of deliberation regarding cyberspace security [33]:

- **Social Level**: Loss of specific values as a result of attacks on informational and technical resources of cyberspace.

- **Informational Level**: Loss of informational resources (data, information, knowledge).
- **Technical Level**: Loss of reliability of technical systems.

IoT systems function in conditions of cyberspace. It is possible to translate the assigned cyberspace security levels into IoT systems security. The results are depicted in Table 2.2.

The areas of challenges toward IoT security indicated in Table 2.2 are interrelated and interdependent. For instance, the technical threats have a direct effect on security within the informational and social aspects. Based on the indicated levels of cyberspace security, it is possible to classify IoT security threats. Within the theory of security sciences, the following types of threats in the aspect of cybersecurity are indicated [33]:

- **Natural**: Resulting from effects of acts of nature.
- **Technical**: Resulting from unreliability of technical systems.
- **Social**: Conscious destructive acts by persons and threats not considered conscious destructive acts by persons.

Based on the indicated general classification, it is possible to detail threats, including the specificity of the multilayer architecture of the IoT. Within the business layer, the key threat are the errors related to business logic, which translates to an impairment of business processes realization and the requirements within the aspect of broadly understood security of the entirety of IoT ecosystem. Errors regarding business logic at the system design stage may result from intentional or unintentional

---

TABLE 2.2
Security Levels of the IoT

Cyberspace Security Level	Security of the IoT
Social level	• Security of patients of medical facilities using IoT systems
	• Security of production systems based on the assumptions of Industry 4.0
	• Security of intelligent transportation and logistics systems
	• Security of autonomous vehicles
	• Security of buildings using intelligent devices
	• Security in agriculture based on IoT systems
	• Other social issues: economic, energetic, ecological
Informational level	• Ensuring confidentiality of informational resources
	• Ensuring availability of informational sets
	• Ensuring integrity of informational resources
	• Ensuring accountability of processed information
Technical level	• Mechanisms of authorization and access control
	• Security of computer networks
	• Physical safety of devices
	• Software vulnerability toward attacks
	• Ensuring reliability of IoT systems

human activity. They determine the security of the other IoT architecture layers. Within the application layer, the key threats consider attacks on service servers and are related to intentional human activity. Classic attacks on the availability of services, such as denial of service (DoS) and distributed denial of service (DDoS), as well as malicious software, errors, bugs, and vulnerabilities in the software, may be indicated. Attacks on service availability may result in paralysis and blocking of the entirety of IoT ecosystem. In the processing layer, the threats consider the availability of services related to processing of data received from the lower layers. This layer considers, among others, attacks on database servers, data warehouses, and other data processing services. Among possible attack vectors, again the DoS and DDoS attacks may be indicated, as well as SQL-injection and wildcard attacks. In the transport layer, the key threats include classic attack vectors on TCP/IP computer networks. Other possible threats are related to the effects of the forces of nature and unaware human activity. Threats in the transport layer consider especially attacks on confidentiality attributes (wiretapping, data theft), availability (preventing data from reaching its destination), and data integrity (data manipulation and modification). In the perception layer, the threats consider IoT sensors and end devices. It is possible to take control of the devices or to install false sensors. Threats of this type are difficult to detect without safe authorization mechanism of devices, and they have an impact on the functioning of the entire IoT ecosystem. A detailed classification of threats for specific IoT architecture layers is found in Table 2.3.

Based on the data from Table 2.3, it can be stated that natural, technical, and social threats diffuse through all IoT architecture layers. Security incidents that may occur in the perception layer, e.g., by taking control over a sensor system, may cause significant consequences for other IoT layers.

2.4 IOT CYBERSECURITY MANAGEMENT MODEL

IoT cybersecurity management is related to rationalizing choice of security measures that provide uninterrupted functioning of the IoT and ensure data protection. Cybersecurity management includes organizational and technical mechanisms within the range of providing information security, which should be adequate to the risk of occurrence of threats [34]. Referring to the systemic analysis, the essence of security management can be characterized, including the systemic case described in (2.2)–(2.6), in which the following values are given [35]:

- $A(t)$ threats, which correspond to the function of a destructive threat;
- $B(t)$ system resistance toward threats, which corresponds to the function of system defensive potential.

These characteristics are the random functions of known probability distribution [35]:

$$F(a, t) = P\{A(t) < a\}, \quad a \geq 0$$

$$G(b, t) = P\{B(t) < b\}, \quad b \geq 0$$

$$t \in T \tag{2.2}$$

TABLE 2.3
IoT Threats

IoT Layers	IoT Threats
Business layer	• Faulty business logic • Faulty implementation of business logic during system designing stage • Lack of adjustment of business logic to legal requirements and standards within the range of information, personal data, and IT systems security protection
Application layer	• Attacks based on social engineering • DoS and DDoS attacks on service servers • Ports scanning • Attacks on login systems and use of password-breaking techniques (dictionary attack, brute-force password attack) • Computer viruses, bugs, and Trojan horses • Exploitation of software errors • Assuming false service • XSS • "Man-in-the-middle"-type attack
Processing layer	• DoS and DDoS attacks on database services servers, data warehouses servers, and servers of other processing services • SQL-injection attacks • SQL wildcard attack
Transport layer	• Classic threats toward computer networks, including passive and active attacks • Classic attack vectors on computer networks covering the specific layers of ISO/OSI model (e.g., ARP spoofing, MAC spoofing, IP spoofing, DNS spoofing, TCP session takeover, DoS-type attacks, routing algorithms attacks) • Ethernet and wireless networks sniffing • Natural threats for network infrastructure resulting from effects of forces of nature (e.g., fire, flood, and low temperatures) • Social threats for network infrastructure resulting from unwitting human activity
Perception layer	• Installation of false end devices and sensors • Overtaking control over sensors or end devices • Wiretapping signals registered by sensors and end devices • Natural threats for sensors and end devices resulting from the effects of forces of nature (e.g., fire, flood, and low temperatures) • Social threats for sensors and end devices resulting from unwitting human activity

Own work.

According to the above indicators, which can define the system security, there is a probability that the effect of a threat on the system will not exceed the acceptable level $a_o \geq 0$, and the system resistance will be larger than the limit value b_o [35]:

$$\beta(t) \equiv \beta(a_o, b_o) = P\{A(t) \leq a_o, B(t) > b_o\} \tag{2.3}$$

Provided statistical independence of the analyzed values, it results in system security assessment indicator [35]:

$$\beta(t) = F\left(a_o, t\right)\left[1 - G\left(b_o, t\right)\right] \tag{2.4}$$

Assuming the desired system security level as $\beta_o > 0$, one can ascertain that in period T, a system is safe if, in a given moment, the following condition is fulfilled [35]:

$$\beta(t) \geq \beta_o, t_o < t \leq t_o + T \tag{2.5}$$

In system security analyses, the use of simplified procedures can also be found, which breaks down to indicating probability [35]:

$$P = p\left(P_s < P_o\right) \tag{2.6}$$

That is, the probability of the generalized resistance (defensive potential) P_s is higher than that of the generalized P_o threat. Referring the above considerations to security of the IoT, the issue of cybersecurity management can be identified with a strategy of optimization of choice of security measures in relation to possible system penetration paths. Thus, level P_s that will maximize β security level can be indicated. IoT cybersecurity management is related to the choice of operating strategies out of a set of acceptable variants, for which the predicted value of effects of threats takes a minimal value, while costs of implementation of security measures will not exceed the acceptable value. In other words, the implementation of solutions facilitating effective IoT cybersecurity management should be preceded by defining limit values for effects of threats and the maximal value of financial resources dedicated for security. Therefore, it is reasonable to adopt a specific method of cyberthreat risk management, as well as a cost calculation method. The issue of the IoT cybersecurity management constitutes a complex and interdisciplinary problem. In the realized research, the following issue areas within the scope of the IoT cybersecurity management, among others, were identified:

- Increase in threats toward IoT and evolution of malicious software for mobile platforms;
- Lack of standardization or limited guidelines regarding the IoT cybersecurity management, e.g., risk management;
- Lack of trust of users of the IoT in the context of delivery of data and safety security;
- User errors and accidental modifications;
- Insufficient devices authentication;
- Lack of safe update mechanism;
- Some IoT devices that use untrusted networks which can increase the risk of threats.
- Challenges within the aspect of access control, e.g., lack of strong password support and lack of two-factor authentication;
- Routers' vulnerabilities toward cyberattacks (see, e.g., [36]);

- Some producers of intelligent devices limit security measures due to production costs (see, e.g., [36]);
- Threats in the transport layer, e.g., MiTM attacks, falsifying network communication, routing path attacks, and DoS attacks;
- Necessity to provide data protection in case of new business processes (see, e.g., [31]);
- Issue of providing security for hybrid systems (see, e.g., [31]);
- Many IoT systems were designed and implemented using various protocols and technologies, which constitute the complex configurations (see, e.g., [31]).

The indicated vulnerabilities and issue areas constitute only a part of the discussed issue. It must be noted that for IoT security, it is necessary to provide protection, among others, on the level of IoT devices, ICT systems, processed data, and users while simultaneously including legal solutions and good practices (Figure 2.3).

The model of the IoT cybersecurity management depicted in Figure 2.3 includes threats toward national cyberspace that may affect IoT data and services. In the architecture of IoT, it is necessary to implement security measures enabling to reach an acceptable level of security. The security measures should include both security requirements for specific IoT model layers, as well as security requirements between layers, and the requirements regarding functioning and maintaining of services (see, e.g., [37]). Security measures significantly minimize risk of occurrence of threats

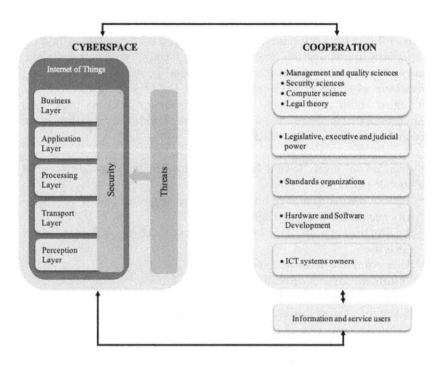

FIGURE 2.3 The IoT Cybersecurity Management Model. (Own work.)

toward IoT data and services security. The Cybersecurity Management Model also includes realizing multisided cooperation of institutions and companies involved in the IoT security management. Moreover, it is necessary for higher education and research facilities to conduct research in the area of IoT cybersecurity improvement. In this aspect, an important matter is to develop mechanisms enabling practical utilization of the results of the research. In the presented model, the legislative, executive, and judicial powers are responsible for law-making, implementation of safe solutions, and supervision over compliance with the law within the discussed range. IoT security institutions, standardization institutions, and owners of ICT systems should develop norms and good practices to provide protection in cyberspace and to provide supervision over security of the IoT. On the other hand, hardware and software producers in the system fulfill a significant role of software development and updating and implementation of IT systems security measures. Information and service users should respect the rules of safe hardware and software utilization. The IoT Cybersecurity Management Model may constitute a general reference model for designing and improving the specific components of security architecture.

2.5 SUMMARY

Developmental tendencies of the IoT and the increased popularization of IoT applications in multiple socioeconomic areas determine a need to manage IoT cybersecurity. With reference to the subject and goal of the research, the following conclusions have been formulated:

- The authors propose to expand the five-layer architecture toward a multicomponent model. This enables to perform a systemic consideration of security, including the key factors of IT systems: requirements, hardware, and software;
- Cybersecurity of the IoT means the ability to provide security of service provision of proper quality and ability to protect legally protected information against unauthorized access and undesired interference. Providing IoT cybersecurity requires systemic approach based on simultaneous development of all the constituent elements of IoT and the users of services. An important element is to provide the information security attributes;
- IoT systems operate in cyberspace in which safety can be analyzed on the social, informational, and technical levels. The aforementioned levels are interrelated and generate many challenges within the context of IoT security. IoT cybersecurity threats on a high level of generality may be classified, among others, according to the source of threats, which enables us to differentiate natural, social, and technical threats;
- In the layer architecture of IoT, there are specific vulnerabilities and the probable threats resulting from them. Examples of threats include errors related to business logic, lack of adjustment of business logic to legal requirements and standards, malicious software, DoS, DDoS, MiTM, XSS (cross-site scripting), SQL-injection attacks, wildcard attack, and spoofing. A separate group of threats comprises human factor vulnerabilities,

i.e., threats resulting from the lack of knowledge or failure to maintain safety procedures;

• Efficiency of IoT cybersecurity management is determined by its resistance toward the occurrence of dangerous situations. The issue of cybersecurity management can be broken down to the optimization of distribution of security measures in relation to probable threats for the purpose of delivery of protection. These assumptions should be realized, including the acceptable limit values of effects of threats and acceptable costs dedicated for security measures.

IoT Cybersecurity Management Model should implement security measures in each layer and security measures between the layers. Moreover, it is recommended that the process of designing efficient solutions is realized with the cooperation of specialist of technical, legal, management and quality, and security sciences. An important element is also mutual cooperation of scientific facilities and companies involved in the management of security of the IoT. The authority of a state, cyberspace security institutions, and normalizing organizations play the main role, respectively, within the aspects of law-making, development of norms and good practices, and supervision over cybersecurity of the IoT. Hardware and software producers participate in the development of software and updates, and implementation of security measures of IT systems. On the other hand, information and service users should respect the rules of utilization of hardware and software.

REFERENCES

1. Sinha, P., Rai, A.K., Bhushan, B. (2019). Information Security Threats and Attacks with Conceivable Counteraction. *2nd International Conference on Intelligent Computing, Instrumentation and Control Technologies (ICICICT)*, pp. 1208–1213. IEEE. Kannur, Kerala, India. doi: 10.1109/ICICICT46008.2019.8993384.
2. Arora, A., Kaur, A., Bhushan, B., Saini, H. (2019). Security Concerns and Future Trends of Internet of Things. *2nd International Conference on Intelligent Computing, Instrumentation and Control Technologies (ICICICT)*, pp. 891–896. IEEE. Kannur, Kerala, India. doi: 10.1109/ICICICT46008.2019.8993222.
3. Europol. (2019). *Cyber-Telecom Crime Report 2019*. Trend Micro Research and Europol's European Cybercrime Centre (EC3). Retrieved from: https://www.europol. europa.eu/publications-documents/cyber-telecom-crime-report-2019 (Accessed 03 January 2020).
4. Sienkiewicz, P. (1995). *Analiza systemowa*. Warsaw, Poland: Bellona.
5. Hoffman, L.J. (1982). *Poufność w systemach informacyjnych*. Warsaw, Poland: WNT.
6. Stokłosa, J., Bilski, T., Pankowski, T. (2010). *Bezpieczeństwo danych w systemach informatycznych*. Poznań, Poland: PWN.
7. Nord, J.H., Koohang, A., Paliszkiewicz, J. (2019). The Internet of Things: Review and Theoretical Framework. *Expert Systems with Applications*. 133, 97–108. doi: 10.1016/j. eswa.2019.05.014.
8. Burhan, M., Rehman, R.A., Khan, B., Kim, B.S. (2018). IoT Elements, Layered Architectures and Security Issues: A Comprehensive Survey. *Sensors (Basel)*. 18(9), 2796. doi: 10.3390/s18092796.
9. Said, O., Masud, M. (2013). Towards Internet of Things: Survey and Future Vision. *International Journal of Computer Networks*. 5(1), 1–17.

10. Liu, J., Li, X., Chen, X., Zhen Y., Zeng, L. (2011). Applications of Internet of Things on Smart Grid in China. *13th International Conference on Advanced Communication Technology (ICACT2011)*, pp. 13–17. IEEE. Seoul.

11. Wu, M., Lu, T. J., Ling, F.Y., Sun, J., Du, H. (2010). Research on the Architecture of Internet of Things. *3rd International Conference on Advanced Computer Theory and Engineering (ICACTE)*, pp. V5-484–V5-487. IEEE. Chengdu. doi: 10.1109/ICACTE.2010.5579493.

12. Darwish, D. (2015). Improved Layered Architecture for Internet of Things. *International Journal of Computing Academic Research*. 4(4), 214–223.

13. Sethi, P., Sarangi, S.R. (2017). Internet of Things: Architectures, Protocols, and Applications. *Journal of Electrical and Computer Engineering*. 2017. doi: 10.1155/2017/9324035.

14. Khan, R., Khan, S.U., Zaheer, R., Khan, S. (2012). Future Internet: The Internet of Things Architecture, Possible Applications and Key Challenges. *10th International Conference on Frontiers of Information Technology (FIT): Proceedings. Institute of Electrical and Electronics Engineers Inc*, pp. 257–260. doi: 10.1109/FIT.2012.53.

15. Ning, H., Wang, Z. (2011). Future Internet of Things Architecture: Like Mankind Neural System or Social Organization Framework? *IEEE Communications Letters*. 15(4), 461–463. doi: 10.1109/LCOMM.2011.022411.110120.

16. Noor, M.B.M., Hassan, W.H. (2019). Current Research on Internet of Things (IoT) Security: A Survey. *Computer Networks*. 148, 283–294. doi: 10.1016/j.comnet.2018.11.025.

17. Soltani, F., Yusoff, M.A. (2012). Concept of Security in the Theoretical Approaches. *Research Journal of International Studies*. 24, 7–17.

18. Degaut, M. (2015). What Is Security? *Revista Brasileira de Inteligência*. 9, 9–28. Retrieved from: https://www.researchgate.net/publication/310495076_What_is_Security (Accessed 20 December 2019).

19. Sienkiewicz, P. (2010). Systems Analysis of Security Management. *Scientific Journals Maritime University of Szczecin*. 24(96), 93–99.

20. Liderman, K. (2002). System bezpieczeństwa teleinformatycznego. *Biuletyn Instytutu Automatyki i Robotyki WAT*. 17, 77–96.

21. Bell, E., LaPadula, L.J. (1973). *Secure Computer Systems: Mathematical Foundations*. Vol. I, Technical Report MTR-2547. McLean, VA: The MITRE Corporation.

22. Pfleeger, Ch., Pfleeger, S., Margulies, J. (2006). *Security in Computing*. Westford, MA: Prentice Hall.

23. Biba, K. (1975). *Integrity Considerations for Secure Computer Systems*. Technical Report MTR-3153. Bedford, MA: The MITRE Corporation.

24. Graham, G.S., Denning, P.J. (1972). Protection: Principles and Practice. *Proceedings of the AFIPS Spring Join Computer Conference*, pp. 417–429. doi: 10.1145/1478873.1478928.

25. Harrison, M.A., Ruzzo, W.A., Ullman, J.D. (1976). Protection in Operating Systems. *Communications of the ACM*. 19(8), 461–471.

26. Shockley, W.R., Heckman, M., Lunt, T.F., Denning, D.E., Schell, R.R. (1990). The SeaView Security Model. *IEEE Transactions on Software Engineering*. 16, 539–607.

27. Fernandez, E.B., Summers, R.C., Wood, C. (1981). *Database Security and Integrity*. Boston, MA: Addison Wesley.

28. Yang, G., Geng, G., Du, J., Liu, Z., Han, H. (2011). Security Threats and Measures for the Internet of Things. *Journal of Tsinghua University*. 51, 1335–1340.

29. Alhassan, M.M., Adjei-Quaye, A. (2017). Information Security in an Organization. *International Journal of Computer*. 24(1), 100–116.

30. Lu, X., Qu, Z., Li, Q., Hui, P. (2015). Privacy Information Security Classification for Internet of Things Based on Internet Data. *International Journal of Distributed Sensor Networks*. 11(8), 1–8. doi: 10.1155/2015/932941.

31. Li, S., Xu, L.D. (2019). *Securing the Internet of Things*. Cambridge, MA: Elsevier.
32. Szczepaniuk, E.K. (2016). *Bezpieczeństwo struktur administracyjnych w warunkach zagrożeń cyberprzestrzeni państwa*. Warsaw, Poland: AON.
33. Sienkiewicz, P. (2012). Bezpieczeństwo cyberprzestrzeni państwa. *Ekonomiczne Problemy Usług*. 88, 802–809.
34. Szczepaniuk, E.K., Szczepaniuk, H., Rokicki, T., Klepacki, B. (2020). Information Security Assessment in Public Administration. *Computers & Security*. 90. doi: 10.1016/j.cose.2019.101709.
35. Sienkiewicz, P. (2015). Podstawy inżynierii systemów bezpieczeństwa. In Sienkiewicz, P. (Ed.). *Inżynieria systemów bezpieczeństwa*, pp. 4–18. Warsaw, Poland: PWE.
36. Avast (2019). Avast Threat Landscape Report. Retrieved from: https://cdn2.hubspot.net/hubfs/486579/Avast_Threat_Landscape_Report_2019.pdf (Accessed 23 December 2019).
37. Li, S., Tryfonas, T., Li, H. (2016). The Internet of Things: A Security Point of View. *Internet Research*. 26(2), 337–359. doi: 10.1108/IntR-07-2014-0173.

3 ROSET

Routing on SENSEnuts Testbeds in IoT Network

Rahul Johari, Reena Tripathi, and Ishveen Kaur
University School of Information,
Communication Technology (USICT),
Guru Gobind Singh Indraprastha University, Sector-16C

Kalpana Gupta
Centre for Development of Advanced Computing(C-DAC)

Rakesh Kumar Singh
IGDTUW

CONTENTS

3.1   INTRODUCTION

In almost every field of IoT (Internet of Things), there is an accumulation of new knowledge. IoT is a device-to-device interfacing, which is affected by user in a direct or indirect way. It involves sensing and analysis of data to make a meaningful interpretation. IoT with the help of innovative technologies predicts the natural disasters such as flood, earthquakes, and cyclone. There are infinite opportunities in IoT. The world becomes more technical with the support of internetworked IoT devices. IoT increases productivity, reduces cost, improves efficiency, and utilizes less resources. It also improves the tracking of devices by using sensors and other communication devices such as Bluetooth, Wireless Fidelity (Wi-Fi) and Global System for Mobile Communications (GSM). Though it makes life easier, IoT security is a main issue because directly or indirectly everyone is dependent on this, and even our personal details also get shared on IoT devices. There are so many published research papers and studies on security and privacy issues on IoT dealing with various aspects around the world.

For the sake of simplicity and clarity, the rest of the paper is organized as follows: In Section 3.2, the applications of WSN have been discussed. In Section 3.3, the problem statement has been discussed. In Section 3.4, the literature survey has been discussed. In section 3.5, the experiment that is performed has been described. In Section 3.6, the algorithm has been discussed. In Section 3.7, the program as well as the steps to execute the program has been discussed. In Section 3.8, the experimental setup has been discussed. In Section 3.9, the conclusion and future work followed by references have been described.

3.1.1   INTRODUCTION TO WIRELESS NETWORKS

Wireless network can be described as a computer network that involves wireless data connection between two different nodes. These nodes don't need any cable connection for the transmission as the communication is over the air(OTA). The main advantage of this types of network is that they assist in end user's mobility as the user, with the help of Wi-Fi connection can connect to network from any location at any point of time. Usually, these types of networks are complex, extremely hostile, and random in nature.

Wireless networks can be classified as Wide Area Network (WAN), Local Area Network (LAN), and Personal Area Network (PAN), which can be further categorized into three subparts, namely Wireless Wide Area Network (WWAN), Wireless Local Area Network (WLAN), and Wireless Personal Area Network (WPAN) respectively. These subparts can be briefly described as follows:

For WWAN, mobile signals are used for data transmission. The cellular service provider maintains WWAN. It has a low performance and mobile access to the Internet. Second, for WLAN and in case of LAN networks, the AM-FM waves are used for transmitting the data. The backbone network uses cable in most of the cases, with one or more wireless access users to the wide area network. The LAN has high performance and mobile extension of wired area networks. Precisely, we can define WLAN as a collection of two or more devices sharing data with air as the transfer medium. Last, WPAN can be defined or stated as a network with short range that may be the same as Bluetooth. It is used for interconnecting devices centered on an individual person's workspace. We can commonly use these networks for connecting compatible devices being used around a central location. We can state that a specific range of about 30 feet is used in PAN networks. PAN has a moderate performance and cable replacement for peripherals.

The wireless network is further classified as wireless MAN, wireless mesh network, cellular network, global area network, space network, and CAN (campus area network).

Install the network drive and make the sharing process easier using wireless networks. Now this networks drive can be the drive of the computer which is hooked up in network hub. The wireless network has a self-organizing capability with another feature of changing the topology of the network whenever a failure or fading of nodes is observed. The sensor nodes provide heavy deployment and combined efforts.

3.1.2   Introduction to Wireless Sensor Networks (WSNs)

WSNs provide sensing service to monitor atmospheric conditions. User can categorize these networks into three types: sensor node, gateway, and a user. The combination of the first two makes the sensor field. Gateway and user are interconnected to each other through the common gateway Internet. Sound surveillance system (SOSUS) invented the first WSN. In the 1950s, it was developed in order to detect Soviet submarines [1].

All nodes are connected in WSNs, and to be more precise, the user can define the nodes in WSN as small computers itself. The WSN is based on the technology of sensor nodes, power sources, microcontroller, small transducer, transceiver, and operating systems. The sensor net uses a protocol stack for WSNs that integrates the power using routing awareness. Further, the data integration is done with the networking protocols along with the communication process in which medium is over the air.

There are various advantages of WSNs. WSN has a simple infrastructure, is cheaper, and spends less energy. Another feature of WSN is that these networks show a good compatibility for external devices and new plug-ins. WSN will increase the usage areas of these networks, but it has some disadvantages too, such as short speed of communication and battery dependency (Figure 3.1).

Currently, it can be said that WSN consists of quality services which are further employed in commercial and industrial applications. WSNs are built in a way that they sense the surrounding activities such as temperature and humidity. Apart from sensing the surrounding activities, these networks can also be used in performing various day-to-day activities such as detecting smart things, discovering the other nodes, and collecting data. WSNs are basically another type of wireless network, which includes hundreds of circulating, lower-power devices called as motes. Further, these types of networks cover huge area loaded with embedded devices that are interconnected to collect, pass, and operate the data to the other operators with the ability to control the capabilities of processing and transferring the data. These nodes connect with other to form data networks, and the sensor node

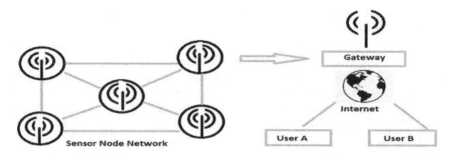

FIGURE 3.1 Wireless sensor networks.

here is the multifunctional wireless device, which saves the energy as well. Motes defined earlier in the industrial field have widespread applications. The sensor nodes come together in cluster, and then, this cluster collects the data from the surroundings. Every network needs some communication media, and similarly, motes also need a communicating medium, which is transceiver in this case. For WSNs, motes can be in number of hundreds/even thousands as compared to the sensor network nodes, which are fewer in number and do not even have a structure.

3.1.3 ARCHITECTURE OF WIRELESS SENSOR NETWORKS

WSN is nothing but an OSI (Open System Interaction) model. The OSI model describes how application will communicate over the network. The main role of OSI model is to communicate between two end points in the network [2] (Figure 3.2).

In the OSI model, a programmer will make use of seven layer functions, whereas in the sensor network, a programmer will make use of only the fifth layer function.

The aforementioned layers can be stated as follows:

a. Application layer;
b. Transport layer;
c. Data link layer;
d. Physical layer.

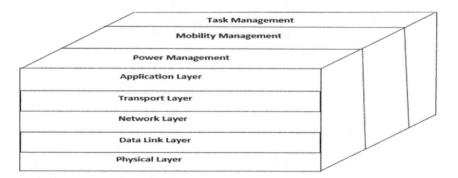

FIGURE 3.2 Architecture of WSN.

It is further divided into three cross-planes, and each plane has its specific task to perform.

a. **Power Management**: It is used to supply power to the sensor nodes for sensing and communication.
b. **Mobility Management**: It maintains a network connection, and checks configuration and reconfiguration of sensor nodes.
c. **Task Management**: It improves energy efficiency and distributes responsibility among sensors [3].

3.2 APPLICATIONS OF WSN

WSNs are used in various fields in everyday objects like touch-sensitive object in electronic devices (Figure 3.3). There are different types of sensors, and each sensor has its own specific role, e.g., seismic, magnetic, thermal, low sampling rate, radar, and infrared. IoT has different domains in which the qualitative work is to be done. IoT brings immense value into our lives. In the field of healthcare, it turns medical system into energetic wellness-based system. The IoT gives a vision of data through analysis, real-time field data, and testing. The government and engineers also need IoT in the field of water management and waste control, as an air pollution observer, and for security and surveillance. The IoT approach works in forest fire prevention also. Everybody knows that the forest fire is a natural disaster; IoT designs devices that are based on automatic feature-embedded system so that the necessary actions are taken before the fire destroys and spreads over a large area. In spite of all these areas, IoT is also used for commercial solutions with the help of Bluetooth, Wi-Fi, and ZigBee. In the present scenario, IoT is mainly used in the field of transportation; it regulates the entire system like traffic controller, for parking area and fuel consumption.

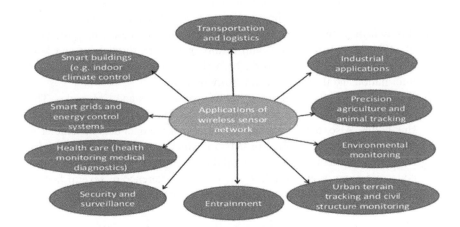

FIGURE 3.3 Applications of WSN.

3.3 PROBLEM STATEMENT

Every application, software, or a methodology has some pros and cons. The same IoT-based WSN methodology has some problems, which can be stated as follows:

1. QoS (quality of service);
2. Bandwidth (congestion);
3. Routing.

Let us try to understand each one of them in brief.

1. **QoS**: With the growing demand of WSN application, QoS can be considered as the most paramount issue for WSN. Guarantee of providing QoS in WSN is the most tough and demanding task claiming to the evidence that assets available for the sensors which are running over the network have different constraints. Also originally, QoS works on some metrics such as jitter, delay, and throughput. Unlike IP networks, sensor networks naturally support service types providing different QoS. The main challenges that occur for QoS using sensor networks are as follows:
   a. **Bandwidth Limitation**: The most typical issue with WSNs is bandwidth limitation. Bandwidth needs to be secured while proceeding with sensor networks and assuring QoS. Burst in traffic can happen due to the combination of real- and non-real-time traffic. To assure and meet the QoS requirements in sensor networks, independent routes can be used. These tasks can be difficult and complex at times due to energy limitation, less resources, and the probable increase in collisions.
   b. **Removal of Iteration**: Sensor networks offer high redundancy for the generated data. Now to eliminate the redundancy, if the traffic is unconstrained, then it is easy to remove. But conducting data aggregation of QoS traffic, the process becomes a bit complex. To make QoS more computational, the user need to have a mixture of sensor level and system rules that will be requisite to accumulate the QoS [4].
2. **Bandwidth (Congestion)**: Bandwidth plays a role in maintaining the QoS of WSNs. Congestion/bandwidth results in reduced throughput and low-power battery mode of network. In WSN, congestion leads to the loss of information and the limited availability of nodes.
3. **Routing**: WSN is a collection of set of sensor nodes and sink nodes. The main role of these nodes is to assemble the data and then send it to the sink nodes. Also, the sensor nodes predict that the data delay character is the infrastructure-less nature of network which is built where these nodes behave like routers that will forward data for other nodes. Further, sensor nodes are more prone to failure due to the interference of several factors such as environmental hazards, energy reduction, and device failure.

Hence if the user need to have operational WSNs, then the clustering and routing algorithm should survive with the network. Hence, in order to have a good optimization, the user need to have a right choice of algorithm. Routing in WSN faces a lot of challenges, which are as follows:

a. **Node Deployment**: Deployment of the sensor networks is done in the area of interest where the presentation of routing protocols is also not hampered. Node can be deployed both manually and physically. If the node is deployed manually, then routing happens through the predetermined path. Sensor nodes are scattered in a spontaneous manner.

b. **Network Topology**: Topology defines a path for routing protocols, and this is needed to be conserved in large node denseness.

c. **Data Aggregation**: This process is done in order to achieve a reduction in redundancy of data transmission. Aggregation is basically a combination of data from dissimilar nodes.

d. **Node Capability**: Determining the application of the sensor node has different roles and capabilities. The roles can be relaying, sensing, and aggregation. Roles, if different, should be performed by different nodes only as if one single node will do multiple roles, the performance will be degraded, and energy of the node will be drained [5] (Figure 3.4).

Routing in WSNs is done using different protocols. These protocols can be divided as follows:

1. **Featured-Based Routing**: The different features such as energy, security, delay, and errors are identified using specific mannerism.

2. **Energy-Efficient Routing**: In [6], the authors have developed a network for routing and also proposed an energy-efficient optimization.

3. **Delay-Less Protocols**: To analyze the delay in networks, many researches were done, which give an output of how much delay is there while transferring the data from sensor nodes. In [6], the authors performed certain experiments, analyzed few issues in routing like a delay in transmission, and proposed a bypassing protocol.

4. **Secure Protocols**: Many experiments show that routing needs to be secure, and for secure routing, the user need appropriate routing strategies and routing techniques.

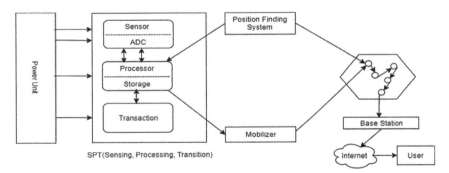

FIGURE 3.4   Components of sensor nodes.

3.4 LITERATURE SURVEY

There are many research data on IoT all around the world. According to many researches, it has been found that the users have a large number of records or cases which have been registered in many countries and upon various issues. Some of them are presented here.

In [2016], the authors [4] discussed smart city application-based models using IoT technology. This audit is suggestive in outlining information about IoT, such as definition, market size, and status of IoT, which has become one of the emerging topics for technology nowadays, and in viewing applicable IoT business samples to help business realities and research institutes participating in related projects build a smart city as a part of future vision.

In [2015] [5], the authors illustrate the view on IoT; according to the mode of communication, it is either between humans or human to device, but the assurance of IoT is a great idea for future aspects in the era of M2M (machine-to-machine) communication. In this paper, the authors provide a wide view of IoT situation and retrospect, mandate of IoT technologies and sensor networks. In [2016], the authors [6] discussed satisfiers and dissatisfiers of smart IoT service and customer attitude. As per the conclusion of their study, users are active drivers of IoT service. Consumers will greet IoT service when the service fulfills their requirements and gives hedonistic value. On the other hand, those who are not using IoT services are not powerful according to the result. Technology and progress play an important role in success.

In [2017], the author(s) [7] focused on the case study of security and privacy issues prevalent in IoT. The survey consisted of four parts. The first part explored the limitations of IoT devices and their solutions. The second part presented the types of IoT attacks. The third part focused on instrument and architectures for authentication and surveillance. The users inspect the security issues that occurred at the separate layers, show the extremity of IoT devices in battery and computing sources, and also discuss about lightweight computing and extension of battery life.

IoT is everywhere in our day-to-day life. In [2016], the authors [4] discuss that people wanted to connect with latest updates which have happened in other parts of the world. They mention various things: The first one is M2M (machine to machine), and the second one is Internet of Everything (IoE). In IoT, chances are endless and innovations are infinite. In a machine learning, surrounding objectives are talkative and data are proceeded to give results [8].

In [2015], the authors [9] include architecture to communicate securely between IoT devices by using datagram transport layer security-positioned certificates with manual authentication. There are almost 25 billion IoT Devices that are going to be linked by 2020, which is a big figure for the existing architecture of Internet with TCP/IP protocols that are adopted in 1980 [10]. The six layers of IoT are as follows:

1. **Coding Layer**: It is the basic building block, which describes the authentication of objects of Internet, where each object has its own identity.
2. **Perception Layer**: This layer describes the physical meaning of each component. To sense the devices, the components are physically connected to each other. Few examples of sensors are RFID (radio frequency

identification) tags and IR sensors. The sensor can sense the humidity from atmosphere, temperature, speed of the object, and location of the object.

3. **Network Layer**: It receives the signal or information in the form of digital signal. Wi-Fi, Bluetooth, GSM, and 3G are some devices that are used with Internet Protocol version 4, Internet Protocol version 6, MQTT (message queuing telemetry transport), and DDS (data distribution service).

4. **Middle Ware Layer**: It works upon the technologies such as cloud computing and ubiquitous computing that ensure a direct connection to the database, the middleware layer, and process the information.

5. **Application Layer**: Application layer is very useful and helpful layer. In this layer, application promotes the development of IoT. Initial applications focus on four main areas: smart cities, agriculture, healthcare, and manufacturing.

6. **Business Layer**: Business layer is responsible for all the benefits and drawbacks of the IoT. It tells about how efficiently user used IoT in day-to-day life. This layer handles the entire IoT system in such a way that it includes business and profit modules, applications, and user's privacy. There are large number of applications that have been already developed because IoT technology needs lots of attention. Due to this, IoT service connects the physical world with economy and efficiently changes consumer's life.

Researchers have found that the authors often discuss the findings of survey data, the user introduces the concept of satisfaction and dissatisfaction, satisfiers are considered as positive drivers of technology, and dissatisfiers are negative drivers of technology.

i. Security and Privacy Issues:

To find a better way to eliminate risks and security issues, a survey was conducted and many research work was done, so as to study the effects on privacy and security of user's requirements.

The analysis can be divided into four segments:

1. Restrictions with IoT devices and overcomes;
2. Types of IoT attacks;
3. Focus on the instrument and architectures;
4. Analysis of security issues in different layers.

People are using IoT in their day-to-day life. IoT devices will collect the user's personal information as well, such as name and account details; to keep in mind all these things, one should take care of security and privacy issues for using IoT devices.

In [2015], the authors [11] proposed the research difficulties and their answers in the field of IoT security, which is the spotlight on security issues, which are further divided into nine different categories:

1. Authentication;
2. Access control;

3. Confidentiality;
4. Privacy;
5. Trust;
6. Secure;
7. Middleware;
8. Mobile security;
9. Policy enforcement.

In [2016] according to [12], the authors proceed with the research as in today's world everyone is familiar with the use of WSN in their day-to-day life. Apart from this, WSNs are even more fruitful in the area of homeland security and battlefield surveillance scenarios. In general terms, the user can say that sensor nodes are static in nature, whereas mobile nodes are deployed based on the needs of the users, and the base station has both static and mobile features. The sensor network works upon the network area and gives the result according to the user's requirement. The WSN's lifetime is affected by the sensor nodes and energy consumption. There are so many technologies that are used for energy saving in sensor nodes, such as energy harvesting, node replacement, and cycle scheduling. On the other hand, the decentralized distribution system comes in light. Nowadays, WSN security is the main issue. There are various characteristics of WSN: The WSN has limited resources, limited memory, low computing power, limited energy, and low bandwidth and communication range. The security protocol affects the performance of an application as the attackers break weak security protocols. It is true that easy and high-security protocols degrade the system performance.

In [2017], the authors [13] observed that as the demand of wireless networks is rapidly increasing, the misuse of wireless network is growing like in cybercriminal activities, data forging, hacking of data, and so on. Due to these securities, wireless network is an important issue for today's world. Attacks such as man-in-the-middle, black hole, and wormhole may affect the system badly because the system has a serious negative impact on data integrity for data monitoring. WSN distinguishes between authorized and unauthorized users. The MAC address and network interface card are used in sensor network.

In [2017], the authors [14] have described how the longevity of WSNs can be increased. Moving forward with the research, it has been found that this can only be achieved with one factor, namely, "clustering". The need to consider this factor is due to the fact that the process of clustering in wireless networks reduces the amount of raw data being transmitted in the entire network. Unfortunately, this process also has some drawbacks that are due to a "global" clustering operation in WSNs. The "global" clustering operations are used in global round-based policy (GRBP). To mitigate this problem of GRBP, the research continued with another approach known as hierarchical clustering task scheduling (HCSP). The major difference or the process being followed for this was that in the entire network of WSN, each cluster is configured only once in each round. Hence due to this fact, the frequency varies across the whole network. This policy helped to diminish the problem of clustering over the entire network of WSN and further calculate the energy consumption for the network.

In [2018], the authors [15] described that WSNs in today's life have become an indispensable part of IoT, which is further being applied in various fields in order to monitor and collect the data from the surroundings. The unique features of WSNs are large scalability, self-organization, dynamic topology, and constraint resources. These features make the sensor networks more prone to attacks. Many researches about WSNs have been done but cannot clearly define about the attacks in WSN. The attacks in WSN are described for the detection along with the security measures in sensor networks. The security data collection for these networks is done in 11 mainstreams, namely, JA, RDA, CNA, DA, SA, BHA, GHA, WA, SHA, HFA, and FA. These attacks can also be detected using different layers of networking as well. It has also been reviewed that these attacks are not only single attacks but also synthesized attacks.

In [2020], the authors [16] started research on IoT in connection with the medical field. An IoT plays an important role in the field of health also; now, the term is defined as an IoHT (Internet of Health Things). In developing countries, cervical cancer is growing as a global health problem among women; using IoHT automated Pap smear cell recognition, the detection of diseases at early stage is possible. The Pap smear concept promotes transfer technique. There are various conventional machine learning techniques such as CNN (convolutional neural network), K-nearest neighbor, logistic regression, Random Forest, and support vector machines. The important way of finding cervical cancer is biopsy, which is a satisfactory method for detection. IoHT is very important for medical researchers and practitioners to diagnose diseases and provide maximum benefits to the women globally. The main advantage of IoHT is that it is an automatic system and gives an accurate result to reduce the computational cost and provide a fast and accurate model for cell-based image classification.

In [2019], the authors [10] described that IoT is basically a mode of transferring data over a network without any human interference. The main advantages of this technology are improved accuracy, minimization of errors from various applications due to the less human interference, low cost, more effective, easy to trace records, easy maintenance of database, reduction in waste, and reduction in loss of data. The IoT has various sectors, and its accuracy and privacy are main issues nowadays. An IoT is a need of everyday life, but its threats and vulnerabilities come in picture. It is mandatory to overcome with these two issues: Vulnerability is basically a flow in the system by which the attacker can grab the system easily and damage the resources and data of the system, and threats take benefits from advantages of vulnerabilities in a negative way.

In 2019, the authors [8] stated that IoT plays various roles in different fields such as healthcare unit, wearable, and farming with the help of different technologies (such as blockchain and WSN). The most important aspect of IoT is to establish a helpful communication between two objects so that a user gets better results for future aspects. The organization gets benefits from the IoT; for the better result, the IoT revamps the consumer experience and increases the efficiency or productivity of the employees. The earning of the employees also increases.

In 2015, the authors [17] coined a new term "distance function" in the field of IoT. The distance function includes data analysis and processing tools or is the core of image processing task; it is used for the underlying assumptions and mathematical constructions so that a user gets a hyperspectral image to maintain the accuracy of hyperspectral image processing result. Euclidean distance of cumulative spectrum

(ECS) is one of the most effective distance functions. The distance function increases image quality (IQ) assessment. The information is based on IQ matrix only. The IQ matrix is further divided into three categories: no reference, reduced reference, and full reference quality matrices. The distance function is most commonly used to measure the number of intersections between two histograms. The transformation cost-based distance is another way to measure dissimilarity between spectra.

In 2019, the authors [18] described the world's most important problem of purity of water. For this big challenge, the IoT sensor plays an important role. The sensors are used in water treatment plant, and in industries, it reduces human efforts, and saves money and time also. One of the kinds of devices along with some sensors and modules was created with Arduino. These modules have Wi-Fi in nature, which sends all the data to the server for record-keeping. These devices monitor the most important factors such as EC (electrical conductivity), pH, dissolved oxygen, color, and turbidity, which affect the quality of the water and help in the treatment process of water.

3.5   EXPERIMENTS PERFORMED

3.5.1   SENSEnuts

SENSEnuts provides software support for developing a device that basically uses a sensor. It focuses on viewing a line data and analysis part of that device. Here, the GUI is used for programming the motes and displaying data from the sensor network. In this experiment, the users focus on a research in WSN domain and implement their algorithms by the real-time implementation. The GUI further consists of three separate programs, namely, device programmer, sense live, and print window. Each program has its own specific feature. Brief instructions and tasks about the three programs are as follows.

Device Programmer: The name itself suggests that user can program a device with a feature provided by SENSEnuts. The programmer programs the nodes according to the rules defined in programs that are further composed in C language with the help of Eclipse IDE. Another function is that it allows the device to decode the MAC address of the node through which device is connected.

Sense Live: It is a visualizing medium that exhibits the data received from the sensor networks. Sense Live output or GUI is featured in two separate sections: The first section can be used to view the current data coming from mote nodes, and the second section creates a database of all the data incoming from the motes in order to save and analyze them later if needed.

Print Window: It is the feature that allows the nodes to display some custom data on the GUI.

3.5.2   SENSEnuts Design Flow

SENSEnuts WSN platform is diverse in nature, and it is programmable to accomplish different tasks. This guide is designed in order to give a quick push to a user working for the first time on SENSE NUTS platform. There is no programming-specific information present in this guide. On the other hand, a user must read this guide thoroughly in order to understand the information in all other guides. This guide covers the following points in exact sequence:

1. Installing the toolchain;
2. Basic hardware setup;
3. Compiling SENSEnuts code;
4. Flashing radio modules;
5. Installing USB gateway drivers;
6. Creating a new project in Eclipse [10].

SENSEnuts is basically a Windows running application used to program on windows operating systems. Also, SENSEnuts radio modules display the data received from the network that is made up of sensor motes. The GUI is a collection of programs that perform a separate function, namely, device programmer, sense live, print window, special command, and network topology. Starting with the GUI, it lands up on the home screen with titles of the programs (Figure 3.5).

3.5.3 DEVICE PROGRAMMER

Device programmer is used as an instrument to program the motes. When a piece of code is written to apply certain set of rules on, or to develop an application on, SENSEnuts platform for compiling that piece of code, it generates a binary (.bin) file. Once the user connects the device with the PC, this device programmer has the feature that helps to print the MAC address on the SENSEnuts GUI. The options available on the GUI are explained as follows:

MAC ADDRESS: This option, when clicked, displays the MAC of the mote attached with the system, which is further attached to the system with USB cable and the SENSEnuts gateway.

File: This option opens the window to select the .bin file with which the radio module has to be programmed.

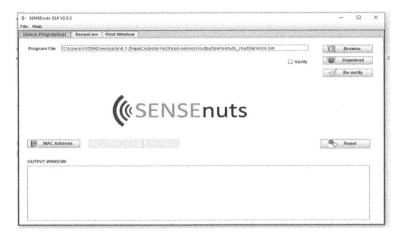

FIGURE 3.5 SENSEnuts GUI snapshot.

a. Browse enables the user to locate and select the .bin file to be programmed saved on the hard disk;

b. Flash initiates the flashing of the mote connected with the computer with the help of USB cable and SENSEnuts gateway.

c. Verify, when selected, compares the file flashed on the device and the file present on the hard disk, and validates if the flashing operation went through without any error.

d. Reverify allows the user to compare the .bin file already present on the mote and the file selected using "Browse" button. This particular option can be used to check if the mote at hand is programmed according to the algorithm written.

3.5.4 SENSE LIVE

Sense Live is an interface that prints the data received from the SENSEnut network in an instinctive way. Only the current or the latest data that are received from the motes that are present in the network are displayed, which further creates database for all these messages or data that are received. Another important feature of SENSE LIVE is that it gives an opportunity to the user to create a new custom interface, which will not be a part of this SENSEnuts platform [19] (Figures 3.6 and 3.7).

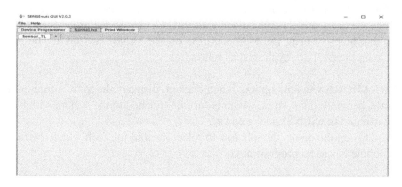

FIGURE 3.6   Sense live window.

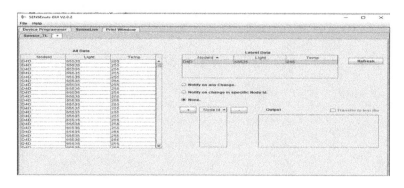

FIGURE 3.7   Sense live tables.

3.5.5 PRINT WINDOW

This feature gives the output or displays the output to the end user which is received from all the motes that are present in the network. The options available in the tool are explained in further sections (Figure 3.8).

3.5.6 LED BLINK

LED Blink is a demo program provided by SENSEnuts software toolkit chain, in which we need to connect the hardware device and perform the requisite function. Led blink programming is done in C language in which different functions are used for different perspectives. We would be proceeding with the understanding of each function being used.

This demo of led blink requires the use of startNode () and void userTaskHandler (uint8 taskType) functions. At the time of initialization in startNode function, we are adding a LED_ON task and then handling it in void userTaskHandler (uint8 task-Type) when timer expires in 100 seconds. Then, a task LED_OFF is added to switch off the LED after 100 seconds. The process is repeated by adding a task LED_ON again when the LED is switched off. This results in repeating LED switch-on and switch-off processes.

3.5.6.1 LED Blink Results Using SENSEnuts GUI

As stated above, LED BLINK demo provided by SENSEnuts allows the hardware device to blink the LED light that is there on the chip. The LED BLINK is used to connect the device with PC, and then, the deployed program starts functioning to make the device active. All the processes start, but the initial step, that is, to start with the functioning of the device programmer in sense live, is to detect the MAC address as soon the hardware device is connected to the PC [20]. Let us understand the steps to allow the Program to run on device.

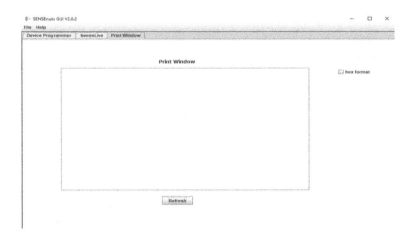

FIGURE 3.8 Print window.

TABLE 3.1
LED Blink Results

S. No	GUI for Experiment	Target Machine
1	SENSEnuts GUI v_6.1.2	Windows 10/7
2	Device attached with SENSEnuts	Windows 10/7

3.5.6.1.1   Experimental Result for LED BLINK
See Table 3.1

Step-1: Open the SENSEnuts GUI.

Result: On opening the SENSEnuts GUI, the user can see the device programmer tab initially with logs box at the bottom and MAC address button with
the fields where MAC address can be seen.

Step-2: Connect the hardware device with PC as soon as one open the GUI.

Result: As soon as the user connects the device with the PC, it should show in
the logs the status as SUCCESS: Device found.

Step-3: Click on MAC address button on the GUI to detect the MAC of the
SENSEnut device.

Result: On clicking on MAC address button, the GUI detects the MAC address of
the device and then it shows in fields next to the button. Also, the user can perform all the steps in the log box as shown in the figures attached after the steps.

Step-4: Load on the SENSEnut GUI by clicking on the browse button and then
selecting the Program from the folder saved in PC.

Result: As the user browses the folder, the location of the folder is visible in
the program file field and then loader starts loading along with the logs visible to the user in the output window. The outputs that are relevant to this
step are as follows:

Success: Bin File written;

Processing: Writing to the device;

Requesting: Write bin file;

Requesting: MAC address.

3.5.6.1.2   Result Using GUI Mode
See Figures 3.9–3.13

3.5.6.2   API-Related Functions
DIO functions (dio.h) are being used while writing program for LED BLINK.

i. void setPin(uint8dioVal):
Description function is to set a digital input/output pin specified by
dioVal. The DIOs are used for some special functions. The list is as follows:

DIO1: Power to SNHTP module DIO2: LED on radio module;

DIO5: critical interrupt from light sensor;

DIO6: PC communication interface;

FIGURE 3.9   Device connection logs after attaching the device with PC.

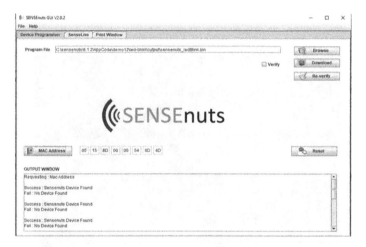

FIGURE 3.10   MAC address detection.

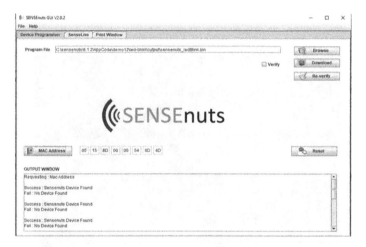

FIGURE 3.11   Logs on running the demo.

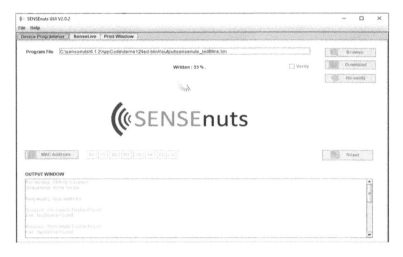

FIGURE 3.12   Running the device after detection.

FIGURE 3.13   Command prompt logs.

DIO7: PC communication interface;

DIO8: Flag pin to update availability of accelerometer data DIO14: clock
    to sensors connected on i2c interface DIO15: data to/from sensors con-
    nected on i2c interface parameters dioVal: the DIO pin to be set (from
    0 to19).

Implementation setPin (1) shall set (logic 1/Vcc) DIO 1 on the processor.

ii. void clearPin (uint8dioVal):

Description clears (unset/logic 0) the DIO pin specified by dioval.
Parameters dioVal: the DIO pin to be cleared (from 0 to 19).

Implementation clearPin (1) shall clear (logic 0/Ground) DIO 1 on the
processor.

iii. initDioInterrupt (uint8dioVal):

Description enables the registry of interrupts from the DIO specified by dioVal. The interrupt is registered from the falling edge of the signal. This call also enables the wake interrupts from the specified DIO to wake the radio module from sleep.

Parameters dioVal: The DIO pin from which the interrupt has to be registered (from 0 to 17).

iv. void ledOn():

Description: Switches on the LED in radio module parameters: NONE.

v. void ledOff():

Description: Switches off the LED in radio module parameters: NONE.

vi. void setInput (uint8dioVal):

Description: Sets the DIO specified by dioVal as the input pin parameters: dioVal: value of DIO to be set as input.

vii. void setoutput (uint8 dioVal):

Description: Sets the DIO specified by dioVal as the output pin parameters: dioVal: value of DIO to be set as output.

3.5.6.3 SENSEnuts Notations

See Table 3.2.

3.6 ALGORITHM

3.6.1 LED Algorithm

Algorithm 1 and Algorithm 2 state that LED blink needs Start Node and Task Handler function The Start Node Added LED_ON task and the handling it to void Task Handler in which two cases are performed in first.

Case LED is ON for 100 sec. When time is expired, LED OFF is added to switch off the LED after 100 sec.

This process is repeated by adding a task LED ON again when the LED is switched off. This process is repeated by adding LED ON and OFF process.

3.6.1.1 Algorithm 1: Start Node

1. LED(ON, 100) // LED is ON for 100 seconds

TABLE 3.2
SENSEnuts Program Notations

ON	0
OFF	1
LED(X,Y)	LED is X for Y seconds; X belongs to (ON, OFF)
SEND_TO_PC	0
addTask(X,Y,Z)	X belongs to USER; Y belongs to SEND_TO_PC; Z is the time in seconds
updateSTLdb(X,Y,Z)	X belongs to user id; Y and Z are the user-defined variables

---

ALGORITHM 2: TASKHANDLER
1. Case 1: if(LED ==ON)
2. LED(OFF, 100)
3. Case 2: if(LED == OFF)
4. LED(ON, 100)
5. Method called for a node receives a datapacket
6. **If** (Interrupt occurs for a criticaldevice)
7. **then** method is called for critical taskhandling
8. Method is called for PC interrupt handling

3.6.2 TEMPERATURE SENSOR

Algorithm 3 states that the user needs to add task function, user task handler, and again add task function. The user initializes temperature, light sensor, and communication part to the personal computer. The add task function adds task in the queue repeatedly. The user task handler adds task when the timer is expired; this function reads temperature and senses the temperature from the environment. The update STLdb function updates PC information in the database. The add task function adds the next task in the program. If the interrupt occurs in a device, then the method is called for critical task handling.

3.6.2.1 Algorithm 3: Temperature
1. Temperature Initialization
2. Light Sensor Initialization
3. Communication port to PCinitialization
4. **addTask** (USER, SEND_TO_PC, 1) // Add task in the queue repeatedly
5. **userTaskHandler** (uint8 tastType) // user added task is handled when the task timer is expired
6. tmp = readtmp ();
7. light = readlux ();
8. **updateSTLdb**(Node_id, light, tmp) // information present in PC is updated indatabase
9. **addTask** (USER, START_TO_PC, 1) // next task isadded
10. Method called for a node receives a data packet
11. **If** (Interrupt occurs for a critical device)
12. **then** method is called for critical task handling
13. Method is called for PC interrupt handling

See Figures 3.14 and 3.15.

3.7 LED BLINK CHANGED PROGRAM

The C file of the program can be changed by the following steps:

1. Open the source folder following the path as:
   C:\SENSE NUTS\6.1.2\AppCode\demo11\led-blink\source;

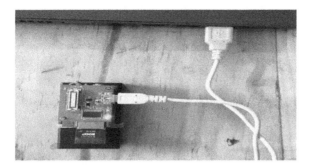

FIGURE 3.14 SENSEnuts device.

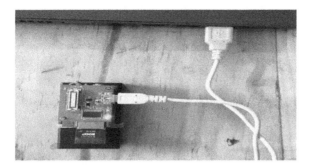

FIGURE 3.15 Temperature sense.

2. Open the C file in code editor and make changes in the function as per your requirements;
3. After making the changes, save the file;
4. Open command window and change the path to the source folder; once the path is changed, code is recompiled again; thereafter, the "make" command is entered in command prompt window and enter is pressed.

On the execution of the command, it shows the generation of a binary file and a bin file. Now open the folder where the changes were done, and run the bin file in SENSEnuts GUI (Figures 3.16–3.18).

3.8 EXPERIMENTAL SETUP

Hardware and software requirements adopted for carrying out the SENSEnuts-based experiments are detailed in Table 3.3.

FIGURE 3.16 SENSEnut GUI.

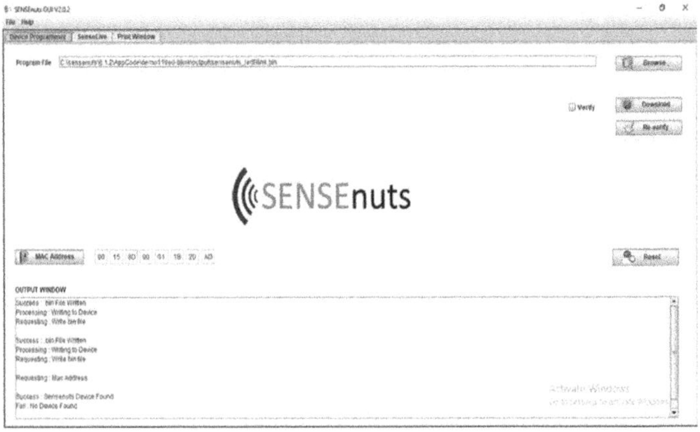

FIGURE 3.17 SENSEnut GUI after changes.

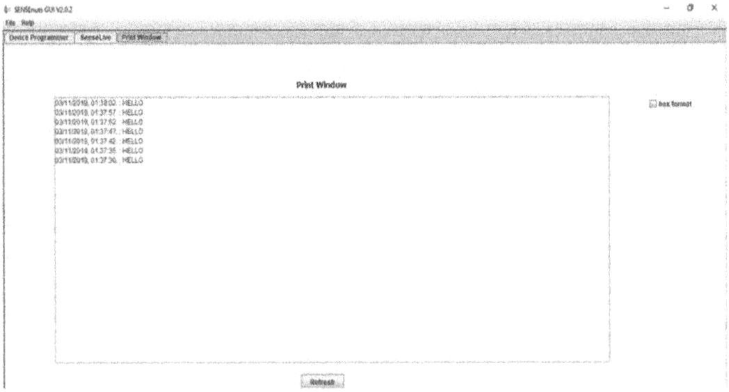

FIGURE 3.18 Output of the changes.

TABLE 3.3
H/W and S/w Specifications

Modules	Speed	Temperature Range	Configuration	Power	Memory
Radio module	1–32 MHZ clock speed	–	2.4GHz IEEE 802.15.4 compliant transceiver 128 bytes receives buffer and 256 bytes transmits buffer Configurable transmit/ receive buffer size	2–3.6 V battery operation	256 KB flash, 32 KB RAM, 4 KB EPROM
USB gateway module	USB to asynchronous serial data transfer interface	–		–	–
Ethernet gateway module	8-Kbyte transmit/receive packet dual port SRAM	-40° to +85°C		–	
Wi-Fi gateway module	Broadcom BCM23362 Single B and 2.4 GHz IEEE 802.11b/g/n 1×1 Wi-Fi transceiver	-30° to +85°C	Active receive : 6.9mA @ 1Mbit/s Active transmit : 12.5mA @ 1Mbit/s	Low-power Wi-Fi networking module Wi-Fi power save: 0.77 mA	Integrated 1 MB flash memory and 128 kB SRAM
2G/3G gateway module	SIM-based solution which helps in receiving and sending data from various sensors communication using 802.15.4 to Internet	–	–	–	–
TL sensor module	Light range 3–64k lux with 16-bit resolution, excellent IR/UV rejection	–25° to 80° With 12-bit resolution		1.5 Microampere shutdown current	

3.9 CONCLUSION AND FUTURE WORK

The sensors used for carrying out the experiment in this chapter have been procured from Eigen Technologies, by utilizing the grant provided under FRGS [Faculty Research Grant Scheme] – 2018 of Guru Gobind Singh Indraprastha University.

As is well known, routing is a very important attribute toward an efficient delivery of data packets/messages in any network. In the current research work, effective and optimized routing using sensors has been done between nodes in an IoT-driven wireless network on the real-time test beds of SENSEnuts nodes introduced by Eigen Technologies, and the results have been positive and encouraging.

In the future, it is proposed to carry out more experiments on the real-time test beds using many more sensors such as GAP sensor (gyro meter, accelerometer, and pressure), ultrasonic sensor module, color sensor, and humidity sensor to be procured from Eigen Technologies and TI (Texas Instruments).

REFERENCES

1. Wang, Qinghua, and Ilangko Bala singham. "Wireless sensor networks-an introduction." *Wireless Sensor Networks: Application-Centric Design* (2010): 1–14.
2. http://www.techrepublic.com/blog/enropeaun-technology/what-the-internet-of-things-means-for-you [Accessed on 02-02-2020].
3. Mbowe, Joseph E., and George S. Oreku. "Quality of service in wireless sensor networks." *Wireless Sensor Network* 6, no. 02 (2014): 19.
4. Byun Jaehak, Sooyeop Kim, Jaehun Sa, Sangphil Kim, Yong-Tae Shin, and Jong-Bae Kim. "Smart city implementation models based on IoT technology." *Advanced Science and Technology Letters* 129, no. 41 (2016): 209–212.
5. Fallah, Monireh, Mostafa Ghobaei Arani, and Mehrdad Maeen. "NASLA: Novel auto scaling approach based on learning automata for web application in cloud computing environment." *International Journal of Computer Applications* 113, no. 2 (2015): 18–23.
6. Lee, Won-jun. "Satisfiers and dissatisfiers of smart IoT service and customer attitude." *Advanced Science and Technology Letters* 126 (2016): 124–127.
7. Tiwana, Amrit, Benn Konsynski, and Ashley A. Bush. "Research commentary—Platform evolution: Coevolution of platform architecture, governance, and environmental dynamics." *Information Systems Research* 21, no. 4 (2010): 675–687.
8. Varshney, Tanishq, Nikhil Sharma, Ila Kaushik, and Bharat Bhushan. "Architectural model of security threats & their countermeasures in IoT." In *2019 International Conference on Computing, Communication, and Intelligent Systems (ICCCIS)*, pp. 424–429. IEEE, 2019.
9. Dos Santos, Giederson Lessa, Vinícius Tavares Guimarães, Guilherme da Cunha Rodrigues, Lisandro Zambenedetti Granville, and Liane Margarida Rockenbach Tarouco. "A DTLS-based security architecture for the Internet of Things." In *2015 IEEE Symposium on Computers and Communication (ISCC)*, pp. 809–815. IEEE, 2015.
10. Arora, Aman, Anureet Kaur, Bharat Bhushan, and Himanshu Saini. "Security concerns and future trends of Internet of Things." In *2019 2nd International Conference on Intelligent Computing, Instrumentation and Control Technologies (ICICICT)*, vol. 1, pp. 891–896. IEEE, 2019.
11. Sicari, Sabrina, Alessandra Rizzardi, Luigi Alfredo Grieco, and Alberto Coen-Porisini. "Security, privacy and trust in Internet of Things: The road ahead." *Computer Networks* 76 (2015): 146–164.

12. Bhushan, Bharat, and Gadadhar Sahoo. "Recent advances in attacks, technical challenges, vulnerabilities and their countermeasures in wireless sensor networks." *Wireless Personal Communications* 98, no. 2 (2018): 2037–2077.

13. Sharma, Mini, Aditya Tandon, Subhashini Narayan, and Bharat Bhushan. "Classification and analysis of security attacks in WSNs and IEEE 802.15. 4 standards: A survey." In *2017 3rd International Conference on Advances in Computing, Communication & Automation (ICACCA)(Fall)*, pp. 1–5. IEEE, 2017.

14. Neamatollahi, Peyman, Saeid Abrishami, Mahmoud Naghibzadeh, Mohammad Hossein Yaghmaee Moghaddam, and Ossama Younis. "Hierarchical clustering-task scheduling policy in cluster-based wireless sensor networks." *IEEE Transactions on Industrial Informatics* 14, no. 5 (2017): 1876–1886.

15. Xie, Haomeng, Zheng Yan, Zhen Yao, and Mohammed Atiquzzaman. "Data collection for security measurement in wireless sensor networks: a survey." *IEEE Internet of Things Journal* 6, no. 2 (2018): 2205–2224.

16. Khamparia, Aditya, Deepak Gupta, Victor Hugo C. de Albuquerque, Arun Kumar Sangaiah, and Rutvij H. Jhaveri. "Internet of health things-driven deep learning system for detection and classification of cervical cells using transfer learning." *The Journal of Supercomputing* (2020): 1–19.

17. Deborah, Hilda, Noël Richard, and Jon Yngve Hardeberg. "A comprehensive evaluation of spectral distance functions and metrics for hyperspectral image processing." *IEEE Journal of Selected Topics in Applied Earth Observations and Remote Sensing* 8, no. 6 (2015): 3224–3234.

18. Aggarwal, Sachin, Rahul Gulati, and Bharat Bhushan. "Monitoring of input and output water quality in treatment of urban waste water using IOT and artificial neural network." In *2019 2nd International Conference on Intelligent Computing, Instrumentation and Control Technologies (ICICICT)*, vol. 1, pp. 897–901. IEEE, 2019.

19. Downloads/6.1.2/docs/Release%20Notes%20V6.1.2.pdf [Pdf Retrieved on 02-02–2020]

20. Downloads/6.1.2/docs/SENSE NUTS%20GUI%20User%20Guide.pdf [Retrieved on 02-02–2020]

4 Machine Learning
Approach towards Security

Vidushi and Manisha Agarwal
Banasthali Vidyapith

CONTENTS

4.1 INTRODUCTION

Wireless sensor networks development is enabled that covers essential areas in the society such as medical and home management, with the advances of wireless communication, growth in technology of hardware, and microelectrical devices [1,2]. Along with the rising in-depth demand of technology, Internet, and automation, a security problem comes in front of society. It is a severe problem and becomes a challenge in front of researchers and scientists. Stealing of data, deletion of data, damage of hardware and software, etc. can be done in few seconds. In the current IT world, Internet becomes mandatory in use, and it integrates with the social life. Growth of new technology completely changes the living life, which means it changes the way to work and think. Internet of Things (IoT) [3] finds the new way of communication.

In the present era, it exhibits a dramatic success and achieves a multitude of different domains like home automation [3]. IoT is a system of connecting different interrelated devices, objects, and living ones via Internet. All of them connect with each other and establish an effective communication by collecting and sharing data. IoT works in almost every industry (like consumer, commercial), and its major applications include smart home, health sector, traffic management in transport department, control of process in manufacturing, smart farming, etc. without human intervention. [4]. Although Internet and IoT both have a number and broad spectrum of benefits ; make the life smooth, easy, and fast; and provide less time to complete work, both still face security problems [5]. They may lose all their potentials and thus cannot provide a trusted environment.

A group of dedicated sensors are used to fulfill the purpose of monitoring and recording the physical environmental condition and collecting the data at the central point referred to as wireless sensor network. This network is used to monitor and control not only general pieces of equipment but also industrial pieces of equipment [6]. As the wireless network popularity is enhancing day by day, paralleling is followed by the security threat [7]. Further, this network application in the domain of industry helps in lowering the failures and improving the overall efficiency [8]. This network is responsible for organizing the data that is collected from different sources. The environmental conditions include sound and humidity. In the wireless sensor network, the transportation of sensor data through the wireless medium depends on the connectivity of the wireless network. The distribution of sensors is in such a way that the monitoring of condition related to the physical environment happens along with the transmission of data to the destination using the wireless medium. The need of the wireless environment directly opens the door for security. While either the data is traveling through the wireless medium or the desired data is captured using sensors, the breach of security can lead to unthinkable and unbelievable problem. The wireless sensor system has numerous, multifunctional, and small sensor nodes. These sensors help in collecting the data from different sensing fields [8]. Due to the possibility of security attacks, the design of routing protocols used in the networks of wireless sensors considers the issues related to design as well as routing challenges [8]. The authors of Ref. [9] discuss SDN (software-defined networking) and NFV (network function virtualization) security mechanisms for IoT. Due to the limited memory supply as well as battery, the conventional security solution cannot be used in case of sensor networks [10]. Through the sensors, enormous massive data can be collected and then can be transmitted from network using any media. This transmission should be safe and secure [11]. The need of security becomes the primary paramount because of the necessity of deployment even in hostile area of sensor networks applications [12]. Due to the emergence of wireless sensor networks in the crucial areas like military investigation, these are the unprotected networks that make them prone to the security attacks [13]. The wireless sensor network can share information or highly critical data using wireless medium. The presence of the wireless medium increases the vulnerability of network attacks like jamming attacks [14]. Security is the main threat for the wireless sensor systems; a clustering system that is based on fuzzy system using BSL (balanced load subcluster) formation can be the solution toward the security threat [15]. Wireless mesh network is the multi-hop and rapidly developed network. This network is also facing the security challenge that needs to be addressed [16]. The wireless mesh network system that has multi-hop network is beneficial in providing the effective broadband service [17]. For balancing the network energy consumption, the selection of aggregate nodes on the location basis needs to be made. This may further bring the security challenges that raise the need of various privacy-preserving schemes that are homomorphic based [18]. The rise in the need of routing protocols is attributable to the dynamic nature of topology, constraints in the resources, and wireless sensor network that is distributed in nature [19].

Localization of the security issues can be related to privacy, intrusion, authentication, malware, and many more. These security issues drastically affect the business as well as personal day-to-day activities. Therefore, it is crucial to deal with this security problem, and finding some prevention is also important. If the data shared between the devices is breached, damaged, or modified, then how powerful destruction can be happened? If the images captured by camera are breached and modified by someone, then a serious authorization and authentication problem will happen. Intrusion detection is very important to preserve the privacy in the development of projects such as smart cities, environment, and farming. Data breaching in the healthcare sector can lead to severe damage and even the loss of life. All these points indicate that advance research in technology is the demand of today's life, but not with the cost of security. To protect the destruction that can lead to increasing or improving the level of security is crucial. This research arranged the paper framework section-wise. Section 4.2 gives the background survey that is related to this research. Section 4.3 briefly introduces the IoT concept. This section also explains the different possible security challenges related to IoT. Section 4.3 summarizes the introduction of machine learning as well as the brief study of learning algorithms. Section 4.4 evaluates and compares the performance of the machine learning models. Finally, Section 4.5 provides the conclusion of this work along with the future work.

4.2 HISTORICAL BACKGROUND AND RELATED SURVEY

In the current era and for the future prospective, security is the major concern. Day by day, technology is increasing with pace. Human is now more dependent on technology, and in the future, this dependence will increase to work fast, easily, accurately, and on time and the repetition of work should be less. But security can be a major obstacle in the way of this technological achievement. Worldwide, security is currently the most important challenge and opportunity for the scientists and researchers. There are many present surveys and researches that are already available and related to security and data or individual privacy. This contribution shows the path for the new discovery and innovation. The crucial work that has been done by researchers and base for the upcoming research is as follows.

Vikas et al. [20] elaborated the security-related problems and security threat sources in detail. Different technologies that can be used to achieve the trust in IoT applications are discussed. The authors of Ref. [21] review the deep learning as well as forensic methods to find out botnets and their IoT environment applications.

In Ref. [22], the work provides malware detection for the real-time implementation using deep learning. Andreas et al. [23] provide the summarized principles of wireless communication with the perspective towards the need for the connectivity between cyber-physical systems and the IoT devices. Ikram et al. [24] have analyzed and surveyed a number of IoT techniques for trust management. The authors of Ref. [25] explained blockchain technique and review the adaption of blockchain for IoT applications called BIoT (blockchain-based IoT). Yang et al. [26] provide the literature survey on the methods of machine learning and deep learning for intrusion detection. The authors of Ref. [27] use deep learning to detect malicious code by converting into gray-scale image. The authors of Ref. [28] make a comparison between different machine learning algorithms to evaluate the intrusion detection using the well-known datasets. The authors of Ref. [29] address the problem of security threats in IoT applications. To classify and detect the object in X-ray baggage security imagery, the authors of Ref. [30] use the concept of transfer learning with deep convolutional neural network. The authors of Ref. [31] summarized IoT device limitations and respective solutions. They also classify the IoT attacks, explain the access control, and also examine the security problems in layers. The authors of Ref. [32] discussed the location-based security issues. In order to improve the performance of IoT networks, the authors of Ref. [33] provide a comprehensive survey of edge computing and compare the edge computing with traditionally available cloud computing. The authors of Ref. [34] discuss the relationship between IoT and fog computing and the issues related to security in fog computing. Ref. [35] highlights and provides a taxonomic analysis for the IoT security issues with respect to the main IoT layers. Problems specific to the position and location of devices are targeted by the authors. The authors of Ref. [36] address the problem of face identification and propose the fog computing-based framework. Min et al. in Ref. [37] proposed the risk prediction algorithm for multimodal disease that is based on the convolutional neural network. In Ref. [38], the authors elaborate the IoT application layer and at the application layer represent the security threats, principles, challenges, and countermeasures against challenges. The authors of Ref. [39] provide a machine learning framework for big data called MLBiD to discuss about its opportunity as well as challenges. Anne et al. [40] mainly survey on IoT middleware-layer security issues and also elaborate the related protocols and concerned security. The advantages and disadvantages of the techniques are also presented in the research. Arsalan et al. in Ref. [41] briefly discussed the vulnerabilities faced by IoT. Mete et al. in Ref. [42] experimentally analyzed that the performance of machine learning algorithms in detecting attack is higher than that of other algorithms in detecting attack. In Ref. [43], the authors survey and discuss about deep learning, and emphasize on the current research, challenges, and future trends of big data. The authors of Ref. [44] proposed a neural network-based multilayer perceptron to detect normal behavior as well as the behavior of the system after attack. Result is also represented graphically by the authors. To detect the different fraudulent access, the authors of Ref. [45] provide a fake detection method for use in different biometric systems. The author of Ref. [46] surveys the machine learning algorithms for identification using the fingerprint. The authors of Ref. [47] discussed about the IoT middleware system.

4.3 IOT BASIC CONCEPT AND RELATED SECURITY TECHNOLOGIES

4.3.1 IoT Basic Concept

It is the way of providing connection between any devices, made by humans via Internet. A giant group (physical devices, objects, and people) share information with each other. The pace of connection of physical objects via Internet is rapidly increasing. It covers almost every field and penetrates speedily in the present industries. All IoT applications that are either currently available or will deploy in coming future need security. Need of security is highly critical. Its critical applications and security concerns are shown in Table 4.1.

The IoT framework [49] can be divided into four essential layers. Diverse techniques or devices like Internet and sensors are used in these layers. Along with facilities, these techniques also bring the risk of security in each layer. IoT layers are shown in Figure 4.1.

- **Sensing Layer** [49,50] directly deals with the end point devices such as mobile phone and computers. This layer receives the input from the devices. It senses the input and, if required, records the received data also. It is also able to connect to the network.

TABLE 4.1
IoT Applications and Security Concern

Application Area	Applications	Security Concern
Smart environment [48]	Detect forest fire, monitor snow level, identify the existence of earthquake, pollution level, etc.	Government concerns areas like earthquake detection, if security breaches may lead to the loss of monetary at high level.
Smart cities [48]	Management of smart traffic, home, etc.	Security problem may arise in these areas, which may lead to privacy concern.
Smart agriculture [48]	Monitor and control of soil moisture, humidity, temperature, etc.	Loss in security may lead to crop loss.
Smart farming [48]	Monitor animals' activities, health, etc.	Steal of animal may happen.
Smart retail [48]	Track the products, intelligent shopping, etc.	Steal of customer personal and monetary information may happen.
Smart metering [48]	Measure and monitor the consumption of electricity, gas, water level, etc.	May cause money loss to both customers and providers.
Smart emergencies [48]	Dangerous gas leakage, measure radiation level, prohibit unauthorized entry, etc.	Serious problems like death can occur in case of breach of security.

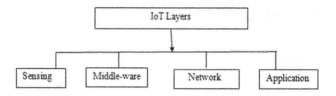

FIGURE 4.1   IoT layers.

- **Network Layer** [49,50] has the responsibility to transmit the data received from sensing layer for processing to the next computation unit. It provides the ubiquitous environmental access to the sensing layer.
- **Middleware Layer** [49,50] is an abstract layer that lies in between the layers of network and application. It also has the capability to provide computation as well as storage facility. To fulfill the application-layer demands, it provides different API.
- **Application Layer** [49,50] is the last layer that deals directly with the users. It provides high-quality services and smart need requested by the customer. Smart cities, environment, grid, meters, home, and emergencies, come under the application layer.

4.3.2   IoT-Related Security Technologies

Different techniques or devices are used for the IoT layers mentioned in Section 4.1. Using these technologies, different security risk and threats are also determined. These security risks cannot be ignored. The ignorance can lead to big problems such as monetary loss, privacy breach, unauthorized access, and ultimately the loss of life. The device or technology used in different layers and the concern related to the security with respect to layers are mentioned in Table 4.2. IoT layers can also be considered the sources of security threats for IoT applications.

TABLE 4.2
Devices/Technologies and Security Threat Possible to IoT Layers

IoT Layer	Device/Technology Used	Security Threats
Sensing layer [50]	Sensors, actuator	Node capture, booting, and side-channel attacks; false data and malicious code injection attack, etc. [50]
Network layer [50]	Internet, Wi-Fi, routing, etc	Phishing site, data transmit, routing attacks, etc. [50].
Middle-ware layer [50]	API, web service, datacenter, cloud, etc.	Flooding attack and malware injection in cloud, SQL injection, etc [50].
Application layer [50]	Smart home, transport, healthcare, etc.	Data theft, access control, distributed denial-of-service (DDOS) attacks, etc. [50].

TABLE 4.3

Technologies Used to Enhance IoT Device Security

S. No.	Technology for IoT Device Security	Technology Description	Goal
1	Deep learning [51]	It is the extension of machine learning, and its idea is based on human brain.	It can be used in almost every area of machine learning and is more effective with nonlinear data.
2	Blockchain [52]	It is a ledger that is publicly shared, distributed, decentralized, and useful in recording transactions among many computers.	In order to provide the security, blockchain provides a secured way, i.e., encryption using hash key to accumulate and transfer the data.
3	Edge computing [53]	It is extension of the cloud computing. Its architecture includes cloud server, some fog nodes, and a number of edge devices.	It reduces the cloud computing burden by doing analyses and computation at the edge nodes and hence improves the efficiency.
4	Fog computing [53]	It is also the extension of cloud computing. To handle a large number of data in real time needs fog computing.	It boosts the overall performance and efficiency by reducing the load of data on cloud.
5	Cloud computing [53]	It provides facilities on the cloud via Internet to store, analyze, and transfer of data between the various connected components.	To provide security, and put a stop to the theft of data.
6	Machine learning [54]	It provides machine the ability of self-learning and improvement with experience.	It is widely used in pattern, image recognition, and intrusion detection.
7	Forensics mechanism [55]	For legal acceptance, it is a way to identify, preserve, analyze, and present evidences digitally.	To investigate the incidents of security in IT sector.
8	Biometric methods [56]	Iris, fingerprint, and face recognition are some different biometric methods, basically used for authentication and authorization.	It provides security by authenticating the right person and restricting the entry of unauthorized ones.

Table 4.2 explains the different possible security threats along with the devices used with the respective IoT layers. Table 4.3 shows some of the well-known technologies and their respective goals that can be useful in enhancing the security level.

4.4 MACHINE LEARNING AND ITS ALGORITHM OVERVIEW

The way of living life of people is changing day by day with the in-depth integration of social life and Internet. The habit of learning and working is changing drastically. Upcoming technology brings the comfort; makes life simple, fast, accurate, easy; and also causes the security breech problem. Security is the current challenge in front of scientists and researchers. It is a vital issue because this severe issue can break the user trust and can lead to the prime issue of loss of technology. As discussed in Section 4.1, a number of different technologies such as cloud computing, edge computing, fog computing, biometric, forensic, machine, and deep learning are available to enhance the security level. This section elaborates the machine learning and its algorithms, and the next section analyzes the performance of machine learning algorithms for intrusion detection using a well-known dataset network intrusion detection.

Presently, in the world of computer science, machine learning is the most prevailing technique. It is successfully applied in pattern recognition, image processing, cybersecurity, and many more fields. It can be successfully applied in intrusion and malware detection, and automatic driving. Its goal is the focus of intelligent system to learn from massive and enormous amount of information in order to achieve high accuracy, less computational time, and low cost. In the coming time, machine learning is the future of intelligent society. It spans in multiple disciplines, which means it is multidisciplinary research area that includes computer, mathematics, and science. The purpose of this learning is to efficiently imitate the learning activities of humans by machines.

As per feedback differences, the tasks related to the machine learning are categorized as reinforcement, unsupervised, and supervised learning. Supervised one has the training samples with their class labels in the model for classification or regression. Decision tree and support vector machine (SVM) are some typical examples of supervised learning. Unsupervised leaning is the learning where training samples have no prior knowledge of category labels. Auto-encoders and clustering are some typical examples of unsupervised learning. Reinforcement learning is the learning where behavior strategies are optimized using try-and-error. This learning is entirely different from the supervised and unsupervised learning.

For intrusion detection, this chapter compares and analyzes different machine learning algorithms based on the accuracy and execution time. The descriptions of the algorithms are as follows.

4.4.1   K Nearest Neighbor (KNN)

It is simple, and easy to understand and implement machine learning algorithm used for classification as well as regression tasks [57]. This algorithm assumes that similar objects are in near proximity. In both classification and regression tasks, the input depends on the value of k. From the feature space, choose k nearest points as input. In classification, output is the class that gets the maximum vote from the k closest neighbors [57]. However, in regression, output is the average of k closest neighbors value. The working of this algorithm depends on the k value and the distance formula. This algorithm starts working with selecting and loading the dataset. The next step is to choose the value of k. Now, to classify the new point, find the distance of new point from each data point in dataset. Based on distance, pick the k closest ones and get their respective labels. In case of regression problem, output is the average of k labels, and in case of classification, output is the label which gets the maximum vote.

Different distance formulas can be used in this algorithm. Some common and frequently used distance formulas are as follows:

1. Euclidean distance function:

$$f(x) = \sqrt{\sum_{i=1}^{k} (x_i + y_i)} \tag{4.1}$$

2. Manhattan distance function:

$$f(x) = |x_i + y_i| \tag{4.2}$$

3. Minkowski distance function:

$$f(x) = \left[\sum_{i=1}^{k} \left((|x_i + y_i|) \right)^q \right]^{\frac{1}{q}} \tag{4.3}$$

where
K: nearest neighbor numbers,
x_i, y_i: the respective coordinate values.

There are some positive as well negative points of this learning model. Some positive points that attract the research to use this algorithm include that it is simple, and easy to understand and implement. In this algorithm, there is no need to make extra effort for model building. In this algorithm, simple distance formulas are used, and no complex calculation is required. It can be used and shows good result in case of classification and regression tasks. Along with some favorable points, there are also negative points while using this learning model: as example increases, it becomes slow and time-consuming. Along with it, it also depends on the k value and which distance formula is used.

4.4.2 NAÏVE BAYES ALGORITHM

It is one of the powerful probabilistic classification techniques of machine learning. It works on the basis of Bayes theorem. It assumes that predictors do not depend on each other [58]. It means that the features present in a dataset are independent of each other, and that is why, it is called naive. Bayes word comes from the Bayes theorem. The main applications of this algorithm are document classification, sentiment analysis, and spam filtration. Bayes theorem is as follows [58]:

$$P(c|x) = \frac{P(x|c)P(c)}{P(x)} \tag{4.4}$$

where
$P(c|x)$: Target class c posterior probability, and x is given attribute,
$P(c)$: Probability of class c,

$P(x)$: Attribute x probability,

P($x|c$): x likelihood probability, where c class is given.

With the assumption that features in the dataset are independent of each other, the basic working principle of this algorithm starts with selecting and loading the dataset. After the selection of dataset, this algorithm constructs the frequency table and prepares the likelihood table. The next step is to use the Bayesian equation to calculate the posterior probability for each class. Finally, the probabilities of each class are compared, and the prediction outcome is the class having the maximum probability.

This algorithm has some advantages: Its use is simple and can be easily understood; even it also shows good result in case of prediction of multiple class problems. Along with mentioned positives, it can also show better result even with less training data and show a good performance even with categorical data. Besides the advantages, there are also some disadvantages: Its working starts with the assumption of predictor's independence; for estimation, Bayes theorem is not a good solution; and this algorithm completely depends on Bayes theorem.

4.4.3 SUPPORT VECTOR MACHINE ALGORITHM

It is supervised, simple, powerful, and popular machine learning algorithm. It shows good result in case of classification but can be used for the challenges, regression, and classification [59]. These features of the dataset are plotted in n-dimensional feature space, where n is the total feature. To separate the classes, more than one hyperplane can be present. However, select the best one which is able to separate well. In case of confusion, select the one with the maximum margin [59]. To find out the maximum margin, popular maximum margin classification algorithm is used. In the case of outliners, this algorithm has the capability to ignore them. In-built function kernel is used to convert low dimension to high dimension if required. For finding the loss, the typical hinge function can be used, which is as follows [59]:

$$l(y) = \max(0, 1 - t.y)$$ (4.5)

where

Y: the output of classifier decision function,

Intended output is $t = \pm 1$.

$l(y)$: loss.

The working of this machine starts with selecting and loading the dataset. The next step is to train the training dataset subset. In the feature space, all features are plotted as coordinates, and the aim is to find the best hyperplane for classes' separation. In the case, if more than one plane possibility is available, then in-built function kernel is used to convert low dimension to high dimension.

The leading points of this learning technique involved are that this model is memory efficient because this algorithm uses the training dataset subset. In the case of clear margin separation, this algorithm shows a good result and high dimensionality. The negative part of this model is that this algorithm degrades its performance in the presence of noise as well as in the case of large dataset. If high time is required to train the dataset, then it doesn't show good result.

4.4.4   DECISION TREE ALGORITHM

Decision tree algorithm is an example of supervised learning and belongs to the machine learning family group. It can efficiently solve both regression and classification problems. In the fashion of supervised learning, it works in simple way, is easily understandable, and is easy to implement. The working principle of this algorithm is based on the decision tree learning [60]. It creates a model to train the data which further predicts the class label. This algorithm uses tree representation for solving problems. Tree nodes can be classified as internal and external nodes [60]. Internal node is used to represent the attribute, and leaf node is used to represent class label. The first step of this algorithm is to select and load the dataset. The second step of this algorithm is to choose the best feature or attribute from the dataset as the root node. The next step is to form the subsets of root after splitting the training data, and the final step is to continue forming the subsets until reaching the leaf node. For the classification of any target that begins from the main root, the results are compared, and this comparison becomes the basis to follow the branch and move towards the next node. Repeat this process unless and until predicting the required label which is represented by leaf [60]. To select the attribute is the challenge for this algorithm. It can take place by using different methods like information gain. Information gain finds the information for every attribute. This complete task is accomplished using two steps. Through the first step, calculate the target entropy, and then by the second step, find out the attributes entropy. Finally, get the information gain by using the following formula:

$$\text{Information gain (IG)} = \text{E(Target)} - \text{E(Target of attribute)} \qquad (4.6)$$

where E denotes the entropy.

Using Equation 4.6, find this gain for each attribute and then finally select the highest gain attribute. The leading points that make this algorithm suitable for the task of classification are as mentioned. This algorithm is simple to learn and is easy to understand. Along with its simplicity, it has the ability of easy implementation that generates the capability of making the rules and further taking decision based on rules. It is beneficial for a person who is either technical or nontechnical, can understand, and further can make decision through tree visualization. The negative consequences of this algorithm are as follows: It has the possibility of having the issue of overfitting and the accuracy of models is diminished in case of testing the model, which can be further taken care by post- and pre-pruning methods. It also shows the comparatively low prediction accuracy, and with multiple class labels, its calculation becomes so complex. Even the information gain with more categories attributes shows a biased response.

4.4.5 RANDOM FOREST ALGORITHM

It works in the fashion of supervised learning that belongs to the machine learning family. The building block and basic model of Random Forest is decision tree [61]. It is made up of several decision trees where trees are independent of each other, and this feature shows its power [61]. Every Random Forest tree predicts the class, and the class with the highest vote is the prediction of model. The basic working principle of this algorithm is the same as the decision tree model but it is much more powerful, is comparatively strong, and shows better result with a maximum time. Initially, select the dataset consisting of features and their values, and load it. Then, to predict the desired model applies different decision trees. The next step is to examine and analyze the result in finding the outcome supported by more number of decision trees. In the final step, the final outcome of this model is the one that is supported by maximum decision trees.

The advantages of this algorithm are as follows: It is easy and simple to understand. Along with simple nature, it shows more efficient result compared to other algorithms and shows less error. Its positivity also includes that all trees present are independent of each other and also have comparatively better accuracy. Besides these advantages, some disadvantages of this model are that it has complex calculation, it is tough to implement, it also has need of powerful features, and it takes more time to predict the result.

4.4.6 LOGISTIC REGRESSION ALGORITHM

It is famous, simple, and understandable machine learning family algorithm [62]. The main purpose of this algorithm is to classify binary task. For prediction, this algorithm uses a linear function [62]. It uses the sigmoid function which is S-shaped curve, whose value varies, or it can map the real value between 0 and 1. Its applications include loan sanction, and spam email. The shape of the function is shown in Figure 4.2.

The equation of this S-shaped sigmoid curve is given below:

$$g(x) = \frac{1}{1 + e^{-x}} \tag{4.7}$$

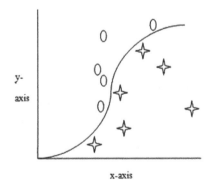

FIGURE 4.2 S-shaped sigmoid function.

Figure 4.2 and Equation 4.7 show the S-shaped sigmoid function. The basic working principle of this algorithm initially selects the dataset and loads it. The next step is to convert the input into binary form. The final step is to make the classification and get the desired result.

The merits of this algorithm include some solid points: It is a simple algorithm, and it has no complex mathematics. Its implementation is easy and well known to estimate the result. There are also some disadvantages of this algorithm: Basically, it is for binary classification only, and it also has the overfitting problem.

4.5 PERFORMANCE OF MACHINE LEARNING MODELS USING NETWORK INTRUSION DATASET FOR INTRUSION DETECTION

Intrusion is the major problem of security and prime tension in the case of security breech. In just few moments, steel of data or deletion of data from computer can happen with single intrusion instance. Intrusion can affect system hardware. Huge monetary losses by intrusion lead to the inferiority of information in cyber world. Hence, detection of intrusion is very critical and crucial, and to prevent it is also essential. A number of its detection techniques are already present, but the main issue is accuracy. Thus, this work applies different machine learning algorithms such as SVM, K-nearest neighbor (KNN), Random Forest, decision tree, logistic regression, and naïve Bayes; these algorithms can be effective in addressing the problem of classification. To evaluate the method of intrusion detection, this research uses the standard network intrusion detection dataset.

4.5.1 NETWORK INTRUSION DETECTION DATASET

This is a standard dataset and shows a benchmark to evaluate the intrusion detection in this chapter. This dataset is the new, latest dataset present on Kaggle. It has an extensive selection of intrusions that is simulated in a network of military environment.

4.5.2 LEARNING MODELS EXPERIMENTAL ANALYSIS

Different learning models are employed to analyze the intrusion detection dataset, and they have their own advantages and disadvantages that have been already discussed in this chapter in the previous section. The complete dataset is divided into training and test. After performing training to the training part of dataset using the particular model, the evaluation of models is made.

4.5.2.1 Evaluation Model

Evaluation models are applied one by one on the training part of the dataset. The analysis of the result of the evaluation model is based on the accuracy parameter and the execution time in second (which means how much time the particular model needs to complete the task). The accuracy and the execution time taken by the machine learning model classifier of the well-known six models are given in Table 4.4.

TABLE 4.4
Model Evaluation Based on Accuracy and Execution Time

	Model Evaluation	
Classifiers	Model Accuracy	Execution Time (in s)
K-nearest neighbor	0.994329	58.141325
Naïve Bayes	0.907961	4.847277
Logistic regression	0.954349	24.828419
Random Forest	0.999943	26.396278
Support vector machine	0.956788	233.677275
Decision tree	0.999943	3.359415

The aforementioned analysis is also shown in Figures 4.3 and 4.4 using graph. Figure 4.3 shows the comparison of machine learning models using the accuracy parameter.

From the observations of Figure 4.3 and Table 4.4, it can be concluded that Random Forest and decision tree models are performing better than the other models.

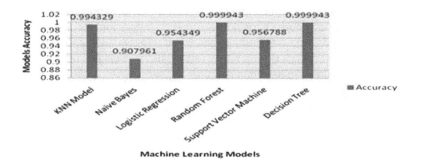

FIGURE 4.3   Evaluation of machine learning models.

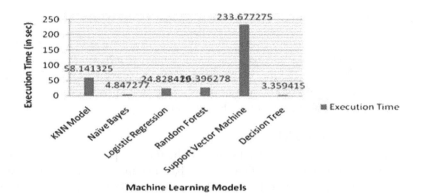

FIGURE 4.4   Model evaluation based on execution time.

Figure 4.4, which is based on Table 4.4, represents the comparison of models with their execution time.

From the observations of Figure 4.4 and Table 4.4, it can be concluded that decision tree models take the least time to complete the execution task. This evaluation model shows that the decision tree is the machine learning model that performs better than the other models and it requires a minimum time.

4.5.2.2 Validation Model

Validation models are applied one by one on the test part of the dataset. The analysis of the result of the validation model is based on the accuracy parameter and the execution time in second (which means how much time the particular model needs to complete the task). The accuracy and the execution time taken by the machine learning model classifier of the well-known six models are given in Table 4.5.

The aforementioned analysis is shown in Figures 4.5 and 4.6 using graph. Figure 4.5 shows the comparison of machine learning models using the accuracy parameter.

TABLE 4.5

Model Validation Based on Accuracy and Execution Time

Validation Model		
Classifiers	**Accuracy**	**Execution Time**
K-nearest neighbor	0.990738	10.168581
Naïve Bayes	0.906589	0.442025
Logistic regression	0.956337	0.848048
Random Forest	0.997089	1.016006
Support vector machine	0.958851	2.901165
Decision tree	0.995766	0.418011

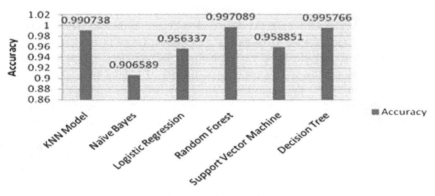

FIGURE 4.5 Validation of machine learning models.

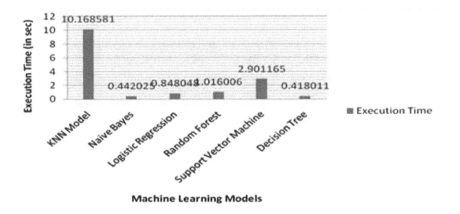

FIGURE 4.6 Model validation based on execution time.

From the observations of Figure 4.5 and Table 4.5, it can be concluded that Random Forest model is performing better than the other models. Figure 4.6, which is based on Table 4.5, represents the comparison of models with their execution time.

From the observations of Figure 4.6 and Table 4.5, it can be concluded that decision tree model takes the least time to complete the execution task. Overall, this evaluation shows that the Random Forest is the machine learning model that performs better than the other models, and Decision Tree model takes a minimum time to complete the task.

4.6   CONCLUSION AND FUTURE WORK

This chapter reviews the literature of different algorithms used for learning the models as well as IoT concept. This research briefly explains different techniques that can be used for enhancing the security of IoT devices. This chapter evaluates various algorithms of machine learning. The focus of this chapter is to detect the intrusion using learning algorithms on network intrusion detection dataset. The experimental work conducted by this study on six well-known learning models shows that Random Forest model yields better result than other models in both the evaluation and validation phases. This experimental work also calculates and shows the execution time taken by the respective learning models. The findings in this study can be further utilized by scientists for upcoming innovative, interesting, and crucial research. In the future, this experiment can be conducted on dataset with more availability of data. To continue this research in the coming time, deep learning techniques and further deep analysis of comparative work can be established. This chapter mentions the security-enhancing techniques other than the machine learning. In the coming future, these techniques can be implemented and analyzed with results of this research.

REFERENCES

1. Bhushan, B., & Sahoo, G. (2018). Recent advances in attacks, technical challenges, vulnerabilities and their countermeasures in wireless sensor networks. *Wireless Personal Communications, 98*(2), 2037–2077.

2. Bhushan, B., & Sahoo, G. (2019). E2SR2: An acknowledgement-based mobile sink routing protocol with rechargeable sensors for wireless sensor networks. *Wireless Networks, 25*(5), 2697–2721.

3. Meneghello, F., Calore, M., Zucchetto, D., Polese, M., & Zanella, A. (2019). Iot: Internet of threats? A survey of practical security vulnerabilities in real IoT devices. *IEEE Internet of Things Journal, 6*(5), 8182–8201.

4. Farooq, M. S., Riaz, S., Abid, A., Abid, K., & Naeem, M. A. (2019). A survey on the role of IoT in agriculture for the implementation of smart farming. *IEEE Access, 7,* 156237–156271.

5. Neshenko, N., Bou-Harb, E., Crichigno, J., Kaddoum, G., & Ghani, N. (2019). Demystifying IoT security: An exhaustive survey on IoT vulnerabilities and a first empirical look on internet-scale IoT exploitations. *IEEE Communications Surveys & Tutorials, 21*(3), 2702–2733.

6. Bhushan, B., & Sahoo, G. (2020). Requirements, protocols, and security challenges in wireless sensor networks: An industrial perspective. In *Handbook of Computer Networks and Cyber Security* (pp. 683–713). Springer, Cham.

7. Lin, Y., & Chang, J. (2019, July). Improving wireless network security based on radio fingerprinting. In *2019 IEEE 19th International Conference on Software Quality, Reliability and Security Companion (QRS-C)* (pp. 375–379). IEEE.

8. Bhushan, B., & Sahoo, G. (2017, July). A comprehensive survey of secure and energy efficient routing protocols and data collection approaches in wireless sensor networks. In *2017 International Conference on Signal Processing and Communication (ICSPC)* (pp. 294–299). IEEE.

9. Burg, A., Chattopadhyay, A., & Lam, K. Y. (2017). Wireless communication and security issues for cyber–physical systems and the Internet-of-Things. *Proceedings of the IEEE, 106*(1), 38–60.

10. Bhushan, B., Sahoo, G., & Rai, A. K. (2017, September). Man-in-the-middle attack in wireless and computer networking—A review. In *2017 3rd International Conference on Advances in Computing, Communication & Automation (ICACCA)(Fall)* (pp. 1–6). IEEE.

11. Karakaya, A., & Akleylek, S. (2018, March). A survey on security threats and authentication approaches in wireless sensor networks. In *2018 6th International Symposium on Digital Forensic and Security (ISDFS)* (pp. 1–4). IEEE.

12. Bhushan, B., & Sahoo, G. (2017, September). Detection and defense mechanisms against wormhole attacks in wireless sensor networks. In *2017 3rd International Conference on Advances in Computing, Communication & Automation (ICACCA)(Fall)* (pp. 1–5). IEEE.

13. Sharma, M., Tandon, A., Narayan, S., & Bhushan, B. (2017, September). Classification and analysis of security attacks in WSNs and IEEE 802.15. 4 standards: A survey. In *2017 3rd International Conference on Advances in Computing, Communication & Automation (ICACCA)(Fall)* (pp. 1–5). IEEE.

14. Jaitly, S., Malhotra, H., & Bhushan, B. (2017, July). Security vulnerabilities and countermeasures against jamming attacks in wireless sensor networks: A survey. In *2017 International Conference on Computer, Communications and Electronics (Comptelix)* (pp. 559–564). IEEE.

15. Bhushan, B., & Sahoo, G. (2020). ISFC-BLS (Intelligent and Secured Fuzzy Clustering Algorithm Using Balanced Load Sub-Cluster Formation) in WSN environment. *Wireless Personal Communications*, *111*, 1667–1694. doi: 10.1007/s11277-019-06948-0.

16. Wang, N., & Wang, H. (2010, March). A security architecture for wireless mesh network. In *2010 International Conference on Challenges in Environmental Science and Computer Engineering* (Vol. 2, pp. 263–266). IEEE.

17. Katayama, M., Mizuno, K., Nakayama, M., & Shimizu, M. (2003, April). A multi-protocol wireless multi-hop network employing a new efficient hybrid routing scheme. In *The 57th IEEE Semiannual Vehicular Technology Conference, 2003. VTC 2003-Spring.* (Vol. 3, pp. 2013–2017). IEEE.

18. Bhushan, B., & Sahoo, G. (2020). Secure location-based aggregator node selection scheme in wireless sensor networks. In *Proceedings of ICETIT 2019* (pp. 21–35). Springer, Cham.

19. Bhushan, B., & Sahoo, G. (2019). Routing protocols in wireless sensor networks. In *Computational Intelligence in Sensor Networks* (pp. 215–248). Springer, Berlin, Heidelberg.

20. Hassija, V., Chamola, V., Saxena, V., Jain, D., Goyal, P., & Sikdar, B. (2019). A survey on IoT security: Application areas, security threats, and solution architectures. *IEEE Access*, *7*, 82721–82743.

21. Koroniotis, N., Moustafa, N., & Sitnikova, E. (2019). Forensics and deep learning mechanisms for botnets in Internet of Things: A survey of challenges and solutions. *IEEE Access*, *7*, 61764–61785.

22. Vinayakumar, R., Alazab, M., Soman, K. P., Poornachandran, P., & Venkatraman, S. (2019). Robust intelligent malware detection using deep learning. *IEEE Access*, *7*, 46717–46738.

23. Farris, I., Taleb, T., Khettab, Y., & Song, J. (2018). A survey on emerging SDN and NFV security mechanisms for IoT systems. *IEEE Communications Surveys & Tutorials*, *21*(1), 812–837.

24. Din, I. U., Guizani, M., Kim, B. S., Hassan, S., & Khan, M. K. (2018). Trust management techniques for the Internet of Things: A survey. *IEEE Access*, *7*, 29763–29787.

25. Fernández-Caramés, T. M., & Fraga-Lamas, P. (2018). A review on the use of block-chain for the Internet of Things. *IEEE Access*, *6*, 32979–33001.

26. Xin, Y., Kong, L., Liu, Z., Chen, Y., Li, Y., Zhu, H., … Wang, C. (2018). Machine learning and deep learning methods for cybersecurity. *IEEE Access*, *6*, 35365–35381.

27. Cui, Z., Xue, F., Cai, X., Cao, Y., Wang, G. G., & Chen, J. (2018). Detection of malicious code variants based on deep learning. *IEEE Transactions on Industrial Informatics*, *14*(7), 3187–3196.

28 Ahmad, I., Basheri, M., Iqbal, M. J., & Rahim, A. (2018). Performance comparison of support vector machine, random forest, and extreme learning machine for intrusion detection. *IEEE Access*, *6*, 33789–33795.

29 Zhou, W., Jia, Y., Peng, A., Zhang, Y., & Liu, P. (2018). The effect of IoT new features on security and privacy: New threats, existing solutions, and challenges yet to be solved. *IEEE Internet of Things Journal*, *6*(2), 1606–1616.

30. Akcay, S., Kundegorski, M. E., Willcocks, C. G., & Breckon, T. P. (2018). Using deep convolutional neural network architectures for object classification and detection within X-ray baggage security imagery. *IEEE Transactions on Information Forensics and Security*, *13*(9), 2203–2215.

31. Yang, Y., Wu, L., Yin, G., Li, L., & Zhao, H. (2017). A survey on security and privacy issues in Internet-of-Things. *IEEE Internet of Things Journal*, *4*(5), 1250–1258.

32. Chen, L., Thombre, S., Järvinen, K., Lohan, E. S., Alén-Savikko, A., Leppäkoski, H. … Lindqvist, J. (2017). Robustness, security and privacy in location-based services for future IoT: A survey. *IEEE Access*, *5*, 8956–8977.

33. Yu, W., Liang, F., He, X., Hatcher, W. G., Lu, C., Lin, J., & Yang, X. (2017). A survey on the edge computing for the Internet of Things. *IEEE Access, 6*, 6900–6919.

34. Lin, J., Yu, W., Zhang, N., Yang, X., Zhang, H., & Zhao, W. (2017). A survey on internet of things: Architecture, enabling technologies, security and privacy, and applications. *IEEE Internet of Things Journal, 4*(5), 1125–1142.

35. Frustaci, M., Pace, P., Aloi, G., & Fortino, G. (2017). Evaluating critical security issues of the IoT world: Present and future challenges. *IEEE Internet of Things Journal, 5*(4), 2483–2495.

36. Hu, P., Ning, H., Qiu, T., Song, H., Wang, Y., & Yao, X. (2017). Security and privacy preservation scheme of face identification and resolution framework using fog computing in Internet of Things. *IEEE Internet of Things Journal, 4*(5), 1143–1155.

37. Chen, M., Hao, Y., Hwang, K., Wang, L., & Wang, L. (2017). Disease prediction by machine learning over big data from healthcare communities. *IEEE Access, 5*, 8869–8879.

38. Swamy, S. N., Jadhav, D., & Kulkarni, N. (2017, February). Security threats in the application layer in IOT applications. In *2017 International Conference on I-SMAC (IoT in Social, Mobile, Analytics and Cloud)(I-SMAC)* (pp. 477–480). IEEE.

39. Zhou, L., Pan, S., Wang, J., & Vasilakos, A. V. (2017). Machine learning on big data: Opportunities and challenges. *Neurocomputing, 237*, 350–361.

40. Ngu, A. H., Gutierrez, M., Metsis, V., Nepal, S., & Sheng, Q. Z. (2016). IoT middleware: A survey on issues and enabling technologies. *IEEE Internet of Things Journal, 4*(1), 1–20.

41. Mosenia, A., & Jha, N. K. (2016). A comprehensive study of security of internet-of-things. *IEEE Transactions on Emerging Topics in Computing, 5*(4), 586–602.

42. Ozay, M., Esnaola, I., Vural, F. T. Y., Kulkarni, S. R., & Poor, H. V. (2015). Machine learning methods for attack detection in the smart grid. *IEEE Transactions on Neural Networks and Learning Systems, 27*(8), 1773–1786.

43. Chen, X. W., & Lin, X. (2014). Big data deep learning: Challenges and perspectives. *IEEE Access, 2*, 514–525.

44. Pavani, K., & Damodaram, A. (2013). Intrusion detection using MLP for MANETs. Third International Conference on Computational Intelligence and Information Technology (CIIT 2013), 440–444. doi: 10.1049/cp.2013.2626.

45. Galbally, J., Marcel, S., & Fierrez, J. (2013). Image quality assessment for fake biometric detection: Application to iris, fingerprint, and face recognition. *IEEE Transactions on Image Processing, 23*(2), 710–724.

46. Awad, A. I. (2012, December). Machine learning techniques for fingerprint identification: A short review. In *International Conference on Advanced Machine Learning Technologies and Applications* (pp. 524–531). Springer, Berlin, Heidelberg.

47. Bandyopadhyay, S., Sengupta, M., Maiti, S., & Dutta, S. (2011). A survey of middleware for internet of things. In *Recent Trends in Wireless and Mobile Networks* (pp. 288–296). Springer, Berlin, Heidelberg.

48. Chen, S., Xu, H., Liu, D., Hu, B., & Wang, H. (2014). A vision of IoT: Applications, challenges, and opportunities with china perspective. *IEEE Internet of Things Journal, 1*(4), 349–359.

49. Kulkarni, R. V., & Venayagamoorthy, G. K. (2009, June). Neural network based secure media access control protocol for wireless sensor networks. In *2009 International Joint Conference on Neural Networks* (pp. 1680–1687). IEEE.

50. Sinha, P., Jha, V. K., Rai, A. K., & Bhushan, B. (2017, July). Security vulnerabilities, attacks and countermeasures in wireless sensor networks at various layers of OSI reference model: A survey. In *2017 International Conference on Signal Processing and Communication (ICSPC)* (pp. 288–293). IEEE.

51. Otoum, S., Kantarci, B., & Mouftah, H. T. (2019). On the feasibility of deep learning in sensor network intrusion detection. *IEEE Networking Letters, 1*(2), 68–71.

52. Syed, T. A., Alzahrani, A., Jan, S., Siddiqui, M. S., Nadeem, A., & Alghamdi, T. (2019). A Comparative analysis of blockchain architecture and its applications: Problems and recommendations. *IEEE Access, 7*, 176838–176869.

53. De Donno, M., Tange, K., & Dragoni, N. (2019). Foundations and evolution of modern computing paradigms: Cloud, IoT, edge, and fog. *IEEE Access, 7*, 150936–150948.

54. Côté, D. (2018). Using machine learning in communication networks. *Journal of Optical Communications and Networking, 10*(10), D100–D109.

55. Koroniotis, N., Moustafa, N., & Sitnikova, E. (2019). Forensics and deep learning mechanisms for botnets in Internet of Things: A survey of challenges and solutions. *IEEE Access, 7*, 61764–61785.

56. Wang, Y., Wan, J., Guo, J., Cheung, Y. M., & Yuen, P. C. (2017). Inference-based similarity search in randomized Montgomery domains for privacy-preserving biometric identification. *IEEE Transactions on Pattern Analysis and Machine Intelligence, 40*(7), 1611–1624.

57. Gao, Y., Zheng, B., Chen, G., Lee, W. C., Lee, K. C., & Li, Q. (2009). Visible reverse k-nearest neighbor query processing in spatial databases. *IEEE Transactions on Knowledge and Data Engineering, 21*(9), 1314–1327.

58. Jiang, L., Zhang, H., & Cai, Z. (2008). A novel Bayes model: Hidden naive Bayes. *IEEE Transactions on Knowledge and Data Engineering, 21*(10), 1361–1371.

59. Martens, D., Baesens, B. B., & Van Gestel, T. (2008). Decompositional rule extraction from support vector machines by active learning. *IEEE Transactions on Knowledge and Data Engineering, 21*(2), 178–191.

60. Chandra, B., & Varghese, P. P. (2008). Fuzzy SLIQ decision tree algorithm. *IEEE Transactions on Systems, Man, and Cybernetics, Part B (Cybernetics), 38*(5), 1294–1301.

61. Wang, X., Wang, S., Liang, J., & Xu, B. (2008, March). Improved phonotactic language identification using random forest language models. In *2008 IEEE International Conference on Acoustics, Speech and Signal Processing* (pp. 4237–4240). IEEE.

62. Lucas, A., Williams, A. T., & Cabrales, P. (2019). Prediction of recovery from severe hemorrhagic shock using logistic regression. *IEEE Journal of Translational Engineering in Health and Medicine, 7*, 1–9.

5 Privacy Preservation with Machine Learning

P. Bhuvaneswari and Nagender
Kumar Suryadevara
University of Hyderabad

CONTENTS

5.1 INTRODUCTION

Privacy-preserved IoT system means where a system can do its operations efficiently without compromise in privacy against an adversary. An adversary can be the cloud service provider (CSP), a dishonest (malicious) node, or a semi-honest node (an associated node leaking information). The following real-time example shows the details of how an IoT environment is easily affected by adversary attacks.

In 2016, DDoS (distributed denial service) attacks [1] at a DNS (domain name server) service provider severely disrupted websites such as Amazon, Twitter, and Facebook. More than 5000 IoT devices in a university campus were used to launch attacks on DNS queries. The IoT devices are targeted by severe malicious attacks due to the lack of standard privacy frameworks and privacy data standardizations.

IoT plays a vital role in the field of healthcare. The IoT-based health system provides timely health monitoring such as pulse, blood pressure, and temperature. This data collected from the patients helps the doctor in making effective diagnosis. At the same time, data is prone to serious identity leakages of the patients. As a real-time example [2], in 2017, Kromtech Security Researchers discovered a data leakage problem on Amazon S3 repository that contained medical reports in the form of weekly blood test results and about 150,000 people were affected by the leak. The privacy leakage included the identity of patients.

So, privacy becomes a significant concern with the rapid deployment of IoT applications. An IoT application (e.g., smart home-assistive living [3] and smart building [4]) includes user's private data such as personal information and sensitive measures to deliver effective and reliable services. But IoT users are concerned about the protection of their private information (e.g., name, email, and medical and financial records) collected by IoT devices. To procure the trust of IoT users, IoT technologies must guarantee an individual's privacy. That is, the control and management of private data should be done according to user's choice. An IoT environment must preserve (i.e., follow their choice as it is) user's privacy specifications during its storage, communication, and processing of data. The privacy frameworks should be able to give its users assurance to maintain privacy preservation. In terms of transparent inferences, i.e., what an IoT device learns about them and how the utilization of user data further improves the accuracy of decision-making in an IoT system. This creates awareness among users to understand and opt for their privacy preferences.

Despite the existing security solutions, such as authentication, confidentiality, access control, and encryption, IoT networks are still prone to security and privacy attacks. IoT systems security differs from traditional systems security [5] due to the following reasons:

- Resource constraints on computational capability, battery life, network bandwidth, and memory capacity. However, the existing security solutions are more resource-intensive;
- Distributed and heterogeneous systems add more difficulties and constraints in their protection;
- Deployment in an unpredictable physical environment may create a scope for new types of physical attacks;
- Each device is connected to the Internet with its IP address. Hence, it is prone to more threats related to the Internet;
- Limited network bandwidth of the IoT network can be easily flooded and attacked;
- An IoT system that uses heterogeneous technologies and protocols must be considered in the proposed security solutions.

Hence, the aforementioned IoT environment challenges should be taken into account while choosing/building the IoT systems security solutions.

Recent works [6–8] focused on IoT system security. The authors of Ref. [6] presented ML use in identifying major vulnerabilities and challenges faced by IoT; the authors of Ref. [7] focused on DL models and their characteristics in IoT applications, and ML algorithms to overcome the popular issues faced by cybersecurity [7]. The recent works reviewed the application of various ML/DL techniques in terms of various datasets, and the techniques that are used to integrate ML in security systems that give good results are emphasized. This chapter presents various types of security attacks and their influence on IoT applications in detail in Sections 5.2–5.4.

An IoT system performs data analytics efficiently as per the application requirements. The data analytics takes user data as input and identifies the required action. To identify the objectives efficiently, a data analytics system (like CSP) collects huge data samples. But the exposure of the user's private data to CSP ends with privacy, which is to be compromised. Further, a more dangerous threat can be possible if the service provider itself acts as a malicious node by revealing data to third parties. For example, in medical diagnosis, users do not need machine learning (ML) algorithms to remember their sensitive information in training set as a specific detail of individual patients. Therefore, recently much work is proposed for security and privacy preservation with ML and deep learning (DL) techniques. Using these techniques, an IoT system can evaluate the classification/recognition of data without disclosing the original data and make an efficient decision.

This chapter aims to provide an understanding of the advantages of implementing ML approaches (GAN (generative adversarial network), LSTM, and SMC (secure multi-party computation) at various levels of the IoT domain to have privacy preservation. In turn, the essence of the ML can further improve the system with privacy preservation. Once the system is learned privacy preservation, the system can adapt to changes if unknown relations adds in the future. We presented a trade-off in Section 5.5 to understand the state of art of ML approaches for privacy preservation. To evaluate the privacy preservation with ML, we conducted experiments using TensorFlow Privacy. The results show that accuracy can be achieved with an acceptable privacy guarantee.

The structure of the remaining sections of this chapter is organized as follows. Section 5.2 recalls various types of attacks in IoT and WSNs. Section 5.3 presents the privacy-related limitations at the IoT smart devices and wireless communications in terms of real-time examples. Section 5.4 provides an introduction to how privacy is incorporated in wireless communication technologies (Wi-Fi, BLE (Bluetooth Low Energy), ZigBee, 6LoWPAN, and LoRaWAN). Section 5.5 presents the privacy-preserved implementations using various AI (artificial intelligence) methods. Then, Section 5.6 provides critical observations of the presented models. Section 5.7 provides the details about the experimental setup and analysis of privacy preservation with ML. Section 5.8 concludes with a summary of ML methods and their adaptability to IoT privacy.

5.2 VARIOUS TYPES OF SECURITY ATTACKS IN IOT AND WSNs

WSNs are helpless against a wide scope of attack types, which may put basic dangers to the security of the systems. Security-related assaults against WSNs can be divided into two fundamental classes: dynamic assaults and latent assaults. In the dynamic assaults class, assailants are commonly covered up (disguised) and either harm the working segments of the system or tap the interchanges connection to gather valuable data. However, the latent assaults can likewise be additionally sorted into listening in, hub breaking down, hub blackout, and traffic investigation types. In the dynamic assaults class, an assailant impacts the capacities and activities of the focused on arrange. The after-effect of this evil impact can be the genuine goal of the assailant and can likewise be distinguished by security instruments (interruption discovery). The remedies for the security attacks on WSN and IoT can be handled in three different stages – Avoidance: This part targets forestalling assaults before they occur. Strategies should have the option to devise measures to guard against the particular sort of attack(s). Interruption avoidance components can oppose outer assailants towards WSNs and IoT, yet they are not explicitly intended to oppose the inner aggressors. Recognition: In an occasion of an assault, if an enemy figures out how to propel the measures taken by the avoidance part, this implies resistance against the assault has fizzled. Right now, security arrangements that are concocted for the discovery part of the related assault would take in-control and work at particularly in recognizing those hubs that are undermined. Alleviation: Last part targets relieving assaults after they occur, for instance, so as to secure system, a safety effort ought to be taken, e.g., "rejecting the influenced hubs in a system" or, on the other hand, "impairing the ports of a system which were utilized during the assault". In this way, all these three parts constitute an entirety security structure and can't be considered independently in guarding WSNs and IoT against different sorts of assaults.

The origin of various types of security attacks in IoT and WSN was elaborated in Ref. [9,10]. In Ref. [9], Meneghello *et al.* identified major vulnerabilities in the IoT domain at information level, access level, and functional level. They classified attacks against IoT devices implementing a specific technology, such as key-related attacks, DoS attacks to the data plane, DoS attacks to the device, replay attacks, and attacks of the privacy of the communications. B. Bhushan and G. Sahoo in Ref. [10] presented an in-depth analysis of various survivability requirements for WSN-related attacks that can affect security. Data confidentiality and integrity-related attacks, power or energy consumption-related attacks, bandwidth consumption and service availability-related attacks, routing-related attacks, identity-related attacks, and privacy-related attacks are considered as the part of their classification. A survey in Ref. [11] presented the possible attacks and their countermeasures in IEEE 802.15.4 standard for WSNs.

However, our work focuses on specific problems in the WSNs and IoT domain related to privacy loss due to the existing known vulnerabilities and security attacks in the WSNs and IoT domain, such as (1) device scan attacks, device tracking, and user activity profiling attacks; (2) key-related attacks; and (3) traffic analysis attacks of the privacy of the IoT applications. Detailed problems and their scenarios are discussed in Sections 5.3 and 5.4.

5.3 LIMITATIONS OF IMPLEMENTING PRIVACY FEATURES TO THE IOT SMART DEVICES

In the IoT environment, attacks have higher impacts on privacy loss than information system security. The attacker might know the victim's personal information and use this information in physical harm like turning off the fire alarm. Thus, it is mandatory to implement the privacy aspects in the IoT smart environment.

- Healthcare applications using WSN are not yet deployed practically due to the cost and complexity involved in managing IoT technologies on the top of health devices [12];
- A sensor is used to measure physical readings like temperature in a room. Readings are either regular or interpretable variations. In the case of a temperature sensor, a sudden change can be observed when a window is opened. An attacker can inference this change and can identify that type and IP addresses of a sensor [13]. Later, this partial information could be used in cracking encryption keys [14];
- An adversary can use MAC address (first 24-bits) to identify the organizationally unique identifier and hence the device manufacturers based on traffic analysis [15,16];
- The use of persistent identifiers becomes a threat to the identity of a node's privacy [16];
- The communication using IoT devices is mostly to specific servers like that device manufacturer. A single flow is sufficient for an attacker to inference activity [17].
- Traffic size reveals the traffic characteristics in the network. This information can, in turn, be exploited to determine the private activities of the home occupants;
- Data integrity attacks on IoT devices make modified data useful for malicious purposes, and hence, privacy is affected;
- Many individuals are reluctant to participate in data collection due to:
  - Data collected by sensors about utility consumption could reveal daily activities of a household, such as when all family members are gone or when someone is taking a shower, which in turn can reveal their identity [18]. Hence, we need a way to collect the privacy-preserved sensor readings.
  - Data collected by health services like weight information is useful in an obesity study. But this could lead to privacy loss of sensitive data of an individual.

    Hence, data collection or data processing of AI techniques must ensure equal priority both in derivable properties and in privacy preservation.
- End-to-end encryption or link-level encryption alone is not a good method for private data protection. This is because
  - End-to-end encryption makes data aggregation difficult at intermediate nodes;

- Any encrypted communication is breakable and private data of a user is violated;
- Though proven encryption methods guarantee data security, it is still prone to privacy attacks by in-network adversaries.
- The privacy preservation approaches using cryptography are not readily applicable to privacy requirements in IoTs. Due to its computational cost, this approach is not suitable for the resource-constrained WSNs.

5.4   ROLE OF WIRELESS COMMUNICATION TECHNOLOGIES IN IOT PRIVACY

Based on the application area, the IoT environment uses different wireless protocols like Wi-Fi, which is used for the devices accessing the Internet. Bluetooth or BLE is used for wearable devices (e.g., smartwatch, blood pressure monitor). ZigBee is used in smart home appliances (e.g., smart lock), and Z-wave protocols are used in electronic appliances (e.g., light and fan). So, the study about the characteristics of wireless technologies can provide a better understanding of the performance of WSN applications. Hence, we discuss the details of IoT wireless technologies such as Wi-Fi, BLE, ZigBee, 6LowPAN, and LoRaWAN in the following.

5.4.1   Wi-Fi

The IEEE 802.11 a/g/n/ac standards are the most popular in IoT personal area networks like smart home networks. The latest standard IEEE 802.11ac can support a throughput of 1 Gbps. The Wi-Fi standards use encryption methods to keep the communication private. The original wireless security protocol, WEP (wired equivalent privacy), makes wireless networks extremely vulnerable to outside threats. Newer Wi-Fi devices, wireless access point, or router supports more secure protocols WPA, WPA2, or WPA3, as described in Figure 5.1. In real-time scenarios, attackers located outside a house with a wireless laptop can identify websites visited by a person using a home network. Though, the home network Wi-Fi is enabled with Wi-Fi encryption modes. It is worth to note that we do not rely on breaking the encryption algorithm in this scenario. In another real-time scenario, where Wi-Fi communication is secured using Wi-Fi Protected Access 2 (WPA2), it has also been vulnerable to attacks (e.g., the Key Reinstallation AttaCK (KRACK)) [19]. WPA2 is a strong encryption method to secure Wi-Fi communication and uses advanced encryption standard (AES) for encryption of data.

The main weakness of WEP is its use of static encryption keys. An eavesdropper can collect a large enough sample of your transmission stream to derive the WEP encryption key, which allows them to decrypt everything being transmitted by you. WPA is more secure than WEP, and it uses an enhanced encryption protocol called TKIP (Temporal Key Integrity Protocol). WPA key generation includes the shared key and network's name (or SSID) for each client. Keys are refreshed to secure from the WEP attackers. WPA2 [20] is a much better version of Wi-Fi secure protocol than WPA for the stronger security against attacks. WPA2 has stronger encryption and authentication mechanisms, AES and counter mode with Cipher

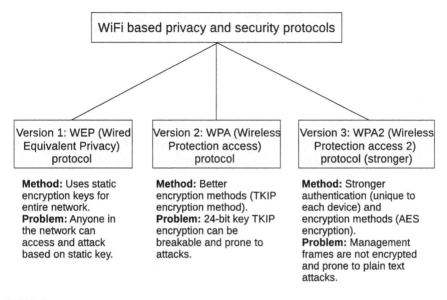

FIGURE 5.1   IEEE Wi-Fi secure protocols support for security and privacy.

Block Chaining Message Authentication Code Protocol (CCMP). WPA2 maintains TKIP for backward compatibility.

However, further security problems persist in Wi-Fi communication. The management frames are sent in clear text without encryption, which makes it easier for the attacker to identify devices present in the IoT environment. For example, home occupancy can be identified with the active devices in the home environment by frequency of Wi-Fi probe request messages [21]. This frequency of probes can be used to accurately track the presence or absence of the device owner at home.

5.4.2   BLE (BLUETOOTH LOW ENERGY)

BLE was merged into the Bluetooth 4.0 onward [22]. BLE standard provides a means for device security and privacy mechanisms through encryption. BLE enables data protection through authentication, confidentiality, and integrity. Especially, data privacy is addressed through a random addressing scheme. BLE prevents the identification of a BLE device by randomly changing MAC addresses so that others can't track the devices based on their addresses.

However, BLE has drawbacks [23]: (1) Even if BLE provides a random addresses scheme, a malicious node can use a replay attack and can identify a BLE device, and (2) there is no provision to access privacy policies of BLE devices. In [23], these two problems are resolved by providing a user-friendly application to users so that a user can express their preferences for the received privacy policies through GATT (The Generic Attribute Profile) services. GATT hosts a service to provide a communication protocol ATT (Attribute Protocol) for BLE devices to exchange all user and profile data over a BLE connection service. Using the application, IoT devices first

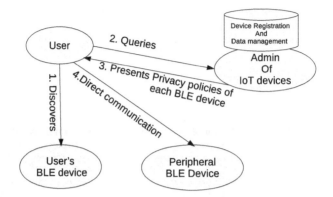

FIGURE 5.2 Overview of user-friendly privacy framework (PrivBat).

obtain privacy policies of the nearby devices, which were added through registration and management interface. BLE's devices information and privacy policies are hosted on a server by the admin of IoT devices, as shown in Figure 5.2. Then, communication starts according to the privacy policies.

Hence, the work mentioned in Ref. [23] proposes a framework that helps to increase user trust in IoT applications with BLE devices. Limitation in this approach is that this framework applies to BLE devices. It can't address legacy of Bluetooth and other wireless technologies. This problem can be solved by introducing a gateway in the network for connecting non-BLE devices. Then, the gateway provides the privacy preference expression GATT services on behalf of non-BLE devices.

5.4.3 ZigBee

ZigBee is a widely used communication protocol with short-range data transfer, such as smart light switches/bulbs, smart meters, and traffic management systems. ZigBee provides privacy and authentication through encryption and authentication using the AES cryptography algorithm in the Counter with CBC–MAC (CCM) mode [9]. Each ZigBee network includes a trust center to authenticate and share encryption keys. Though the messages are encrypted with the AES algorithm, the keys are transferred as plaintext and hence prone to plaintext attacks. Besides, ZigBee shares key among all the devices in the same network. So a malicious node listening in the network range can easily attack. ZLL (ZigBee Light Link) standard [24] is another variant of ZigBee. It distributes key only to certified manufacturers; ZLL still fails in key distribution against attacks [9].

Since ZigBee is the most used technology in IoT smart homes, e.g., lights, Morgner et al. [25] investigated the smart light system and found that different types of attacks are possible on these devices that are connected to a network using ZigBee.

Further, research proposed the methods on enhancements to cryptography key distributions for ZigBee network communication [9]. However, tools like KillerBee make it possible to sniff traffic in the ZigBee network and can perform attacks.

5.4.4   6LowPAN

6LowPAN (IPv6 over Low-Power Wireless Personal Area Networks) [14,26] protocol is predominantly used for short-range distance communications along with ZigBee and BLE. IPSec is used to provide authentication and encryption, which is on the top of the 6LoWPAN header. But Stateless Address Autoconfiguration (SLAAC) of IPv6 addressing mechanism uses device MAC address, which makes easier to learn the capabilities of that node through IP sniff and hence creates privacy threat [27,28]. Lack of privacy and security mechanisms at the 6LoW-PAN link layer makes this communication network vulnerable to a vast number of attacks. Therefore, in 6LoWPAN networks, gateways at each network need to provide authentication and encryption methods.

5.4.5   LoRaWAN

LoRaWAN (Long-Range Wide Area Network) [29] is a new wireless communication protocol for long-range distance (up to 14 km coverage) communication, like smart city applications. Payload messages are encrypted with AES-128-bit key, and encryption keys (NwkSKey and AppSKey) are stored in the end device [30]. Since all the keys are stored in the end devices, a side-channel attack can be possible to recover the keys [31]. LoRaWAN network server is another possibility to store the keys. However, a nonsecure service running on the server may easily work as a door to access the LoRaWAN network [31].

5.4.6   OBSERVATIONS

The abovementioned wireless technologies (Sections 5.4.1–5.4.5) are used in securing privacy during communications by incorporating various security mechanisms for various wireless communication protocols. So, each wireless protocol discussed in Sections 5.4.1–5.4.5 is vulnerable to various attacks and hence results in privacy leakages. Vulnerability is attributable to the design specifications of a communication protocol and the nature of wireless radios. Therefore, it is important that IoT devices and its wireless communication network in an IoT environment should thoroughly evaluate various scenarios against attacks before the deployment of IoT services/applications on smart devices.

So, we propose that extensive testing on the IoT environment can utilize the benefit of ML. ML and DL techniques are well known for accurate anomaly detection and classification of attacks. The IoT system testing can be done either as an offline process or as a preprocess (even before actual services start real-time functioning of the actual system). AI-based IOT scanner [32] and WADAC [33] can be used as a testing tool that is implemented for the anomaly detection and classification of attacks.

Figure 5.3 shows IEEE 802.11 link-layer frame format [34], and we can observe that the frame header fields are in plaintext with payload encrypted. The header information such as source and destinations (MAC addresses, frame size, and type) is again used in most of the attacks, which can be utilized as data features required for ML algorithms [32,33].

Frame Control	Duration	Sender address	Dest. address	Reciever AP address	seq. control	Transmitter AP address	QoS control	HT control	encrypted payload	CRC-32 checksum
16-bits	16-bits	48-bits	48-bits	48-bits	16-bits	48-bits	16-bits	16-bits	variable	32-bits

FIGURE 5.3   IEEE 802.11ac MAC data frame format with encrypted network-layer load.

IoT scanner [32] is a traffic analyzer that uses simple sent-to-received bytes ratio to identify the type of devices that are active in the network and communication. IoT scanner works for the IoT networks connected with Wi-Fi, BLE, and ZigBee. It scans the network passively and identifies active devices based on the analysis of sent-to-received traffic (i.e., MAC address of frames). So we propose that this can be used as a simple testing for IoT privacy leakage in terms of device identity.

Another tool is WADAC [33] that detects and classifies attacks based on traffic features (MACs, frame size, type, etc.). In this work, anomaly (attacks) detection uses unsupervised learning because the training data is unavailable for unknown anomalies. So WADAC uses an unsupervised deep auto encoder algorithm to detect anomalies. Once there is an anomaly detected, the types of attack are classified based on the traffic features. So we propose that WADAC can be used to identify possible attacks that can create harm to user privacy before the actual functioning of IoT network communication.

5.5   PRIVACY PRESERVATION WITH MACHINE LEARNING

ML algorithms work by studying a lot of data and updating their parameters to process the relationships in that data. Unfortunately, ML algorithms do not learn to ignore personal specifics by default.

So recent research concentrated on privacy-preserving data analytics by masking the original data using ML and DL mechanisms. Differential privacy (DP), MPC, GAN, and long short-term memory (LTSM) networks are some of the recent active research proposals.

5.5.1   DIFFERENTIAL PRIVACY

DP is a statistical and ML framework that was designed to protect privacy, which was invented by Cynthia Dwork, Frank McSherry, Kobbi Nissim, and Adam Smith [35]. The basic idea is to add noise to the input behavior in order to provide privacy. DP mathematically guarantees that anyone seeing its result will essentially make the same inference about any individual's private information, as shown in Figure 5.4. Whether or not that individual's private information is included in the input to the analysis [35]. It provides a mathematically provable guarantee of privacy protection since it can mask all original samples.

Mathematical statistics-based methods such as top coding, bottom coding, data swapping, adding random noise, and multiple imputation methods have the following limitations.

Limitations of simple masking techniques/challenges in data masking are as follows:

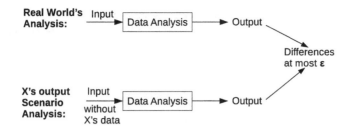

FIGURE 5.4 Differential privacy.

1. If the masking mechanism fails even for a small amount of data, it will result in the risk of privacy leak. Small unchanged data could be the sensitive personal information.
2. Or even if unchanged data is nonsensitive, there is a possibility that the adversary could infer the sensitive data from nonsensitive data. Is it suitable for various types of data (e.g., numerical data, images, text, and video)?
3. The addition of too much noise can lead to a loss of accuracy in analysis. So DL-based mechanisms are recently proposed as an alternative to data masking for privacy preservation during data processing analysis.

5.5.2 GENERATIVE ADVERSARIAL NET (GAN)

GAN [36] is a mechanism of generating simulation data, which has the same statistical properties as the actual data. A basic GAN-generated simulation data is completely different from raw data. A GAN model consists of two components:

1. A generator that generates simulation data and 2. a discriminator (D) to identify whether data is original or not. So the training process finishes once the discriminator can't identify the generated data as the original (e.g., if the probability of D is equal to 0.5, then discrimination is difficult). Multiple types of GAN models are proposed in recent research such as deep convolutional GAN, least-squares GAN, and Wasserstein GAN.

In [37], the authors used the WGAN (Wasserstein GAN) model and proved theoretically that their model with stochastic gradient descent (SGD) optimizer helps to improve the quality of generated data when applied on numerical data. To test the effectiveness of WGAN, WGAN is compared with GAN encoder–decoder and DP. The summary of comparisons using experiment result analysis is as follows: (a) All three models can preserve data privacy by masking the entire original data; (b) the generated data is different from the original data in all three compared models; (c) but only WGAN statistical characteristics are the same as the original dataset and the statistical characteristics of the remaining two models are different from the actual dataset. The experiment uses simulated datasets (for the census and the environmental simulation for training of model and testing).

Usage and Privacy Trade-Off: WGAN models are preserving privacy through generated data, and the generated data can be successfully used as a training dataset for ML analysis.

In [37,38], the authors observed that encoder–decoder using the GAN model has a drawback of lossy compression, which creates a difficulty in constructing data back while decoding. So this is a helpful observation for validating DL encoder–decoder models.

5.5.3 A DEEP LEARNING APPROACH FOR PRIVACY PRESERVATION IN AMBIENT ASSISTED LIVING (AAL)

This paper [39] focuses on a LSTM encoder–decoder, which is a recurrent neural network (RNN) DL algorithm. LSTM is used to create different encoded views for AAL's data based on the privacy level of information. The end user is given access at the privacy level. This mechanism uses DL's ability (i.e., using multiple hidden layers) to transform the original data into higher-level abstract expressions (Figure 5.5). This neural network model consists of three types of privacy operations that are encoded at the sender of application and decoded at the receiver of application: (1) Data from each entry can be completely revealed to the receiver, (2) generalized (semantically changes the value), or (3) deleted. Here in this assisted living model, receivers are a family member, doctor, caregiver, and researcher. Each will have separate predefined views and a different decoder output based on their privacy level of access. For example, output sentence for "John has Alzheimer" after deletion and generalization operations is "* has Dementia". The encoder–decoder model using LSTM ensures that a right decoder can only decode the corresponding view-related information. Hence, user's privacy is preserved. The experiment uses simulated datasets (for AAL environment) and simulation for training of the LSTM model. Experiment results show that the average error rate in the decoder is 1.5 characters for an entry with a length of 160 characters. The LSTM network with 256 hidden layers achieves a learning rate of 0.0004. Hence, learning accuracy is better, but the error rate may affect the reconstruction of sensitive data.

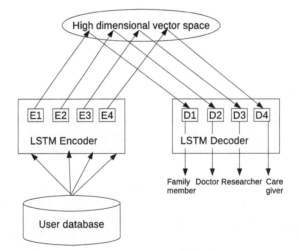

FIGURE 5.5   Assisted living system using multilayer LSTM DL framework.

Usage and Privacy Trade-Off: Both parameters can be realized in LSTM-based encoder and decoder. Where encoding user information provides privacy, the utility is achieved at the receiver by reconstructing this encoded information based on the privacy level of a receiver.

5.5.4 PRIVACY-PRESERVING DEEP LEARNING USING SECURE MULTI-PARTY COMPUTATION (SMC)

Collaborative learning [40] improves the accuracy of the neural network model for multiple parties without sharing input datasets for a given objective. SMC [41] uses the parallel nature of the SGD algorithm for this collaborative learning, so that individual participants can train independently with their datasets and then share some selected small subsets of key parameters for collaborative training.

Selective sharing of parameters is the main key feature of the SMC framework. The choice of SGD optimizer in neural network training enhances the benefit of parallelization and improves the performance of the training process by interleaving shared parameters and updated local parameters. Additionally, SMC uses DP to mitigate the privacy loss due to shared parameters. Besides, DP is applied to avoid privacy loss in choosing which parameters to share as well as to avoid privacy loss due to shared parameter values.

The result says [41] that the SMC works at a low-performance cost in comparison with lightweight cryptographic techniques (secure MPC or homomorphic encryption). So it is suitable for DL models. Experiments are evaluated on practical systems that used benchmark datasets (images of handwritten SVHN dataset and house numbers MNIST dataset). The convergence of learning is faster with multi-party participation than individual party stand-alone learning. Accurate results can be achieved as the parameter sharing rate (0.001 to 0.1) increases; accuracy is reaching the percentage of 99.14 for MNIST and 93.12 for SVHN.

The benefits of the SMC framework are as follows: (a) It addresses privacy threats associated with DL through DP, (b) it enables participants to control their training model dataset, and (c) the learning updates are directly available with the data owners. Also, SMC is the first work proposed [41] for privacy-preserved ML approach where it brings the benefits of DL to privacy preservation.

Usage and Privacy Trade-Off: Here, a better utility is achieved since a participant's learning accuracy can be improvised by benefiting from other models without losing the data privacy.

5.6 TRADE-OFF BETWEEN THE PRIVACY, USAGE, AND EFFICIENCY OF PRESENTED MODELS

Table 5.1 summarizes the pros and cons encountered in privacy preservation and usage of the learning model as a trade-off between the privacy, usage, and efficiency measures. Here, the efficiency is given as a learning speed of AI frameworks presented in Section 4. Values mentioned in efficiency column are presented from their respective research papers [37,39,41].

TABLE 5.1

Trade-off between the Privacy, Usage, and Efficiency for WGAN, LSTM, and SMC Models

Framework	Privacy (Leakproof Data Processing)	Usage (Learning-Relevant Parameters)	Efficiency (Speed of Learning)
WGAN [37]	Privacy preservation is achieved through data masking for all the original samples with the GAN-generated data	Meets the learning objective requirements since the generated data samples are reflecting features as the original data is	Difference between the original and generated data is compared using the measures, ARR, and HD values (5.5679, 0.009 for census and 17.1973, 0.134 for PM2.5 datasets [37]). With smaller score values, WGAN results in better performance
LSTM [39]	Privacy preservation through encoder–decoder sometimes may create lossy compression, which creates a difficulty in constructing data back [38]	Partially meets the learning objective since the decoder cannot fully learn the features of raw data with the average error rate in decoder, which is 1.5/160 characters per entry	In case the decoder cannot fully learn the mandatory features from generated data, then it also cannot transform back into real data. But the given learning rate of 0.0004 in [39] shows good performance
SMC [41]	Partially achievable since there are proven privacy threats available [42]. The collaborative learning approach has the basic drawback in the security of training sets	Meets the learning objective requirements. As multiple parties share their parameters, it helps access to better parameter, and learning in turn can achieve better results	Convergence of learning is faster with multi-party than stand-alone learning. Hence, accurate results can be achieved as the parameter sharing rate (0.001–0.1) increases; accuracy is reaching the percentage of 99.14 for MNIST and 93.12 for SVHN [41].

In this section, we have chosen three recently proposed DL frameworks [37,39,41] for privacy preservation. So each one uses different DL mechanisms such as GAN, CNN (convolutional neural network), and multilayer perceptron (MLP) networks. We focused on complete coverage of data for privacy preservation; preserved data must not lose its role in enhancing data processing goals and then their efficiency in processing speed. As given in Table 5.1, the GAN framework meets all three requirements. However, SMC and LSTM suffer from their fundamental design loopholes although its accuracy results are satisfying.

The headers of Table 5.1 represent the following:

Privacy: Unrevealing the original data during training.
Usage: Including training data set that can meet training objective.
Accuracy: Correctly identifying the objective of the neural network model (e.g., classification/detection accuracy).
Efficiency: Speed of learning.

5.7   EXPERIMENTAL SETUP AND ANALYSIS

To evaluate the privacy preservation with ML, we conducted experiments using TensorFlow Privacy [43]. TensorFlow Privacy library supports DP optimizer to train ML models. It uses differentially private stochastic gradient descent (DPSGD) optimizer as given in [44] on the standard MNIST dataset [45]. The MNIST handwritten digit recognition dataset contains 60,000 training examples where each example is a 28×28 size gray-level image.

The DP can be expressed using two computed values (ε, δ).

This pair of privacy value represents the following:

- ε: Epsilon gives a bound on the probability of output that can increase by including or removing a single training example. It can be a small constant value between 0 and 1 for privacy guarantees. However, epsilon up to 10 may still mean good practical privacy.
- δ: Delta gives a bound on the probability of change in model behavior. This is usually set to a very small number (e.g., $1e^{-5}$). Thumb rule to set delta is by taking a value less than the inverse of the total training data size (i.e., 60,000 for MNIST), which is $1e^{-5}$. We followed the approach given in [43] to compute epsilon value by keeping the delta value fixed. We measure the accuracy to find the impact of the privacy-preserved model.

For the training of DL models, a CNN on MNIST with the DPSGD is used in the experiments, and they are executed on the TensorFlow software library for ML. Our experiments show 96.2% training accuracy with privacy values

$$(\varepsilon, \delta) = (2.68, 10^{-5}).$$

As the number of epochs increases, the accuracy of the CNN model with DPSGD increases and the corresponding ε increases.

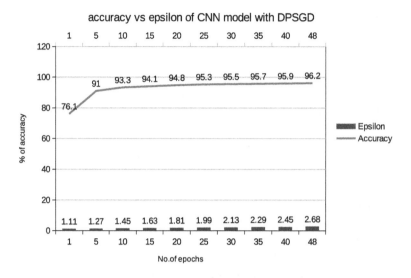

FIGURE 5.6   Training accuracy for privacy-preserved CNN model with DPSGD.

The graph in Figure 5.6 shows the DP model with acceptable privacy preservation. The experimental details can be observed as given in the graph. From the graph, the accuracy and a corresponding ε value are computed for each epoch up to 48 epochs with a δ value fixed. Our experiment result shows 94.1%, 95.3%, and 96.2% test set accuracy for epochs 15, 25, and 48 with DPSGD, respectively. Epsilon values appear less than 3 (1.63, 1.99, and 2.68) at an acceptable level, which meets theoretical guarantees provided by DP.

5.8   CONCLUSION

The IoT environment should provide a balance between the benefits and the values that give people the freedom of control over their personal information with the benefits of smart services. In this chapter, we have provided the intricacies persisting in the implementation of AI methods to have a trust mechanism for the IoT theme applications.

The dynamism of the IoT framework at the sensing systems, the communication mechanisms, and the data processing levels have distinct characteristics that enable us to have appropriate privacy preservation model. As ML is "data-hungry", it can adversely influence data privacy for effective realization, and they are also necessities for making sense and significance in how to have the privacy for IoT data.

In the near future, we plan to conduct a detailed implementation of the privacy mechanism for collaborative learning in terms of distributed and federated learning so that the most proficient approach to impart the privately learned models can be used by different entities without releasing sensitive data. The exploration will also be done for learning the distribution from a dataset "privately".

REFERENCES

1. Verizon, 2016. Data Breach Investigations Report. [online]. Available: https://conferences. law.stanford.edu/cyberday/wp-content/uploads/sites/10/2016/10/2b_Verizon_Data-Breach- Investigations-Report_2016_Report_en_xg.pdf (accessed October 20, 2019).

2. Kromtech, 2017, October 10. Patient Home Monitoring Service Leaks Private Medical Data Online. [online]. Available: https://www.kromtech.com/blog/security-center/ patient-home-monitoring-service-leaks-private-medical-data-online (accessed October 20, 2019).

3. N. K. Suryadevara, A. Negi, and S. R. Rudraraju, 2019. A Smart Home Assistive Living Framework Using Fog Computing for Audio and Lighting Stimulation. In: Satapathy, S., Raju, K., Shyamala, K., Krishna, D., and Favorskaya, M. (eds) *Advances in Decision Sciences, Image Processing, Security and Computer Vision. ICETE 2019. Learning and Analytics in Intelligent Systems*, vol. 3. Springer, Cham. doi: 10.1007/978-3-030-24322-7_47.

4. S. Sarkar, R. Wankar, S. N. Srirama, and N. K. Suryadevara, 2020. Serverless Management of Sensing Systems for Fog Computing Framework, *IEEE Sensors Journal*, vol. 20, no. 3, pp. 1564–1572. doi: 10.1109/JSEN.2019.2939182, http://ieeexplore.ieee.org/stamp/stamp.jsp?tp=&arnumber=8822951&isnumber=8962361.

5. N. Chaabouni, M. Mosbah, A. Zemmari, C. Sauvignac, and P. Faruki, 2019. Network Intrusion Detection for IoT Security Based on Learning Techniques, *IEEE Communications Surveys & Tutorials*, vol. 21, no. 3, pp. 2671–2701. doi: 10.1109/COMST.2019.2896380, http://ieeexplore.ieee.org/stamp/stamp.jsp?tp=&arnumber=8629941&isnumber=8809933.

6. M. Jindal, J. Gupta, and B. Bhushan, 2019. Machine Learning Methods for IoT and their Future Applications, *2019 International Conference on Computing, Communication, and Intelligent Systems (ICCCIS)*, pp. 430–434. doi: 10.1109/ICCCIS48478.2019.8974551, http://ieeexplore.ieee.org/stamp/stamp.jsp?tp=&arnumber=8974551&isnumber=8974457.

7. R. Tiwari, N. Sharma, I. Kaushik, A. Tiwari, and B. Bhushan, 2019. Evolution of IoT & Data Analytics Using Deep Learning, *2019 International Conference on Computing, Communication, and Intelligent Systems (ICCCIS)*, pp. 418–423. doi: 10.1109/ICC-CIS48478.2019.8974481, http://ieeexplore.ieee.org/stamp/stamp.jsp?tp=&arnumber=8974481&isnumber=8974457.

8. S. Soni and B. Bhushan, 2019. Use of Machine Learning Algorithms for Designing Efficient Cyber Security Solutions, *2019 2nd International Conference on Intelligent Computing, Instrumentation and Control Technologies (ICICICT)*, pp. 1496–1501. doi: 10.1109/ICICICT46008.2019.8993253, http://ieeexplore.ieee.org/stamp/stamp.jsp?tp=&arnumber=8993253&isnumber=8993111.

9. F. Meneghello, M. Calore, D. Zucchetto, M. Polese, and A. Zanella, 2019, October. IoT: Internet of Threats? A Survey of Practical Security Vulnerabilities in Real IoT Devices, *IEEE Internet of Things Journal*, vol. 6, no. 5, pp. 8182–8201. doi: 10.1109/JIOT.2019.2935189, http://ieeexplore.ieee.org/stamp/stamp.jsp?tp=&arnumber=8796409&isnumber=8863548.

10. B. Bhushan and G. Sahoo, 2018. Recent Advances in Attacks, Technical Challenges, Vulnerabilities and their Countermeasures in Wireless Sensor Networks, *Wireless Personal Communications*, vol. 98, no. 2, pp. 2037–2077. doi: 10.1007/s11277-017-4962-0.

11. M. Sharma, A. Tandon, S. Narayan, and B. Bhushan, 2017. Classification and Analysis of Security Attacks in WSNs and IEEE 802.15.4 Standards: A Survey, *2017 3rd International Conference on Advances in Computing, Communication & Automation*, pp. 1–5. doi: 10.1109/ICAC-CAF.2017.8344727, http://ieeexplore.ieee.org/stamp/stamp.jsp?tp=&arnumber=8344727&isnumber=8344648.

12. H. Zhu, A. S. Chang, R. S. Kalawsky, K. F. Tsang, G. P. Hancke, L. L. Bello, and W. K. Ling, 2017. Review of State-of-the-Art Wireless Technologies and Applications in Smart Cities, *IECON 2017 – 43rd Annual Conference of the IEEE Industrial Electronics Society*, pp. 6187–6192. doi: 10.1109/IECON.2017.8217074, http://ieeexplore.ieee.org/stamp/stamp.jsp?tp=&arnumber=8217074&isnumber=8216002.

13. F. Nizzi, T. Pecorella, F. Esposito, L. Pierucci and R. Fantacci, 2019, April. IoT Security via Address Shuffling: The Easy Way, *IEEE Internet of Things Journal*, vol. 6, no. 2, pp. 3764–3774. doi: 10.1109/JIOT.2019.2892003, http://ieeexplore.ieee.org/stamp/stamp.jsp?tp=&arnumber=8606197&isnumber=8709863.

14. L. Brilli, T. Pecorella, L. Pierucci, and R. Fantacci, 2016. A Novel 6LoWPAN-ND Extension to Enhance Privacy in IEEE 802.15.4 Networks, *2016 IEEE Global Communications Conference (GLOBECOM)*, pp. 1–6. doi: 10.1109/GLOCOM.2016.7841523, http://ieeexplore.ieee.org/stamp/stamp.jsp?tp=&arnumber=7841523&isnumber=7841475.

15. M. Vanhoef, C. Matte, M. Cunche, L. S. Cardoso, and F. Piessens, 2016. Why MAC Address Randomization Is Not Enough: An Analysis of Wi-Fi Network Discovery Mechanisms, *Proceedings of the 11th ACM on Asia Conference on Computer and Communications Security (ASIA CCS '16)*. Association for Computing Machinery, New York, NY, pp. 413–424. doi: 10.1145/2897845.2897883.

16. L. Demir, M. Cunche, and C. Lauradoux, 2014. Analysing the Privacy Policies of Wi-Fi Trackers, *Proceedings of the 2014 Workshop on Physical Analytics (WPA '14)*. Association for Computing Machinery, New York, NY, pp. 39–44. doi: 10.1145/2611264.2611266.

17. H. Guo and J. Heidemann, 2018. IP-Based IoT Device Detection, *Proceedings of the 2018 Workshop on IoT Security and Privacy*. Association for Computing Machinery, New York, NY, pp. 36–42. doi: 10.1145/3229565.3229572.

18. Y. Hong, W. M. Liu, and L. Wang, 2017, September. Privacy Preserving Smart Meter Streaming against Information Leakage of Appliance Status, *IEEE Transactions on Information Forensics and Security*, vol. 12, no. 9, pp. 2227–2241. doi: 10.1109/TIFS.2017.2704904, http://ieeexplore.ieee.org/stamp/stamp.jsp?tp=&arnumber=7929354&isnumber=7948894.

19. J. Karasek, 2018, June 13. Security 101: Protecting Wi-Fi Networks against Hacking and Eavesdropping. [online]. Available: https://www.trendmicro.com/vinfo/us/security/news/cybercrime-and-digital-threats/security-101-protecting-wi-fi-networks-against-hacking-and-eavesdropping (accessed December 2, 2019).

20. A. Paul, 2006. Benefits and Vulnerabilities of Wi-Fi Protected Access 2 (WPA2). INFS 612, pp. 1–6. https://dl.irstu.com/wp-content/uploads/Download/Education/Book/Network/Network%20Security/WEP-WPA-Article/Benefits%20and%20Vulnerabilities%20of%20Wi-Fi%20Protected%20Access%202%20(WPA2).pdf (accessed December 2, 2019).

21. A. Di Luzio, A. Mei, and J. Stefa, 2016. Mind Your Probes: Deanonymization of large Crowds through Smartphone WiFi Probe Requests, *IEEE INFOCOM 2016 – The 35th Annual IEEE International Conference on Computer Communications*, pp. 1–9. doi: 10.1109/IN-FOCOM.2016.7524459, http://ieeexplore.ieee.org/stamp/stamp.jsp?tp=&arnumber=7524459&isnumber=7524326.

22. Bluetooth Core Specification 4.0, 2010, June. [Vol 0], Bluetooth SIG, Kirkland, WA. https://www.bluetooth.org/docman/handlers/downloaddoc.ashx?doc_id=456433 (accessed December 2, 2019).

23. S. Cha, M. Chuang, K. Yeh, Z. Huang, and C. Su, 2018. A User-Friendly Privacy Framework for Users to Achieve Consents with Nearby BLE Devices, *IEEE Access*, vol. 6, pp. 20779–20787. doi: 10.1109/ACCESS.2018.2820716, http://ieeexplore.ieee.org/stamp/stamp.jsp?tp=&arnumber=8327825&isnumber=8274985.

24. ZigBee Specification, 2015. ZigBee Document 05-3474-21, ZigBee Alliance, August 5, Davis, CA. https://zigbeealliance.org/wp-content/uploads/2019/11/docs-05-3474-21-0csg-zigbee-specification.pdf (accessed December 2, 2019).

25. P. Morgner, S. Mattejat, Z. Benenson, C. Müller, and F. Armknecht, 2017. Insecure to the Touch: Attacking ZigBee 3.0 via Touchlink Commissioning, *Proceedings of the 10th ACM Conference on Security and Privacy in Wireless and Mobile Networks (WiSec '17)*. Association for Computing Machinery, New York, NY, pp. 230–240. doi: 10.1145/3098243.3098254.

26. IEEE Standard for Telecommunications and Information Exchange between Systems – LAN/MAN Specific Requirements – Part 15: Wireless Medium Access Control (MAC) and Physical Layer (PHY) Specifications for Low Rate Wireless Personal Area Networks (WPAN), IEEE Std 802.15.4-2003, pp. 1–680, 1 October 2003. doi: 10.1109/IEEESTD.2003.94389, http://ieeexplore.ieee.org/stamp/stamp.jsp?tp=&arnumber=1237559&isnumber=27762.

27. J. Hui and P. Thubert, 2011, September. Compression Format for IPv6 Datagrams over IEEE 802.15.4-Based Networks. Internet Engineering Task Force, Fremont, CA, USA, RFC 6282, pp. 1–24. [Online]. Available: https://www.rfc-editor.org/rfc/rfc6282.txt (accessed November 27, 2019).

28. J. S. Atkinson, M. Rio, J. E. Mitchell, and G. Matich, 2014. Your WiFi Is Leaking: Ignoring Encryption, Using Histograms to Remotely Detect Skype Traffic, *2014 IEEE Military Communications Conference*, pp. 40–45. doi: 10.1109/MILCOM.2014.16, http://ieeexplore.ieee.org/stamp/stamp.jsp?tp=&arnumber=6956735&isnumber=6956719.

29. LoRaWAN 1.1 Specification, 2016, May. LoRa Alliance, Inc. Fremont, CA. https://lora-alliance.org/sites/default/files/2018-05/lorawan1.0.1final_05apr2016_1099_1.pdf.

30. K. Tsai, F. Leu, I. You, S. Chang, S. Hu, and H. Park, 2019. Low-Power AES Data Encryption Architecture for a LoRaWAN, *IEEE Access*, vol. 7, pp. 146348–146357. doi: 10.1109/ACCESS.2019.2941972, http://ieeexplore.ieee.org/stamp/stamp.jsp?tp=&arnumber=8840874&isnumber=8600701.

31. X. Yang, E. Karampatzakis, C. Doerr, and F. Kuipers, 2018. Security Vulnerabilities in LoRaWAN. *2018 IEEE/ACM Third International Conference on Internet-of-Things Design and Implementation (IoTDI)*, pp. 129–140. doi: 10.1109/IoTDI.2018.00022, http://ieeexplore.ieee.org/stamp/stamp.jsp?tp=&arnumber=8366983&isnumber=8366961.

32. S. Siby, R. R. Maiti, and N. O. Tippenhauer, 2017. IoTScanner: Detecting Privacy Threats in IoT Neighborhoods, *Proceedings of the 3rd ACM International Workshop on IoT Privacy, Trust, and Security (IoTPTS '17)*. Association for Computing Machinery, New York, NY, pp. 23–30. doi: 10.1145/3055245.3055253.

33. Ragav Sridharan, R. R. Maiti, and N. O. Tippenhauer, 2018. WADAC: Privacy-Preserving Anomaly Detection and Attack Classification on Wireless Traffic, *Proceedings of the 11th ACM Conference on Security & Privacy in Wireless and Mobile Networks (WiSec '18)*. Association for Computing Machinery, New York, NY, pp. 51–62. doi: 10.1145/3212480.3212495.

34. IEEE Standard for Information technology – Telecommunications and Information Exchange between Systems Local and Metropolitan Area Networks – Specific Requirements – Part 11: Wireless LAN Medium Access Control (MAC) and Physical Layer (PHY) Specifications–Amendment 4: Enhancements for Very High Throughput for Operation in Bands below 6 GHz, IEEE Std 802.11ac-2013 (Amendment to IEEE Std 802.11-2012, as Amended by IEEE Std 802.11ae-2012, IEEE Std 802.11aa-2012, and IEEE Std 802.11ad-2012), pp.1–425, 18 December 2013. doi: 10.1109/IEEESTD.2013.6687187, http://ieeexplore.ieee.org/stamp/stamp.jsp?tp=&arnumber=6687187&isnumber=6687186.

35. C. Dwork and A. Roth, 2014. The Algorithmic Foundations of Differential Privacy, *Foundations and Trends in Theoretical Computer Science*, vol. 9, no. 3–4, pp. 211–407. doi: 10.1561/0400000042.

36. I. Goodfellow, J. Pouget-Abadie, M. Mirza, B. Xu, D. Warde-Farley, S. Ozair, A. Courville, and Y. Bengio, 2014. Generative Adversarial Nets, *Advances in Neural Information Processing Systems*, pp. 2672–2680. http://papers.nips.cc/paper/5423-generative-adversarial-nets.pdf.

37. L. Wei, P. Meng, Y. Hong, and X. Cui, 2020. Using Deep Learning to Preserve Data Confidentiality, *Applied Intelligence*, vol. 50, pp. 341–353. doi: 10.1007/s10489-019-01515-3.

38. X. Chen, D. P. Kingma, T. Salimans, Y. Duan, P. Dhariwal, J. Schulman, I. Sutskever, and P. Abbeel, 2016. Variational Lossy Autoencoder. arXiv preprint arXiv: 1611.02731, https://arxiv.org/abs/1611.02731.

39. I. Psychoula, E. Merdivan, D. Singh, L. Chen, F. Chen, S. Hanke, J. Kropf, A. Holzinger, and M. Geist, 2018. A Deep Learning Approach for Privacy Preservation in Assisted Living, *2018 IEEE International Conference on Pervasive Computing and Communications Workshops (PerCom Workshops)*, pp. 710–715. doi: 10.1109/PERCOMW.2018.8480247, http://ieeexplore.ieee.org/stamp/stamp.jsp?tp=&arnumber=8480247&isnumber=8480085.

40. L. Jiang, R. Tan, X. Lou, and G. Lin, 2019. On Lightweight Privacy-Preserving Collaborative Learning for Internet-of-Things Objects, *Proceedings of the International Conference on Internet of Things Design and Implementation (IoTDI '19)*. Association for Computing Machinery, New York, NY, pp. 70–81. doi: 10.1145/3302505.3310070.

41. R. Shokri and V. Shmatikov, 2015. Privacy-Preserving Deep Learning, *Proceedings of the 22nd ACM SIGSAC Conference on Computer and Communications Security (CCS '15)*. Association for Computing Machinery, New York, NY, pp. 1310–1321. doi: 10.1145/2810103.2813687.

42. B. Hitaj, G. Ateniese, and F. Perez-Cruz, 2017. Deep Models under the GAN: Information Leakage from Collaborative Deep Learning, *Proceedings of the 2017 ACM SIGSAC Conference on Computer and Communications Security (CCS '17)*. Association for Computing Machinery, New York, NY, pp. 603–618. doi: 10.1145/3133956.3134012.

43. Google et al., 2018. TensorFlow Privacy. https://github.com/tensorflow/privacy (accessed January 20, 2020).

44. M. Abadi, A. Chu, I. Goodfellow, H. B. McMahan, Ilya Mironov, Kunal Talwar, and Li Zhang, 2016. Deep Learning with Differential Privacy, *Proceedings of the 2016 ACM SIGSAC Conference on Computer and Communications Security (CCS '16)*. Association for Computing Machinery, New York, NY, pp. 308–318. doi: 10.1145/2976749.2978318.

45. Y. Lecun, L. Bottou, Y. Bengio, and P. Haffner, 1998, November. Gradient-Based Learning Applied to Document Recognition, *Proceedings of the IEEE*, vol. 86, no. 11, pp. 2278–2324. doi: 10.1109/5.726791, http://ieeexplore.ieee.org/stamp/stamp.jsp?tp=&arnumber=726791&isnumber=15641.

6 Machine Learning-Based Internet-of-Things Security Schemes

Deepak Kumar Sharma,
Shrid Pant, and Mehul Sharma
Netaji Subhas University of Technology

CONTENTS

6.1 INTRODUCTION

The Internet of Things (IoT) has the power to connect billions of smart devices functioning with minimal human interference. As one of the most rapidly growing domains in the history of computing, IoT systems play an integral part in improving numerous real-life applications. This has, however, also brought several security and privacy issues to attention. Traditional schemes have failed to provide efficient solutions for security issues in IoTs, but many machine learning (ML) techniques have succeeded.

This chapter includes the reviews of various ML schemes for IoT security and the experimental analysis of the models. The ML-based security schemes have been known to provide superior levels of accuracy and precision in classification problems of numerous fields. Section 6.4 discusses the application of various ML models for the classification of different IoT attack scenarios. It has been established that the ML-based security schemes, including neural networks and logistic regression, can provide the near-perfect levels of accuracy and precision with enough training time and data.

This chapter has been logically organized into five sections. This part, Section 6.1, introduces the various concepts pertaining to the IoTs, ML schemes, and their interrelationships. The next part, Section 6.2, describes the principal disciplines of ML-based IoT security schemes that include IoT architectures, threats, and challenges; ML-based schemes on malware detection; authentication; secure offloading and access control; and results and analysis, challenges, and applications of these ML-based schemes. Section 6.3 discusses the comparative analysis of ML-based schemes with other methods, viz., proxy re-encryption (PRE) and physically unclonable functions (PUFs) based on IoT security. Section 6.4 thoroughly characterizes the conclusion and future works of ML-based IoT security schemes. Last section includes the references for this chapter.

6.1.1 INTRODUCTION TO THE INTERNET OF THINGS (IoT)

Over the past several years, the IoT technology has been dawning a new era of innovation, influencing the academia and industry alike. Simply, IoT is the networked interconnection of numerous objects possessing the ability to communicate without human-to-machine or human-to-human interactivity. An increasingly distributed network of objects interacting with each other and humans has been created due to the ubiquity of IoT, and this has opened tremendous opportunities and is promising to improve lives.

While the Internet of Things was coined in 1999 by Kevin Ashton in reference to supply chain management (SCM), its present meaning has been inclusive to encompass utilities, transport, healthcare, and many other sectors. Although the technology and means have evolved to a great degree, the underlying aim of allowing machines to sense and transfer information without the help of humans has stayed the same. The open wireless technologies, including Wi-Fi, Bluetooth, Mobile Data (3G/4G/5G), and radio frequency identification, (RFID), have helped IoT mature and integrate almost every part of our daily lives to the Internet. These technologies have enabled IoTs to have successful applications in personal and home, mobile, enterprise, and utilities. However, numerous road blocks, particularly with respect to security, lie ahead for powerful and secure IoT [1–4].

6.1.2 CONTEMPORARY ADVANCEMENTS IN MACHINE LEARNING (ML)

ML is a powerful method that has been successfully employed in the medical, finance, marketing, security, and other fields. It equips devices to perform tasks without explicit instructions to do so, by employing inference and pattern recognition instead.

ML has developed drastically over the last decade, transitioning from laboratory curiosity to applied engineering in the fundamentals of numerous fields. In the medical field, e.g., it has been used in the classification of arrhythmia using electrocardium data and in the automated assessment of movement disorders such as dystonia and Parkinson's disease by analyzing sensor data of specialized wearables. In fraud detection, classification algorithms such as XGBoost classifier, support vector classifier, K-nearest neighbors (K-NN), and logistic regression have been implemented for credit/debit card transactions, insurance claims, tax return claims, etc. In addition to these, ML algorithms, including unsupervised learning, supervised learning, reinforcement learning, and semi-supervised learning, have ensured a reliable and consistent functioning for self-driving cars, advertisement recommendation, image/speech recognition, and many other applications. The fresh momentum in ML has been a result of factors such as powerful and affordable computing, large volumes of organized datasets, and efficient and affordable data storage. Today, ML is implemented in research and industry to work on specialized tasks with little or no human intervention [5–7].

6.1.3 MOTIVATIONS FOR THE USE OF MACHINE LEARNING IN THE INTERNET OF THINGS

IoT devices, including RFIDs and wireless sensor networks (WSNs), require user information to provide certain services and applications. The user information must be protected from attacks such as denial of service, eavesdropping, jamming, and spoofing to ensure data privacy. Although IoT devices with limited resources are restricted to perform intensive security tasks under large input data, most of the currently available solutions generate computational-intensive load for the IoT devices. This makes the devices vulnerable to attacks. Further, an IoT device might even generate false alarms due to the lack of information, and might not be able to detect an attack in time or accurately estimate the current state of the network due to restrained means.

IoT access control, malware detection, authentication, and secure offloading are various techniques that can protect the IoT devices from miscellaneous attacks. Traditional schemes might not be applicable to all IoT devices due to their restrictive computational, memory, and battery resources. ML methods such as reinforcement learning, supervised learning, and unsupervised learning can be utilized to resolve the security issues of IoT devices, as they can help determine the key security parameters in dynamic networks [8,9].

6.1.4 INTRODUCTION TO MACHINE LEARNING-BASED INTERNET OF THINGS SECURITY SCHEMES

IoT integrates physical devices with applications and networks, which makes it vital for the user to have data privacy and security. Security is a challenging task for IoT devices due to their constricted resources. IoT devices have a defensive policy that uses ML and smart attacks to determine the principle criteria in the security protocols. This comes at a bargain with dynamic networks that are an integral part of the IoT network. The different kinds of ML techniques are used for various threats. Supervised learning techniques are useful in detecting network intrusion, spoofing attacks, and malware detection, while unsupervised learning techniques detect DoS attacks. Regression learning techniques are extremely useful in improving the authentication process and also anti-jamming offloading.

Support vector machines (SVMs), naive Bayes, neural networks, and K-NN are the supervised learning algorithms that can be employed in IoT devices for security purposes. Various attacks on the system can be detected and prevented using these algorithms. Furthermore, unsupervised learning techniques like multivariate correlation analysis can be used in a similar way. An advantage of using unsupervised learning techniques is that it does not require information classified into categories, unlike supervised learning techniques. In addition to this, Q-learning, Dyna-Q, post-decision state (PDS), and deep Q-network (DQN) are various reinforcement learning techniques that use trial and error to help an IoT device select security protocols and main criteria against various attacks [10–12].

6.1.5 NOTATIONS AND TERMINOLOGIES

The notations/symbols mentioned below ensure better readability and clarity. Notations that have been most used, throughout this chapter, are defined in Table 6.1.

6.2 PRINCIPAL DISCIPLINES OF ML-BASED IOT SECURITY SCHEMES

The most widely employed ML algorithms in network security include unsupervised learning and supervised learning. The following subsections discuss the principal disciplines of the IoTs and then the ML-based security schemes. A thematic taxonomy of ML-based security schemes for IoT security is presented in Figure 6.1.

6.2.1 INTERNET OF THINGS ARCHITECTURE

IoT is continuously changing and evolving. Thus, the IoT needs to consider factors such as interoperability, scalability, and reliability. While there are different architectures proposed in the IoTs, generally the structure of IoT is divided into the following five layers:

TABLE 6.1
Notations and Terminologies

Notations/Symbols	Definitions
IoT	Internet of Things
ML	Machine learning
RL	Reinforcement learning
MLP	Multilayer perceptron
CNN	Convolutional neural network
DNN	Deep neural network
SVM	Support vector machines
KNN	K-nearest neighbor
RFID	Radio frequency identification
UIT	Unique Identifier Technology
WSN	Wireless sensor network
DoS	Denial of service
RSSI	Received signal strength indicator
PUF	Physical unclonable function
PRE	Proxy re-encryption
DAG	Directed acyclic graph
L-BFGS	Limited-memory Broyden–Fletcher–Goldfarb–Shanno

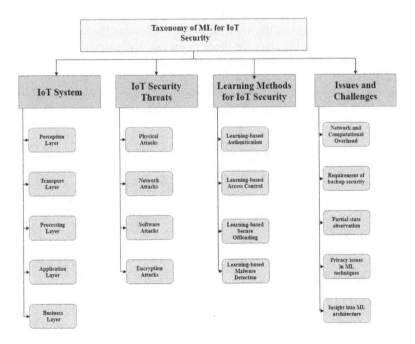

FIGURE 6.1 Taxonomy of ML for IoT security.

1. **Perception Layer (Device/Sensor Layer)**: This layer handles the clustering and detection of object-specific data by the sensor devices. The data may be related to acceleration, motion, location, orientation, temperature, and humidity, on grounds of the types of sensor. The sensors may be infrared sensors, RFID reader, or 2-D barcode, depending upon the method of object identification.
2. **Transport Layer**: This layer is accountable for relaying the data between the perception layer and the processing layer. Transmission mediums, wired or wireless, may be applied using technologies such as Wi-Fi, Bluetooth, 2G/3G/4G, and Infrared, based on the sensor devices.
3. **Processing Layer**: This layer stocks, examines, and processes large datasets coming from the transport layer. It employs technologies, including databases, big data processing, and cloud computing, to achieve its purpose.
4. **Application Layer**: The aim of this layer is to dispense the necessary operations or services by employing the obtained data from the network layer.
5. **Business Layer**: This layer is accountable for administering the IoT system, as a whole. Users' privacy, profit and business models, and applications are all managed under the business layer.

While the five-layer architecture is adequate in itself and has been implemented in a number of systems, there are multiple architecture systems available, such as three-layer architecture and service-oriented architecture. The three-layer architecture is a traditional model based on the main idea of IoT instead of the finer aspects, while the service-oriented architecture has been developed to facilitate services between the network layer and the application layer for data services as well as operations [2,13–17].

Figure 6.2 represents the basic architectures of IoTs. Label A represents the three-layer architecture, whereas Label B represents the five-layer architecture.

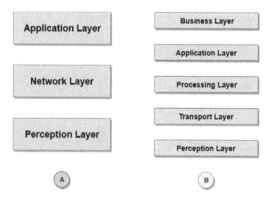

FIGURE 6.2 Basic architectures of IoT systems.

6.2.2 Internet of Things Security Threats and Challenges

IoT devices include cloud computing, RFIDs, and wireless networks to collect user information and provide services. Due to these reasons, there are various threats and challenges to the safety and privacy of the data collected, which may be sensitive. The device needs to provide the user with data confidentiality, integrity, and availability [18].

Security threats on IoT devices can be classified as follows:

1. **Physical Attacks**: The attacks on the hardware of the device, which requires the attacker to be in contact with the device, are known as physical attacks. These include the following:
   a. **Node Tampering**: If the sensor of a device is damaged or removed by an attacker with the intent of stealing information or cryptographic keys, it is known as node tampering.
   b. **Node Jamming**: This attack is caused when the attacker uses radio frequencies to jam the wireless sensor nodes that stop the communication to any other node. This can lead to DoS attack.
   c. **Malicious Node Interjection**: A malicious node can be added to the network of the IoT, which then affects the passage of data between various nodes.
   d. **Physical Damage**: The network of IoT devices is damaged in this sort of attack. It is distinct from node tampering as the attack is on the availability of service and not for data.
   e. **Sleep Deprivation Attack**: The node is kept awake for long periods of time, which drains out its battery and thus makes it to shut down.
   f. **Malicious Code Interjection**: Malicious script is physically introduced into the node to obtain access to the node or the complete network.
2. **Networks Attacks**: Any attack on the wireless network with which the device is connected comes under this category. This type of attack can be undertaken remotely. These include the following attacks:
   a. **Spoofing**: RFID signals can be spoofed to read or record data transmission using the RFID tag. This can then be used by the attacker to send fallacious data in the network and eventually gain access to the system.
   b. **Sinkhole Attack**: The attacker creates a node and directs all the information to it, thereby causing a breach of data confidentiality and denying services to other nodes.
   c. **Denial of Service**: The attacker can increase the traffic of the network, resulting in the denial of service to every node as the network can't handle the volume of information.
   d. **Routing Information Attacks**: Routing information is altered or deleted, which causes the confusions in the network along with inability in communication.

3. **Software Attacks**:- These are some of the biggest risks to the user information as they use malicious scripts to steal or tamper with the data. These attacks include the following:
   a. **Phishing Attack**: The attacker gains the credentials of the user and steals sensitive information on the IoT device.
   b. **Virus, Worms, Spyware, and TrojanHorse**: Malicious software can be used to infect the system, thus making it unchallenging for a hacker to gain hold of sensitive data or deny service to the nodes.
   c. **Denial of Service**: A distributed DoS attack can be executed through the application layer of the network. The attacker will then affect every user on the IoT network. The attack can block users from accessing the services of the network along with giving access to the attacker to the data stored.
4. **Encryption Attacks**: Attack on the encryption done on the system is known as encryption attack. This attack includes the following:
   a. **Side Channel Attacks**: The encryption key used by the device to encrypt user information is retrieved using various techniques.
   b. **Cryptanalysis Attack**: The attacker uses various cryptanalysis techniques to break the encryption on the data [18–22].

6.2.3 MACHINE LEARNING-BASED SCHEMES

In the following texts, various ML-based schemes for improving the security of IoT devices are discussed. Table 6.2 summarizes the ML-based security schemes and their performance evaluation for different examples of attacks on IoT devices.

6.2.3.1 Learning-Based Authentication

Several IoT architectures have been discussed in Section 6.2.1. A general architecture for IoT consists of three primary layers: the application layer, the network layer, and the physical/perception layer. The safety issues pertaining to each of these layers, in terms of authentication, have been thoroughly discussed in Section 6.2.2. The focus, now, will be to resolve those issues.

Many systems that employ unique identifiers have implemented IoT for certain services. The response of such systems is initiated upon authenticating the unique identifier. The uniqueness of ID, processing capabilities, storage, and small size have allowed the Unique Identifier Technologies (UITs) to enable IoT realization. As IoTs become engaged in large-scale endeavors, the integrity and safety of the systems have to be established. This is achieved through authentication methodologies. Since IoT devices have limited resources available, the conventional authentication methodologies are not always applicable. One of the most prominent challenges remains to be the application of distributed, lightweight, and impenetrable authentication schemes to IoT devices. Spoofing and eavesdropping are common attacks on IoT devices that can be resolved by applying authentication techniques. The aforementioned problems can be overcome through ML-based authentication schemes.

Q-learning, a model-free RL algorithm, is applied as a common authentication technique among the sender and receiver nodes in IoTs. In Q-learning and similar

TABLE 6.2
ML-Based IoT Security Schemes

Attacks	ML Technique	Security Scheme	Remarks
Eavesdropping	Authentication	Nonparametric Bayesian [11]	Secrecy data rate
		Q-learning [23]	Proximity passing rate
Spoofing	Authentication	SVM [24]	False alarm rate
		Distributed Frank–Wolfe [25]	Detection accuracy
		Q-learning	Mass detection rate
		Dyna-Q [26]	Classification accuracy
		DNN [12]	
Denial of	Access control	Q-learning, multivariate correlation	Root-mean error
service	Secure offloading	analysis [28], neural network [27]	Detection accuracy
Intrusion	Access control	KNN [29]	Detection rate
		SVM	Classification accuracy
		Neural networks	Root-mean error
		Naive Bayes	
Malware	Access control	KNN	Detection accuracy
	Malware detection	Random forest [30]	Classification accuracy
		Q-learning	True positive rate
		Dyna-Q	Detection latency
Jamming	Secure offloading	DQN [31]	Energy consumption
		Q-learning	

other algorithms, the agents attempt to learn an optimal strategy from their history of interactivity with their surroundings. The algorithm computes the quality of state–action combination as:

$$Q : S * A \rightarrow R \tag{6.1}$$

In Equation (6.1), "S" represents the states, "A" represents the actions, and "R" is the relation obtained from them. As the Q-table is updated, we apply the following equation for the calculation of new Q-values:

$$Q(s_t, a_t) \leftarrow (1 - \alpha) \cdot Q(s_t, a_t) + \alpha(r_t + \gamma) \cdot \max Q(s_{t+1}, a) \tag{6.2}$$

In Equation (6.2), α is the learning rate and r_t is the prize collected while progressing from state s_t to new state s_{t+1}.

Q-learning can be highly beneficial in dynamic unknown environments. So an authentication game is employed where the attacker aims at increasing its illegal advantage and the receiver aims at detecting these attacks. A simple hypothesis test is applied based on the channel response for the attack detections. Packets are labeled on the grounds of the uniqueness of the channel responses. The repeated interactions between the nodes form the zero-sum spoofing game allowing detection through Q-learning. Dyna-Q learning, which extends Q-learning by integrating a world model, is also applied to improve the training efficiency.

Other ML-based techniques include distributed Frank–Wolfe, SVM, nonparametric Bayesian, DNN (deep neural network), and incremental aggregated gradient. To reduce network overhead and improve accuracy, the distributed Frank–Wolfe and incremental aggregated gradient are applied by utilizing the RSSIs (received signal strength indicators) obtained by various landmarks. On the other hand, techniques to authenticate IoT devices/services through proximity-based authentication assess the package arrival time intervals and the RSSIs to identify the attacks out of the propinquity range. The performances of these learning-based authentication schemes depend on their area of use as every technique has its own pros and cons [8,7,12,26,32,33–35].

The general flow of ML-based IoT schemes is illustrated in Figure 6.3 for reference.

6.2.3.2 Learning-Based Access Control

When a connection is already entrusted, access control decides the acceptance of a new connection. A service call will be accepted if the bandwidth in the communication channel is available. It will be discarded or put on wait-list if there isn't enough bandwidth in the communication channel. In IoTs, each of the networks possesses different characteristics. While IoTs provide restricted service bandwidth, they have short transmission delays. Due to numerous constraints on IoT devices, degraded performance in crucial security measures, including intrusion detection, has been found. Many other challenges that come with designing access control for IoTs in

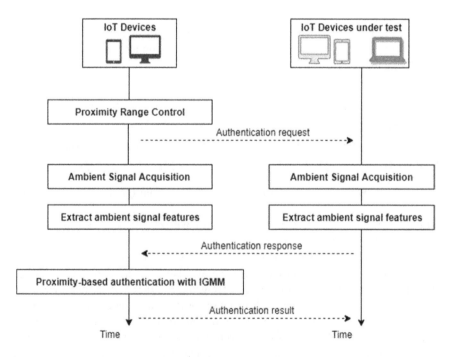

FIGURE 6.3 Depiction of a general ML-based authentication scheme for IoT system.

heterogeneous networks have led to restrictions on the safe use of IoT devices. Thus, various learning-based access control techniques are employed to create lightweight access control algorithms.

SVM, neural networks, and K-NNs are some of ML techniques that are employed for intrusion and malware attacks against access control. A multivariate correlation analysis model is utilized to extract geometric correlations between the network congestion characteristics in order to detect DoS attacks. Similarly, the application of K-NN to resolve the issues pertaining to unsupervised outlier detection for WSNs is shown in a few outlier detection schemes. Furthermore, for the identification of various attacks in smart grids and Internet traffic, supervised learning techniques like SVM are utilized. For instance, a traffic-flooding attack detection mechanism based on the SVM hierarchical structure is proposed. Another protocol, multilayer perceptron (MLP), computes the suspicion factor for indicating the presence of a DoS attack by employing neural networks with multiple neurons in the hidden layer for training the connection weights of the MLP. Experimental simulation results presented in the aforementioned ML techniques prove that the accuracy of detection of the attacks against access control is significantly higher than that using other models, while keeping the computations required by IoT devices to a minimum [24,28,29,36,,37].

6.2.3.3 Learning-Based Secure Offloading in IoT

ML is a beneficial method of providing secure offloading in IoT devices as the threat may be launched from various layers and can be of various types. The offloading solution provided by ML techniques is similar to Markov decision process (MDP) with finite states and addresses jamming, rogue devices, and smart attacks. Reinforcement learning techniques can be applied in changing radio environments, while Q learning is a suitable solution for low computational complexity.

Spoofing and jamming attacks can be prevented using Q-learning-based offloading using a Q function, the anticipated discounted extended period benefit for each action–state pair, and previous anti-jamming offloading along with the current state of the device. To improve the extended period award and to evade being confined in local favorable strategy, there is a compromise between exploration and exploitation, which improves the spoofing rate and jamming rate by 30% and 8%, respectively, as compared with a benchmark strategy. Q function can also be used in another Q learning-based scheme where the IoT device can access the cloud while being oblivious to the interference or jamming model used. This strategy can improve the benefit by 53.8% when analyzed with the standard strategy. Another Q-based scheme uses the spectrum occupancy to attain the ideal sub-band to defy jamming and interference from other radio devices, increasing the jamming cost by 44.3%. DQN improves the decision-making process of the IoT devices by helping them choose the radio frequency faster using convolutional neural networks (CNNs), but requires higher memory and computation resources. This scheme is a breakthrough as it decreases the learning time of the IoT devices significantly (66.7%), thus allowing for faster computation [8,38].

ML-based secure offloading for IoT devices is illustrated in Figure 6.4.

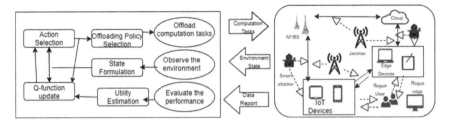

FIGURE 6.4   Depiction of ML-based secure offloading in IoT.

6.2.3.4 Learning-Based IoT Malware Detection

Malware detection in IoT devices can be done by ML algorithms that assess the runtime performances of the applications. An IoT device can use reinforcement techniques, without being aware of malware present in the device, to find the ideal proportion of the app traces that should be offloaded to the cloud. The devices will require faster computation speed, much more powerful security services, and a large malware database to detect the malware. To compute the optimal offloading policy, various schemes that use different ML techniques have been presented.

Q learning can be applied to find the ideal offloading proportion without the use of radio bandwidth model and trace generation of the adjoining IoT devices. The malware detection using Q learning not only improves the accuracy of detection by 40% but also reduces the latency by 15%. Furthermore, Dyana-Q-based scheme utilizes the Dyna architecture to assimilate knowledge from the device and the network. It employs real shielding and virtual experiences, and not only improves the accuracy by 18% as compared to Q learning scheme but also reduces the latency by 30%. Another scheme, known as the PDS-based malware detection scheme, employs the known network data and attack information, and utilizes the channel model to improve the accuracy of detection by 25% as compared to the Dyna-Q-based scheme. This scheme also uses Q learning to study the unknown state space and improves the exploration efficiency to get the most accurate results [12]. Figure 6.5 depicts the ML-based malware detection for IoT devices.

6.2.4 Results and Analysis

In this subsection, the simulation setup, results, and thorough analysis are discussed. ML algorithms are compared with each other in a multiclass classification problem of IoT attack identification. The algorithms are trained and evaluated in the Microsoft Azure Machine Learning Studio [55] to obtain reliable and accurate results in order to facilitate further analysis.

6.2.4.1 Simulation Setup and Implementation

The Bot-IoT dataset [54] was selected for the training and testing of various ML algorithms. It was selected due to its realistic test bed configuration, realistic traffic, diverse attack scenarios, and new generated features. Out of all the available features, the following were utilized for the simulation.

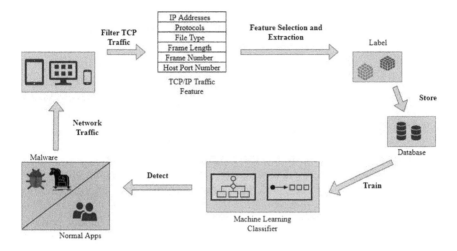

FIGURE 6.5   ML-based malware detection illustrated for IoT devices.

TABLE 6.3
Features and Descriptions

Feature	Description
State number	Numerical representation of feature state
N IN Conn P SrcIP	Number of inbound connections per source IP
N IN Conn P DstIP	Number of inbound connections per destination IP
Drate	Destination-to-source byte count
Stddev	Standard deviation of aggregated records
Mean	Average duration of aggregated records
Min	Minimum duration of aggregated records
Max	Maximum duration of aggregated records
Srate	Source-to-destination packets per second
Seq	Argus sequence number

Table 6.3 includes all the features utilized in the ML models and their brief description. These particular features were selected because they had the largest cumulative correlation coefficient and joint entropy. Employing the aforementioned features, the following attacks were classified:

1. **Denial of Service**: It is an attack mechanism that aims to disturb a particular service, thereby causing that service to be unavailable to the authorized users. In DoS attacks, the targeted resource is flooded by one system. On the other hand, in distributed denial of service (DDoS) attacks, the targeted resource is flooded by multiple systems.
2. **Information Theft**: It is an attack category that tries to weaken the security of a system in order to acquire confidential data. Based on their target of

attacks, information theft may be keylogging or data theft. In keylogging, a remote host is compromised for recording the keystrokes to steal confidential data.

The aforementioned category of attacks were used as labels. If there were no attacks, the classifiers were instructed to label them as "normal" systems. In order to classify these attacks, supervised ML algorithms were trained. The following were the trained algorithms:

1. **Multiclass Neural Network**: It is a system of interconnected layers with the capability of high-grade classification and pattern recognition. It can quickly adapt to varying input data, so there is little or no requirement to redesign the output criteria.
2. **Multiclass Decision Forest**: It is a collective model that rapidly builds numerous decision trees, which is learning from the set of input data. It has the ability to perform integrated classification and feature selection.
3. **Multiclass Logistic Regression**: It generalizes the logistic regression method to multiclass problems for systematic classification.
4. **Multiclass Decision Jungle** [56]: It employs directed acyclic graphs (DAG) to make classification. It is a recent extension to decision forests with lower memory footprint and higher accuracy.

These multiclass algorithms are chosen for their well-known classification problem application. The full specifications of the aforementioned algorithms used for building the models are detailed in Tables 6.4–6.7. The neural network is chosen for the fully connected multilayer feed forward neural network, which is described in Table 6.4. The decision forest that constructs eight decision trees to facilitate the model is given in Table 6.5. Logistic regression is designed to prevent overfitting and converge to an optimal solution, which is given in Table 6.6. Similarly, decision jungle that is detailed with the guidelines, including the decision DAGs, is given in Table 6.7. The specifications for the respective models are chosen based on heuristics to obtain the optimum results.

TABLE 6.4
Multiclass Neural Network Specifications

Parameter	Value/Remarks
Type of normalizer	Min–max normalizer
Hidden-layer specification	Fully connected case
Number of hidden nodes	8
The initial learning weights diameter	0.1
Learning rate	0.5
Number of learning iterations	1

TABLE 6.5
Multiclass Decision Forest Specifications

Parameter	Value
Minimum no. of samples per leaf node	1
No. of random splits per node	128
Maximum depth of the decision trees	32
Number of decision trees	8

TABLE 6.6
Multiclass Logistic Regression Specifications

Parameter	Value
L1 regularization weight	1
L2 regularization weight	1
Memory size for L-BFGS	20
Optimization tolerance	1E–07

TABLE 6.7
Multiclass Decision Jungle Specifications

Parameter	Value
Number of optimization steps per decision DAG layer	2048
Maximum depth of the decision DAGs	32
Maximum width of the decision DAGs	128
Number of decision DAGs	8

6.2.4.2 Simulation Results

The obtained simulation results have been graphically and analytically analyzed based on various parameters. The comparison of decision forest, logistic regression, multiclass neural network, and decision jungle is shown in Figures 6.6–6.9. The following parameters have been utilized for comparison:

1. **Overall Accuracy**: It is defined as the ratio of the total number of correctly predicted records to the total number of records to predict.
2. **Average Accuracy**: It is the average of the accuracy of each individual class.
3. **Precision**: It is the ratio of the true outcomes over all the positive outcomes.
4. **Recall**: It is the fraction of all the correct outcomes returned by the model.

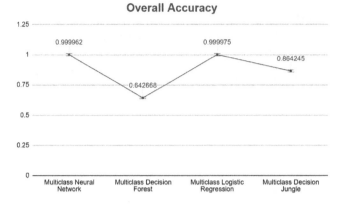

FIGURE 6.6 Overall accuracy of various models.

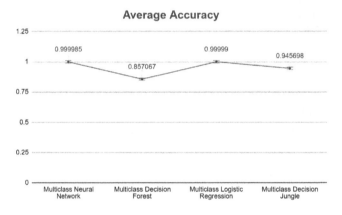

FIGURE 6.7 Average accuracy of various models.

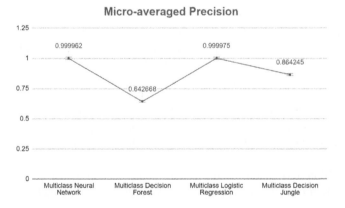

FIGURE 6.8 Micro-averaged precision of various models.

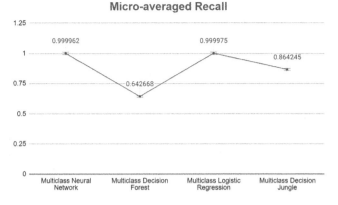

FIGURE 6.9 Micro-averaged recall of various models.

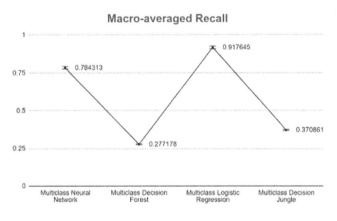

FIGURE 6.10 Macro-averaged recall of various models.

In micro-averaged method, the values from all the classes are aggregated to obtain the average metric. On the other hand, in macro-averaged method, the relevant metric for each class is independently computed and then averaged together.

A comparison of various supervised learning algorithms is illustrated in Figures 6.6–6.10.

The comparison of the overall accuracy of the models is shown in Figure 6.6. The multiclass logistic regression model outperforms all other models. The multiclass neural network model is slightly below the multiclass logistic regression model, followed by the multiclass decision jungle and, finally, multiclass decision forest.

The average accuracy of the four models is compared, which is shown in Figure 6.7. Again, the multiclass logistic regression model outperforms all other models. And the multiclass neural network model is slightly below the multiclass logistic regression model, followed by the multiclass decision jungle and, finally, the multiclass decision forest model.

Micro-averaged precisions are compared in Figure 6.8. In this case, the trained multiclass logistic regression outperforms the other models. The multiclass neural network and the multiclass decision jungle models are slightly below the multiclass logistic regression, followed by, finally, the multiclass decision forest.

The comparison of the micro-averaged recall and the macro-averaged recall is, respectively, shown in Figures 6.9 and 6.10. In case of micro-averaged recall, the multiclass logistic regression model performs better than all other models. It is immediately followed by the multiclass neural network and the multiclass decision jungle models. The decision forest model performs the poorest among them. Again, in the case of macro-averaged recall, the multiclass logistic regression model performs the best followed by the multiclass neural network model and then lastly by multiclass decision jungle and the multiclass decision forest models.

6.2.5 CHALLENGES IN ML-BASED SECURITY SCHEMES

The attack mechanisms and learning-based security schemes have been identified to enhance the IoT privacy and security. Although ML-based security schemes have numerous advantages, there are still some fundamental challenges in certain practical applications [39,40], which are as follows:

1. **Network and Computational Overhead**: To enforce the many prevailing ML-based security schemes, large amounts of training datasets requiring complicated data extraction processes are needed. These have resulted in high communication costs and intense computational needs, and led to major issues, especially for IoT systems without edge computing or cloud-based servers.
2. **Requirement of Backup Security**: Due to the lack of training data, bad feature extraction, or oversampling, both unsupervised and supervised learning techniques might not identify attacks. A network disaster might be caused by the IoT systems at early stages of learning in some RL techniques aiming to achieve the optimal solution. Hence, to obtain a reliable and secure IoT system, backup security solutions are needed along with ML-based security schemes.
3. **Partial State Observation**: The IoT systems generally have issues approximating the attack state and network perfectly. And all this while, they are expected to prevent security disasters on account of bad policy at early stages of learning. This creates many problems that the IoT systems need to tolerate.
4. **Privacy Issues in ML Techniques**: While ML algorithms might leak data, studies have also shown that privacy-preserving ML techniques are open to dominant attacks. Decentralized or even distributed methodologies are incapable of maintaining the training set perfectly private.
5. **Insight into ML Architecture**: ML techniques, unlike traditional methods of solving by programmatically instructing, learn from experience. However, a theory to assess exactly the process of ML execution based on its structure is not available yet. Such theories can facilitate resource allocation and determine the number of layers or quantity of data in the model.

6.2.6   APPLICATION OF IoT EMPLOYING ML-BASED SECURITY SCHEMES

The IoT devices utilizing ML-based schemes for security and privacy have been implemented in various fields, which are as follows:

1. **Agriculture**: The IoT devices can be used to automate the agriculture sector, thus making it smarter and better. The sensors can collect data on the temperature, humidity, moisture, etc., thus enabling real-time monitoring of the fields. This data can be analyzed and used to run the real-time measures. The irrigation of plants, and use of fertilizers and pesticides, etc. can be performed based on the data assessed by the sensors. The monitoring of soil constituents and water quality can help the farmers make an informed choice to the type of vegetation they want to grow.

2. **Healthcare**: IoT devices are currently being used in the majority of the healthcare sectors to provide prompt treatment to patients. To help in recording and observing the patients to ensure that warnings are sent to the concerned healthcare professionals in critical situations, the Internet of Medical Things (IoMT) has been established. Even though it has made the healthcare sector more organized and synchronized, the security in the healthcare IoT devices is complex. The devices need to be flexible enough to ensure that healthcare personnel can access all the patient information in case of an emergency while also being secure so that the data cannot be stolen by attackers. The sensors used in IoT are being used in smartphones and smartwatches to monitor all the daily activities such as the number of steps, sleep analysis, and running distance. IoT devices can be further used to advance the healthcare system and provide a wide range of applications.

3. **Governance**: IoT sensors can facilitate in forming smart governance as the authorities will be provided with data ranging from weather to security. IoT sensors can replace the conventional monitoring systems to generate a huge amount of information. This would be a breakthrough in the government sector as the IoT systems would be able to overcome the limitations of the current systems, thereby providing an opportunity for an optimal decision which could be made considering numerous perspectives.

4. **Transportation**: IoT devices can be used to make transport systems more smart. Data collected from CCTVs, GPS, and weather sensors can be analyzed to help manage traffic by providing users with a smart choice. This can help solve the daily traffic problem in the big cities. Furthermore, the systematic analysis of data related to smart transport can inherently improve shipment itinerary, road safety, and delivery time.

5. **Supply Chain**: IoT devices can be used to make the supply chain process smarter and more flexible. IoT-embedded sensors can be used to track goods throughout production and transportation by the supervisors. The data collected through the process can then be used to improve customer service by making appropriate decisions involving the machine uptime or the transportation methods used.

6. **Smart Homes**: Home devices incorporated with IoT constitute a smart home. The devices can respond to the surrounding changes, along with user requirements. The sensors can collect data on internal and external environments to automatically control the devices in order to suit the user based on their preferences.

7. **Prediction of Natural Disasters**: The IoT systems can be used to collect the real-time information about the environment and weather conditions of an area. The data can be analyzed to predict the occurrence of landslides or flash floods in disaster-prone areas. This could be a breakthrough and would help save thousands of lives [41–47].

6.3 COMPARATIVE ANALYSIS WITH OTHER METHODS

In this section, ML-based security schemes are compared to other methodologies, viz., PRE- and PUF-based security schemes.

6.3.1 PROXY RE-ENCRYPTION-BASED IoT SECURITY SCHEMES

PRE-based schemes include cryptosystems for enabling proxies or third party to modify the cipher text that has been encrypted for one party, so that the cipher texts may be decrypted by another. PREs are similar to the conventional symmetric or asymmetric encryption algorithms, with the addition of these functions: delegation and transitivity. Delegation allows the key holders to create a re-encryption key based on their secret keys and the delegated user's key. Transitivity, on the other hand, ensures the re-encryption of the cipher text unlimited times. Generally, PRE-based schemes consist of five algorithms: KeyGen, RKGen, Encrypt, Re-encrypt, and Decrypt.

1. **KeyGen**: It produces user's public–private keys and is conducted by a trusted entity.
2. **RKGen**: It produces a re-encryption key RKPkey1 → Pkey2 from public keys Pkey1 and Pkey2, and a private key Kkey1.
3. **Encrypt**: It encrypts the message M using a public key Pkey.
4. **Reencrypt**: It generates a cipher text which may be decrypted with the private counterpart of Pkey2. A cipher text, CPkey1, and re-encryption key, RKPkey1 → Pkey2, are taken as input.
5. **Decrypt**: It generates the message M from cipher text CPkey1 and decryption key Kkey.

Many schemes employ PRE for various security issues in IoTs. For data management based on proxy re-encryption, PRE provides a distinct advantage due to its efficiency. Every node in the IoT network will perform encryption to produce n re-encryption keys, which are sent to the proxy server. Cipher texts are, then, produced by the server, which allows the other nodes to decrypt texts. This drastically reduces the encryption calculation burden for the individual nodes. Therefore, unlike the traditional schemes where the increase in the number of devices rapidly raises the computational time, PER-based schemes have linear order of growths. Figures 6.11 and 6.12 compares and contrasts the encrypted communication process of individual nodes.

Many other PER-based schemes work for IoTs under different properties. Collusion resistance, exponentiation-free, noninteractivity, pairing-free, directionality, and multiple uses are some defining properties for PER schemes. To resolve issues pertaining to data hiding and sharing in cloud IoT environments, conditional proxy re-encryption is utilized. Therefore, PER-based security schemes can be seen as beneficial for IoT devices. However, these schemes have nontransferable and unidirectional issues. While the operational time in these schemes is better than that in traditional methods, it is still high. Hence, PER-based methodologies have serious efficiency and security problems [48–50].

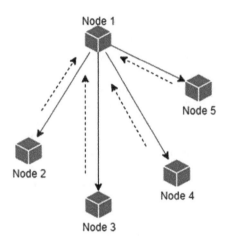

FIGURE 6.11 Management and sharing of data in existing networks.

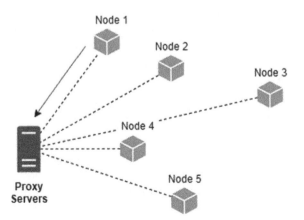

FIGURE 6.12 Management and sharing of data in PER.

FIGURE 6.13   Working of sensor PUF.

6.3.2   PHYSICAL UNCLONABLE FUNCTIONS (PUF)-BASED IoT SECURITY SCHEMES

PUFs provide a digital fingerprint to physical devices such as a microprocessor or an integrated circuit. PUFs dispense an alternative approach to the IoT devices against various threats and challenges. As these provide a "digital fingerprint" in the physical form, every Integrated Circuits (IC) can be uniquely identified. PUFs use challenge-response functions in which the output is dependent on both the input and the tangible microstructure of the device. As there is immanent variability in the manufacturing process of ICs, PUFs can successfully use the differences to employ various challenge-response functions.

PUF accomplishes two goals: providing a process for volatile secrets and delivering a distinct identity to each IoT. Volatile cryptics are embedded in the composition of the IC during manufacturing and are unavailable digitally. These are useful features in providing a powerful technique for authentication. Along with this, PUFs provide reliable security against cloning and physical attacks as creating an impeccable copy of the PUF is impossible. PUFs are insusceptible to timing attacks as well because of their isochronous nature and challenge-response function. PUFs have a low-energy footprint, making them an appealing option for self-trust in IoT systems. These features help PUFs make the IoT systems safe and secure against various attacks.

PUFs provide an attractive alternative approach to security in IoT systems due to their distinctive characteristics. The devices using PUFs do not need to store secrets in themselves, and the manufacturing process of ICs gives a unique and secure component to the system. The PUF-based security systems are efficient and low cost, shielding the IoT systems from various threats.

Figure 6.13 graphically depicts the working of sensor PUFs. Figure 6.13 represents the sensor PUF that generates the output based on both the challenge and physical quality unlike the traditional PUF, which only considers the challenge bits before making a response [51–53].

6.4   CONCLUSION

ML and its applications in various fields have brought a revolution in the way problems are resolved. Since ML-based techniques require little or no human interference, they have been extensively used in recent years to solve many real-world issues. Recently, the emergence of low-computation-cost algorithms, availability of big data, and the evolution of new algorithms have driven considerable advancements of the learning algorithms. Improvements in recurrent neural networks (RNNs), CNNs, deep auto encoders, deep belief networks (DBSs), and other DL techniques have displayed immense potential for advancements in the IoT field. Furthermore, RL-based techniques, including extensions of deep Q networks such as prioritized experience replay, double Q-learning, and continuous control with deep RL, have been studied for use. Aside from purely ML-based techniques, other technologies have been integrated with ML to provide for better implementation.

While the growing popularity of IoTs has highlighted numerous issues pertaining to its security, the availability of limited resources and the dynamic nature of the network have proven traditional security protocols to be obsolete. Hence, ML-based schemes are being preferred over the conventional schemes. Although ML techniques, too, face some challenges, they have shown immense potential to overcome fundamental problems in IoTs. Likewise, there has been a development in various other methodologies, including proxy re-encryption- and physical unclonable function-based security schemes, alleviating numerous challenges faced by IoTs. However, ML techniques are evidently superior in terms of operational time, security, and efficiency.

REFERENCES

1. Luigi Atzori, Antonio Iera, Giacomo Morabito, "The Internet of Things: A Survey", *Computer Networks* (2010), Volume 54, Issue 15, Pages 2787–2805.
2. Jayavardhana Gubbi, Rajkumar Buyya, Slaven Marusic, Marimuthu Palaniswami, "Internet of Things (IoT): A Vision, Architectural Elements, and Future Directions", *Future Generation Computer Systems* (2013), Volume 29, Issue 7, Pages 1645–1660.
3. Franco Cicirelli, Antonio Guerrieri, Carlo Mastroianni, Giandomenico Spezzano, Andrea Vinci, "The Internet of Things for Smart Urban Ecosystems", Springer International Publishing, Cham (2019), Edition 1, Pages 95–124.
4. Louis Coetzee, Johan Eksteen, "The Internet of Things – Promise for the Future? An Introduction", *IST-Africa Conference Proceedings* (2011), 11–13 May 2011, Gaborone, Botswana.
5. Mohammed Ali Al-Garadi, Amr Mohamed, Abdulla Al-Ali, Xiaojiang Du, Mohsen Guizani, "A Survey of Machine and Deep Learning Methods for Internet of Things (IoT) Security", arXiv: 1807.11023.
6. Muhammad Shafique, Theocharis Theocharides, Christos-Savvas Bouganis, Muhammad Abdullah Hanif, Faiq Khalid, Rehan Hafız, Semeen Rehman, "An Overview of Next-Generation Architectures for Machine Learning: Roadmap, Opportunities and Challenges in the IoT Era", *Design, Automation & Test in Europe Conference & Exhibition* (2018), 19–23 March 2018, Dresden, Germany.
7. P. Mohamed Shakeel, S. Baskar, V. R. Sarma Dhulipala, Sukumar Mishra, Mustafa Musa Jaber, "Maintaining Security and Privacy in Health Care System Using Learning Based Deep-Q-Networks", *Journal of Medical Systems* (2018). doi: 10.1007/s10916-018-1045-z.
8. Liang Xiao, Xiaoyue Wan, Xiaozhen Lu, Yanyong Zhang, Di Wu, "IoT Security Techniques Based on Machine Learning: How Do IoT Devices Use AI to Enhance Security?", *IEEE Signal Processing Magazine* (2018), Volume 35, Issue 5, Pages 41–49.
9. Mohammad Abu Alsheikh, Shaowei Lin, Dusit Niyato, Hwee-Pink Tan, "Machine Learning in Wireless Sensor Networks: Algorithms, Strategies, and Applications", arXiv: 1405.4463.
10. Liang Xiao, Weihua Zhuang, Sheng Zhou, Cailian Chen, "Competing Mobile Network Game: Embracing Anti-Jamming and Jamming Strategies with Reinforcement Learning," *Proceedings of the IEEE Conference on Communication and Network Security* (2013), National Harbor, MD, Pages 28–36.

11. Liang Xiao, Qiben Yan, Wenjing Lou, Guiquan Chen, Y. Thomas Hou, "Proximity-Based Security Techniques for Mobile Users in Wireless Networks", *IEEE Transactions on Information Forensics and Security* (2013), Volume 8, Issue 12, Pages 2089–2100.

12. Cong Shi, Jian Liu, Hongbo Liu, Yingying Chen, "Smart User Authentication through Actuation of Daily Activities Leveraging WiFi-Enabled IoT", *Mobihoc '17* (2017), Article No. 5, Pages 1–10.

13. Rafiullah Khan, Sarmad Ullah Khan, Rafaqat Zaheer, Shahid Khan, "Future Internet: The Internet of Things Architecture, Possible Applications and Key Challenges", *10th International Conference on Frontiers of Information Technology* (2012), 17–19 December 2012, Islamabad, India.

14. Dhananjay Singh, Gaurav Tripathi, Antonio J. Jara, "A Survey of Internet-of-Things: Future Vision, Architecture, Challenges and Services", *IEEE World Forum on Internet of Things (WF-IoT)* (2014), Volume 1, Seoul.

15. Lu Tan, Neng Wang, "Future Internet: The Internet of Things", *3rd International Conference on Advanced Computer Theory and Engineering (ICACTE)* (2010), 20–22 August 2010, Chengdu, China.

16. Miao Wu, Ting-Jie Lu, Fei-Yang Ling, Jing Sun, Hui-Ying Du, "Research on the Architecture of Internet of Things", *3rd International Conference on Advanced Computer Theory and Engineering (ICACTE)* (2010), 20–22 August 2010, Chengdu, China.

17. Zhi-Kai Zhang, Michael Cheng Yi Cho, Chia-Wei Wang, Chia-Wei Hsu, Chong-Kuan Chen, Shiuhpyng Shieh, "IoT Security: Ongoing Challenges and Research Opportunities", *IEEE 7th International Conference on Service-Oriented Computing and Applications* (2014), 17–19 November 2014, Matsue, Japan.

18. Ioannis Andrea, Chrysostomos Chrysostomou, George Hadjichristofi, "Internet of Things: Security Vulnerabilities and Challenges", *IEEE Symposium on Computers and Communication (ISCC)* (2015), 6–9 July 2015, Larnaca, Cyprus.

19. Fan Wu, Lili Xu, Saru Kumari, Xiong Li, Jian Shen, Kim-Kwang Raymond Choo, Mohammad Wazid, Ashok Kumar Das, "An Efficient Authentication and Key Agreement Scheme for Multi-Gateway Wireless Sensor Networks in IoT Deployment", *Journal of Network and Computer Applications Archive* (2017), Volume 89, Issue C, Pages 72–85.

20. Fagen Li, Jiaojiao Hong, Anyembe Andrew Omala, "Efficient Certificateless Access Control for Industrial Internet of Things", *Future Generation Computer Systems* (2016). doi: 10.1016/j.future.2016.12.036.

21. Martin Henzea, Lars Hermerschmidt, Daniel Kerpen, Roger Häußling, Bernhard Rumpe, Klaus Wehrle, "A Comprehensive Approach to Privacy in the Cloud-based Internet of Things", *Future Generation Computer Systems* (2015), Volume 56, Pages 701–718. doi: 10.1016/j.future.2015.09.016.

22. Malek Harbawi, Asaf Varol, "An Improved Digital Evidence Acquisition Model for the Internet of Things Forensic I: A Theoretical Framework", *5th International Symposium on Digital Forensic and Security (ISDFS)* (2017), Pages 1–6, IEEE, Tirgu Mures, Romania.

23. Liang Xiao, Caixia Xie, Tianhua Chen, Huaiyu Dai, H. Vincent Poor, "A Mobile Offloading Game against Smart Attacks", *IEEE Access* (2016), Volume 4, Pages 2281–2291.

24. Mete Ozay, Iñaki Esnaola, Fatos Tunay Yarman Vural, Sanjeev R. Kulkarni, H. Vincent Poor, "Machine Learning Methods for Attack Detection in the Smart Grid", *IEEE Transactions on Neural Networks and Learning Systems* (2016), Volume 27, Issue 8, Pages 1773–1786.

25. Liang Xiao, Xiaoyue Wan, Zhu Han, "PHY-Layer Authentication with Multiple Landmarks with Reduced Overhead", *IEEE Transactions on Wireless Communications* (2018), Volume 17, Issue 3, Pages 1676–1687.

26. Liang Xiao, Yan Li, Guolong Liu, Qiangda Li, Weihua Zhuang, "Spoofing Detection with Reinforcement Learning in Wireless Networks", *IEEE Global Communications Conference (GLOBECOM)* (2015), San Diego, CA. doi: 10.1109/GLOCOM.2015.7417078.

27. Anna L. Buczak, Erhan Guven, "A Survey of Data Mining and Machine Learning Methods for Cyber Security Intrusion Detection", *IEEE Communications Surveys & Tutorials* (2016), Volume 18, Issue 2, Pages 1153–1176.

28. Zhiyuan Tan, Aruna Jamdagni, Xiangjian He, Priyadarsi Nanda, Ren Ping Liu, "A System for Denial-of-Service Attack Detection Based on Multivariate Correlation Analysis", *IEEE Transactions on Parallel and Distributed Systems* (2014), Volume 25, Issue 2, Pages 447–456.

29. Joel Branch, Boleslaw Szymanski, Chris Giannella, Ran Wolff, Hillol Kargupta, "In-Network Outlier Detection in Wireless Sensor Networks", *26th IEEE International Conference on Distributed Computing Systems (ICDCS '06)* (2006), 4–7 July 2006, Lisboa, Portugal.

30. Fairuz Amalina Narudin, Ali Feizollah, Nor Badrul Anuar, Abdullah Gani, "Evaluation of Machine Learning Classifiers for Mobile Malware Detection", *Soft Computing* (2016), Volume 20, Issue 1, Pages 343–357.

31. Guoan Han, Liang Xiao, H. Vincent Poor, "Two-Dimensional Anti-Jamming Communication Based on Deep Reinforcement Learning", *IEEE International Conference on Acoustics, Speech and Signal Processing (ICASSP)* (2017), 5–9 March 2017, New Orleans, LA.

32. Aidin Ferdowsi, Walid Saad, "Deep Learning-Based Dynamic Watermarking for Secure Signal Authentication in the Internet of Things", May 2018, Pages 1–6. doi: 10.1109/ICC.2018.8422728.

33. Rajshekhar Das, Akshay Gadre, Shanghang Zhang, Swarun Kumar, Jose M. F. Moura, "A Deep Learning Approach to IoT Authentication", *IEEE International Conference on Communications (ICC)* (2018), 20–24 May 2018, Kansas City, MO.

34. Mohammed El-hajj, Ahmad Fadlallah, Maroun Chamoun, Ahmed Serhrouchni, "A Survey of Internet of Things (IoT) Authentication Schemes", *Sensors* (2019), Volume 19, March 2019, doi: 10.3390/s19051141.

35. Pratikkumar Desai, Amit Sheth, Pramod Anantharam, "Semantic Gateway as a Service Architecture for IoT Interoperability", *IEEE International Conference on Mobile Services* (2015), 27 June–2 July 2015, New York, NY.

36. Raghavendra V. Kulkarni, Ganesh K. Venayagamoorthy, "Neural Network Based Secure Media Access Control Protocol for Wireless Sensor Networks", *International Joint Conference on Neural Networks* (2009), 14–19 June 2009, Atlanta, GA.

37. Jaehak Yu, Hansung Lee, Myung-Sup Kim, Daihee Park, "Traffic Flooding Attack Detection with SNMP MIB Using SVM", *Computer Communications* (2008), Volume 31, Issue 17, Pages 4212–4219.

38. Rodrigo Roman, Javier Lopez, Masahiro Mambo, "Mobile Edge Computing, Fog et al.: A Survey and Analysis of Security Threats and Challenges", *Future Generation Computer Systems* (2018), Volume 78, Pages 680–698.

39. Md. Mahmud Hossain, Maziar Fotouhi, Ragib Hasan, "Towards an Analysis of Security Issues, Challenges, and Open Problems in the Internet of Things", *IEEE World Congress on Services* (2015), 27 June–2 July 2015, New York, NY.

40. Mohammad Abu Alsheikh, Shaowei Lin, Dusit Niyato, Hwee-Pink Tan, "Machine Learning in Wireless Sensor Networks: Algorithms, Strategies, and Applications", *IEEE Communications Surveys & Tutorials* (2014), Volume 16, Issue 4, Pages 1996–2018.

41. Sara Amendola, Rossella Lodato, Sabina Manzari, Cecilia Occhiuzzi, Gaetano Marrocco, "RFID Technology for IoT-Based Personal Healthcare in Smart Spaces", *IEEE Internet of Things Journal* (2014), Volume 1, Issue 2, Pages 144–152.

42. Wen-Tsai Sung, Yen-Chun Chiang, "Improved Particle Swarm Optimization Algorithm for Android Medical Care IOT Using Modified Parameters", *Journal of Medical Systems* (2012), Volume 36, Issue 6, Pages 3755–3763.

43. S. M. R. Islam, D. Kwak, M. H. Kabir, M. Hossain, K. S. Kwak, "The Internet of Things for Health Care: A Comprehensive Survey." *IEEE Access* (2015), Volume 3, Pages 678–708.

44. C. Camara, P. Peris-Lopez, J. E. Tapiador, "Security and Privacy Issues in Implantable Medical Devices: A Comprehensive Survey", *Journal of Biomedical Informatics* (2015), Volume 55, Pages 272–289.

45. George Dimitrakopoulos, "Intelligent Transportation Systems Based on Internet-Connected Vehicles: Fundamental Research Areas and Challenges", *11th International Conference on ITS Telecommunications* (2011), 23–25 August 2011, Pages 145–151, St. Petersburg, Russia.

46. Tanmay Baranwal, Pushpendra Kumar Pateriya Nitika, "Development of IoT Based Smart Security and Monitoring Devices for Agriculture", *6th International Conference – Cloud System and Big Data Engineering (Confluence)* (2016), 14–15 January 2016, Pages 597–602, Noida, India.

47. Diane J. Cook, Aaron S. Crandall, Brian L. Thomas, Narayanan C. Krishnan, "CASAS: A Smart Home in a Box", *Computer* (2013), Volume 46, Issue 7, Pages 62–69.

48. SuHyun Kim, ImYeong Lee, "IoT Device Security Based on Proxy Re-encryption", *Ambient Intelligence and Humanized Computing* (2018), Volume 9, Issue 4, Pages 1267–1273.

49. Ahsan Manzoor, Madhsanka Liyanage, An Braeke, Salil S. Kanhere, Mika Ylianttila, "Blockchain Based Proxy Re-Encryption Scheme for Secure IoT Data Sharing", *IEEE International Conference on Blockchain and Cryptocurrency (ICBC)* (2019), 14–17 May 2019, Seoul, South Korea.

50. Muhammad Baqer Mollah, Md. Abul Kalam Azad, Athanasios Vasilakos, "Secure Data Sharing and Searching at the Edge of Cloud-Assisted Internet of Things", *IEEE Cloud Computing* (2017), Volume 4, Issue 1, Pages 34–42.

51. Muhammad Naveed Aman, Kee Chaing Chua, Biplab Sikdar, "Mutual Authentication in IoT Systems Using Physical Unclonable Functions", *IEEE Internet of Things Journal* (2017), Volume 4, Issue 5, Pages 1327–1340.

52. John Ross Wallrabenstein, "Practical and Secure IoT Device Authentication Using Physical Unclonable Functions", *IEEE 4th International Conference on Future Internet of Things and Cloud (FiCloud)* (2016), 22–24 August 2016, Vienna, Austria.

53. Arun Kanuparthi, Ramesh Karri, Sateesh Addepalli, "Hardware and Embedded Security in the Context of Internet of Things", *2013 ACM workshop on Security, Privacy & Dependability for Cyber Vehicles*, Pages 61–64, Berlin, Germany.

54. Nickolaos Koroniotis, Nour Moustafa, Elena Sitnikova, Benjamin Turnbull, "Towards the Development of Realistic Botnet Dataset in the Internet of Things for Network Forensic Analytics: Bot-IoT Dataset", *Future Generation Computer Systems* (2019), Volume 100, Pages 779–796.

55. Roger Barga, Valentine Fontama, Wee Hyong Tok, "Introducing Microsoft Azure Machine Learning", *Predictive Analytics with Microsoft Azure Machine Learning* (2015), pp. 21–43. Apress, Berkeley, CA. doi: 10.1007/978-1-4842-1200-4_2.

56. Jamie Shotton, Toby Sharp, Pushmeet Kohli, Sebastian Nowozin, John Winn, Antonio Criminisi, "Decision Jungles: Compact and Rich Models for Classification", *Advances in Neural Information Processing Systems* (2013), pp. 234–242.

7 Intrusion Detection of SCADA System Using Machine Learning Techniques
A Study

Koyela Chakrabarti
West Bengal University of Technology

CONTENTS

7.1 INTRODUCTION: BACKGROUND AND CONTRIBUTIONS

A study on the recent trend of cyber-attacks has revealed that SCADA (supervisory control and data acquisition) systems and scientific instruments have been targeted the most [1]. A majority of these attacks have been based on exploiting the vulnerabilities of the websites [2]. According to Ref. [3], around 46 websites have been targeted in energy sector, 23 in transport, and 31 in dam and water. A security researcher in Kaspersky suggests that the security of an industrial computer is breached using a malware distributed through Internet or removable media device [4]. But there is a difference between a common IT security goal and that of a SCADA system. Apart from the obvious security concern regarding confidentiality, integrity, and availability (CIA) as in any network, a SCADA system additionally needs to act in a timely manner and also ensures that security compromised in one part of the system should be isolated and does not corrupt the system as a whole. The SCADA system thus needs a secured access control to ensure that when one security goal is breached, it does not affect the other security parameters. For example, say, after compromising the integrity of the system, if the attacker makes a device to malfunction, the availability of the system might get hampered. Due to the fast evolving nature of the network attacks, the rule or signature-based intrusion detection system (IDS) fails to detect an attack. The learning factor of the machine learning (ML) algorithms helps to detect an unusual activity in the network, which is a usual consequence to an attempt to disrupt the network.

This chapter discusses the different security goals of a SCADA system. This chapter states vulnerabilities of the communication protocols used in IIoT. This chapter gives a short account of various types of attacks based on the security goals they compromise and how ML algorithms can be used to detect those attacks. The advantage that ML algorithm gives over the rule-based ones is stated, and the challenges the ML algorithms pose regarding training the data are also mentioned. A short summary of the existing IDS algorithms based on ML along with a comparison table of the algorithms is given. A short experiment is conducted where a temperature sensor node is installed that sends data to a central computer acting as the supervisory node. When the temperature level rises above 35°C, the central node sends a command to a buzzer node. Using network tools, the normal data is collected. Then, man-in-the-middle (MITM) attack is carried out, and the network data are again recorded. An imbalanced dataset having less than 1% attack data, as is generally observed in real world, is used to train five types of ML algorithms, namely, SVM (support vector machine), naive Bayes (NB), k-nearest neighbor (K-NN), Random Forest (RF), and decision tree (DT). The efficiency of the different classifiers on classifying the test data is compared according to different ML classification efficiency parameters discussed in this chapter.

This chapter is divided into the following sections: Section 7.2 following introduction gives an overview of SCADA system. Section 7.3 discusses the security goals of a SCADA system. Section 7.4 gives a short account of the transmission protocols of IoT used in IIoT and the associated vulnerabilities of each protocol. Section 7.5 gives a short account of the various attacks to the SCADA system. Section 7.6 gives different ML classifiers. Section 7.7 explains the parameters that are used to determine how effective an ML classification is. Section 7.8 explains the security breaches and the consequent changes or anomalies in the network that are used by classifiers. Section 7.9 reviews some of the existing works in ML-based SCADA IDS and tabulates a comparative analysis of the algorithms. Section 7.10 explains the experimental setup conducted. Section 7.11 explains the analysis of the effectiveness of classification of the filters based on a few parameters as stated in Section 7.7, and Section 7.12 finally provides the conclusion of this chapter.

7.2 SCADA SYSTEM OVERVIEW

SCADA is the heart of the industrial control system (ICS). It is the software package that acts as a central control system to monitor the entire area and control accordingly [5]. Working of SCADA is mostly automated, needing human intervention to handle only the critical events. It employs controller for different network interfaces, the input and output sensors and actuators, and communication software. There is a remote terminal unit (RTU) comprising programmable logic controllers (PLCs) with some preset values according to the type of control system. Say, the temperature of the manufacturing unit of a certain pharmaceutical plant is maintained at 40°C, and some action needs to be taken if the temperature rises above it. An overview of the SCADA architecture is shown in Figure 7.1.

SCADA consists of two layers: (1) data server layer (DSL) and (2) clint layer (CL).

1. **DSL**: Data servers connect to PLC directly or through a network or a bus to communicate with the field devices. The PLCs collect the raw signal data from the sensors, convert it into digital data, and send that data to the server. The server processes the data and sends a command to the PLC connected to the relay. The PLC then applies the required electrical signal to the actuator.
2. **CL**: This layer provides the human–machine interface (HMI). The HMI is presented as a graphical user interface for the human controller to access the different control units like PLCs when needed.

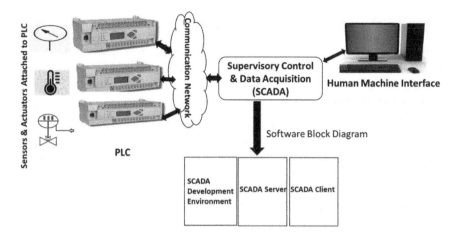

FIGURE 7.1   An overview of SCADA architecture.

7.3   SCADA SYSTEM SECURITY GOALS

The SCADA system apart from CIA goals, which is a prerequisite to any cyber system, imposes some additional factors. According to the importance, the security goals of SCADA are given in the following.

7.3.1   TIME SENSITIVITY

SCADA system software works as a real-time operating system (RTOS) with hard deadline. The system works as a sequence of a series of data input and command provided to the relay. Therefore, any latency in one stage of the system can have a cascading effect of delay in all the subsequent stages, thus resulting in a total system failure eventually. The data input to any stage in a system hence has a lifespan, after which the data no longer serves the purpose. So the system needs protection against latency to send the data and also repeatedly send the same data from or to a field device.

7.3.2   AVAILABILITY

A SCADA system comprises a series of interrelated and interconnected physical devices such as sensors, actuators, wireless and wired communication media, and computational nodes. The process is sequential and continuous; therefore, failure of any of the aforementioned part will stall the entire control system.

7.3.3   INTEGRITY

This refers to the integrity of data and message. First, the data that is received or sent must not be tampered. This includes the integrity of payload, source and destination addresses, sequence of data frames (if any), and timestamp of the data packet passing through the network. Modifying header or payload contents, or deletion or repeatedly sending same data over network might disrupt the system. For example, if the destination address is modified, the control signal for one relay will reach the other one and might bring down a certain part of the system.

7.3.4 CONFIDENTIALITY

It is the property that safeguards unauthorized access to the system. In SCADA system, confidentiality is needed when data is passed into a control algorithm. Additionally, some common cyber information such as passwords or encryption keys need to be guarded against side channel attacks.

7.3.5 GRACEFUL DEGRADATION

This refers to the isolation of the part of the system which has been attacked. The interfacing of the different modules will be such that on detection of an attack event, the compromised data should be isolated in the region/module where the attack has happened rather than propagating to the other parts and affecting the whole system.

7.4 DATA TRANSMISSION PROTOCOLS USED IN IIOT

This section gives a brief description of the most commonly used communication protocols in IoT that are utilized in IIoT for data exchange along with the security flaws in each protocol. The vulnerabilities of different IIoT protocols along with the associated network layer are given in Table 7.1.

7.4.1 MODBUS

Modbus is the most commonly used open source protocol in electronic industries since its inception from 1979. This is a serial communication protocol and works as master slave. The HMI computer here works as the Master and PLC as the slave. Master here retrieves information from the slave. There can be a maximum of 254 devices connected to one data link. The transmission over a Modbus link is continuous. Each slave device has two tables to store the raw analog signal and two for digital data. The data transmitted is not encrypted. Therefore, message confidentiality is not provided and the data packet can be easily sniffed. Also, there might be data insertion and unauthorized command injection.

TABLE 7.1
Vulnerability in Different IIoT Communication Protocols

Protocol	Associated Network Layer	Vulnerability
Modbus	Application	Confidentiality, integrity
BACNet	Physical/datalink	Confidentiality, availability
DNP3	Datalink, transport, application	Authentication
MQTT	Transport	Authentication

7.4.2 BACNet

Building Automation Controls Network is a communication protocol that works in a number of physical/datalink layer such as Ethernet, ARCNET, point-to-point over RS-232, BACnet/IP, ZigBee, and Master/Slave token passing over RS-485. In SCADA systems, mostly Master/Slave modes of communication are carried out. Data confidentiality is not protected in this protocol. Therefore, it is vulnerable to reconnaissance attack. The newer version of the BACNet provides Advanced Encryption Standard (AES) and Data Encryption Standard (DES) encryption schemes. But the encryption adds latency due to the communication overhead and delays, which is not suitable for an RTOS environment with hard deadline. So the newly added encryption scheme is not implemented in industries. The PLC/HMI scan cycle time is some order of milliseconds, which is not sufficient for the encryption/decryption scheme to work. The protocol is also susceptible to DoS attacks that compromise the availability of the system.

7.4.3 DNP3

The protocol works on the top of a TCP/IP or serial bus connection and is composed of link, transport, and application layers. The protocol is more efficient, robust, and hence reliable compared to the aforementioned two protocols, and also provides time synchronization with the RTU. But the protocol does not provide security features such as authentication and encryption or access control. For the lack of these security features, the messages are vulnerable to DoS attacks, spoofing, or eavesdropping. In 2012 IEEE standardization, public key infrastructure (PKI) has been included. But due to time latency and complexity of PKI, the feature is not practically implementable in IIoT systems.

7.4.4 MQTT

This is a lightweight publish/subscribe messaging protocol suitable for IoT connectivity. The protocol is used for remote sensing and control of the IoT devices in ICS systems. The topology works as a client-broker model. The client when sends some data becomes the publisher. The client that requests for a data from the broker becomes the subscriber. The broker sits between the publisher and the subscribers. The system is extremely scalable where a single broker can deal up to thousands of clients. Whenever a new client wants to join the network it needs to register with the broker by sending their ID and password over the network. But MQTT does not provide any encryption, therefore the data is sent as plaintext across the network. Also, if an intruder manages to steal the credentials of a client, he/she will also get access to the data of other clients in the system. MQTT packets can be secured using transport layer security/ secured socket shell (TLS/SSL) while sending across Transmission Control Protocol/ Internet Protocol suite (TCP/IP). But this makes the system too complex to be suitable for use in IIoT context. However, it supports a light weight Message Authentication Code (MAC) function using hashing. Clients can utilize this feature for verifying and signing data if they know the secret key.

7.5 DIFFERENT ATTACKS IN IIOT

In most of the industries, the integration of IoT with ICS has been done on the existing network infrastructure so SCADA is also vulnerable to the common IT security attacks. Some of the prominent attacks that the SCADA system has faced so far according to the security goal it has breached are given in the following. Figure 7.2 shows a bar chart that has been prepared according to a security report of Dell in 2015 with respect to SCADA that shows the different types of attacks the system faces. A brief description of some of the real-life major attacks along with their timeline on SCADA systems reported so far is provided in Table 7.2.

7.5.1 INTEGRITY

According to the Dell, on 2015 security report as mentioned above, most of the attacks have been targeted to compromise the integrity of the SCADA. The prominent attacks that have targeted to disrupt the integrity have been briefly described in the following.

7.5.1.1 Buffer Overflow

This is the attack aimed at exploiting the weakness in coding of software. This error generally occurs when the buffer tries to hold more data than it can handle. The data from buffer overflows to adjacent storage, thus corrupting the memory space. In the worst case, it creates the entry point for a cyber-attack. Usually, languages such as C and C++ lack the necessary type checking or checking the size of arrays, and also lack the input value checking or runtime boundary value checking. This is one of the most prevalent attacks to a SCADA system since the operating system

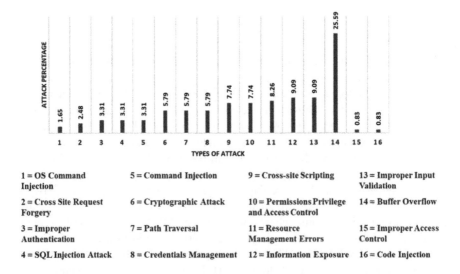

1 = OS Command Injection	5 = Command Injection	9 = Cross-site Scripting	13 = Improper Input Validation
2 = Cross Site Request Forgery	6 = Cryptographic Attack	10 = Permissions Privilege and Access Control	14 = Buffer Overflow
3 = Improper Authentication	7 = Path Traversal	11 = Resource Management Errors	15 = Improper Access Control
4 = SQL Injection Attack	8 = Credentials Management	12 = Information Exposure	16 = Code Injection

FIGURE 7.2 Vulnerability to various attack vectors in SCADA according to "2015 Dell Security Annual Threat Report".

TABLE 7.2

Some of the Prominent Attacks of SCADA Reported

Year	Targeted Industry	Attack Description	Consequence
2000	Queensland Sewage Treatment Plant	A former employee belonging to the software development team gained access and controlled the SCADA.	Pumping of around 8,00,000 liters of raw human waste into a nearby resort ground and river destroying plants and wildlife.
2008	Baku-Tbilisi-Ceyhan oil pipeline of Turkey	A terrorist organization illegally accessed SCADA and disrupted the system by disabling alarm and changing the pipeline pressure reading.	A pipeline explosion that leads to spilling over 30,000 barrels of oil over an area below which was a water aquifer.
2009	A South California Coast Oil Company	The monitoring and pipeline leak detection system was illegally accessed and tampered by a dissatisfied employee of the organization.	The monitoring module was disabled leaving the coastline vulnerable to disasters from a leaking pipeline.
2010	14 industrial sites based in Iran	A self-replicating Stuxnet worm was introduced to the ICS using an USB flash drive.	Spying of the operational information to gain access to the system and attack the centrifuges and disrupt them.
2014	A German steel plant	Social engineering and spear phishing attacks were used to access the production network	Production machines and control components were manipulated to prevent shutting down of the blast furnace, thereby causing physical damages.

used in ICS is mostly based on C, and for the need of constant monitoring, the SCADA workstations are usually not rebooted. This leads to the accumulation of memory fragmentation, thus leading to an overflow condition. The attack can lead to the manipulation of control data sent to the relays.

7.5.1.2 Improper Input Validation

This is another software error that does not provide a proper input validation technique. The attacker in this case provides an incorrect input to a control algorithm and manipulates the functioning of the system. SCADA systems being deterministic in nature are not usually checked for these vulnerabilities in the software.

7.5.1.3 SQL Code Injection

This attack takes the advantage of the system that sends the user data entered in web forms to the database as-is without sanitizing it. Thus, an attacker can send unexpected SQL command into the forms and fetch information directly from the server. This attack can also be used to gain unauthorized access to the system. For SCADA system, it can gain access to data server of the SCADA and manipulate with authentication and control data inputs.

7.5.2 AVAILABILITY

The availability of attacks to the system is also brought about by buffer overflow attacks as explained in 7.5.1.1. Apart from that, the attack most prevalent in ICS to compromise the availability is DoS attack as explained below with respect to SCADA.

7.5.2.1 Denial of Service (DoS)

The attack is intended to consume the resources of a system so that it fails to provide service to its intended users. The attack takes place by either flooding or crashing the services of the server node. In SCADA system, usually flooding attack is more commonly launched. The attacker sends too many get/post requests or fake or SYN (synchronize) packets to the target. This congests the network link and, as a result, introduces a latency to send command to the intended PLC or the monitoring system. Therefore, it greatly hampers the availability of the system.

7.5.3 CONFIDENTIALITY

The main type of attack that falls into this category is the reconnaissance attacks. In this attack, the attacker gathers knowledge about the network. In wireless local area network (WLAN), data is received by anybody within the signal range. Therefore, it is easy to analyze the network data to carry out the attack. Below is a short description of the different means of gathering information for carrying out the reconnaissance attack.

7.5.3.1 Sniffing

This refers to the eavesdropping of packets sent over the network. This is practically impossible to detect in WLAN. There are a number of sniffing tools available commercially or by written free driver, such as Prismdump, Ethereal, Tcpdump, and Wildpackets. These tools capture the packet, analyze it, and display useful information.

7.5.3.2 Phishing

In this attack, the attacker masquerades as a trusted entity and fools the victim to open a link sent via email or text message. The victim is redirected to the wrong website and his/her confidential information like login credentials is stolen. In SCADA system, often an employee of the organization is targeted, and through him/her, the malware is introduced into the network which next infects it. According to Ref. [6] based on a study conducting spear phishing attack on employees of SCADA system, around 26% of such attacks have been successful.

In SCADA system, the intruder gathers information about network such as address of the hosts, the IP addresses, connected devices, and the security policies adopted by the system. Next the vulnerabilities of the network are identified and the vulnerable devices are attacked. The network interface between HMI and PLC or PLC and I/O is passively monitored using sniffers without injecting any traffic. This attack is thus very difficult to detect. Although the attacker is silently monitoring without attacking the system, the information gained can be exploited to infect the network.

7.5.4 AUTHENTICATION

Authentication attacks in SCADA are mostly carried out for information gathering. Mostly, these attacks are passive and are hard to detect. Preventive measures are effective in stopping authentication attacks. The prominent attacks that are listed as authentication attacks are briefed as follows.

7.5.4.1 Man-in-the-Middle Attack

This attack is based on eavesdropping a communication where an external agent or the attacker places itself between two legitimate nodes and impersonates itself as the other while communication. The attacker tampers with the messages sent to and from the nodes. In this case, the attacker posing as PLC might send a malicious command to the relay, or the data sent to HMI from PLC might be changed. In this type of attack, the syntax will be valid, so it is better to use encryption techniques to safeguard the system.

7.5.4.2 Unauthenticated Access

This unauthenticated access is gained in mainly two ways in SCADA: The first way is by carrying out phishing attack, by sending malicious mails, and by provoking the unsuspecting user who is a part of the IIoT network to share his/her login credentials. The second way is by exploiting the reluctance of the legitimate user of the system to change password regularly or any other human error that accidentally gives away the credentials. The password can be guessed using key logging or Brute force methods. This attack lets the attacker to fetch some information from the system. But the attack can be more severe if the attacker is able to gain root access to the system and thus able to traverse the directories present.

7.5.5 AUTHORIZATION

Authorization attacks are mostly carried out by entities that already have a legitimate access to the SCADA system. Usually, these attacks are hard to detect, and a constant monitoring of unusual access of parts of system by a user might help detect the attack. But for that, a log containing the information of the activity of each individual user is to be stored properly for training the algorithm to detect a deviation from the normal behavior. The prominent attacks that fall into this category are as follows.

7.5.5.1 Backdoor

Backdoors are secret entrance that bypasses the authentication schemes to access the system. Usually in SCADA, sometimes vendors or manufacturers leave a backdoor to access the system for maintenance or updation of product or service leaving the system susceptible to attack. Backdoor may be introduced by hidden parameters such as login information; old users who have left the organization but their digital access permissions to the system have not been removed, flawed hardening of operating system or database or application server. The backdoor attack once introduced is very difficult to detect. Usually, backdoor attacks are mostly insider attacks. The impact of this attack can be dangerous since the attacker can fully access the system.

7.5.5.2 Directory Traversal

The attack is the unauthorized access to the directory with restricted access and runs commands remotely in the root directory of the web server. This is an HTTP attack, e.g., by manipulating the view field of the HTTP GET URL, where one can fetch a confidential file. This attack takes place due to poor listing of directories or improper or no validation or filtering of the user input.

7.6 DIFFERENT MACHINE LEARNING CLASSIFIERS USED IN IDS

IDS is a security technique that detects vulnerability attack of a computer or an application. IDS is passive in nature and constantly monitors the data traffic without altering the traffic flow itself. There are two main types of IDS in the network as follows:

i. **Signature Based**: A signature-based IDS monitors the traffic to find the specific pattern or signature as predefined by the vendor. Therefore, the type fails to recognize a new type of attack. This IDS needs to be constantly updated with new release from vendor. With fast-evolving attack patterns, it is not a practical solution to use a rule- or signature-based IDS anymore.

ii. **Anomaly Based**: As a solution to the above-mentioned limitation of the signature-based IDS, an anomaly-based IDS was developed to detect unknown attack pattern. ML techniques are used to first develop a model system free of attack and then train the IDS to learn the hardware and software configuration. Any change is compared against the developed model to detect intrusion. The model is more general in nature but sometimes may falsely report a deviation as an intrusion, i.e., a false positive (FP). It is important to choose a proper feature selection algorithm so that the IDS can effectively classify an anomaly as an attack.

As the trend of cyber-attacks is fast changing, it is very difficult to use a signature-based detection scheme since it works only on some prefixed parameters. Therefore, an algorithm that can also evolve with the changing nature of attack is the need of the hour. As discussed earlier, ML can be trained to detect a change from the usual condition without actually defining the type of change. This ability to detect an anomaly is used in the context of SCADA in order to detect predict the changes in the pattern of data flow or data access in the interfacing network. For example, a user is trying to access a data or part of system which he/she usually does not, or in other words, a user is trying to do an unauthorized access. Again, too many HTTP SYN requests coming from a particular network indicating a DoS attack can also be classified as an anomaly, and hence, the attack can be detected. Based on the study of the IDS designed by researchers in IIoT, a number of classifiers are usually used in conjunction. In this section, the prominent classifiers in ML commonly used are discussed. Figure 7.3 illustrates the different types of classifiers discussed here.

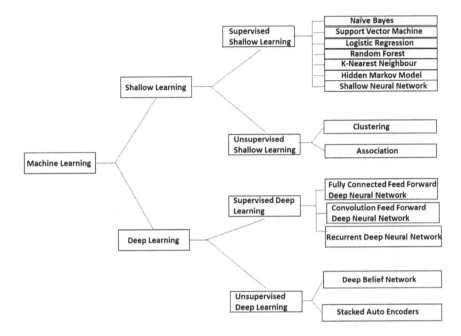

FIGURE 7.3   Classification of different machine learning classifiers.

7.6.1   SHALLOW LEARNING (SL) CLASSIFIERS

This is the traditional ML approach of feature extraction specific to a model and then training the algorithm on that features to later find out anomalies in the input test data. The classifiers in this category based on their type are described shortly in this subsection.

7.6.1.1   Supervised SL

The training data provided to the filters are labeled; i.e., an input maps to a particular output. The popular classifiers in this category used in SCADA IDS are listed as follows:

A. **Naïve Bayes (NB)**: This is a family of simplified algorithm derived from the Bayesian probability model. Here, each feature makes an equal and independent contribution to the final outcome. The probability of an attribute is unaffected by the probability of another one. For n number of attributes, 2n! assumptions are made by the classifiers. The conditional independence assumption regarding the features does not match with the real-world scenario. So the NB is sometimes improvised to explicitly state the dependencies among the attributes.

B. **Logistic Regression (LR)**: This is a set of classifiers that adopt a discriminative model. Like NB, this also assumes that input features are independent of each other. The performance of LR learning model is heavily dependent on the size of the training set.

C. **Random Forest (RF)**: This is a set of classifiers that efficiently run on large datasets. It is capable of handling a large number of input variables and gives a proper estimation of importance of each variable in classification. RF is computationally less exhaustive than other tree ensemble algorithms like boosting. The classifier is capable of generating an unbiased estimate in case of generalizing error. It is robust in handling noise and outliers, and can compute closeness between pairs of cases used to locate outliers.

D. **Support Vector Machine (SVM)**: SVM is a discriminative classifier that separates the different sets of data by drawing a hyperplane or decision plane between them. It is a supervised algorithm that can be used for both regression and classification problems. Each labeled data point is represented as an n-dimensional coordinate vector. The classifier works for both linear and nonlinear data. For nonlinear data, it usually converts it to a higher dimension, works out the hyperplane, and converts it back to the original dimension. So vectors in multidimensional space can also be classified using SVM.

E. **Non-nested generalized exemplar (Nnge)**: This is a data mining technique that computes the distance between the input data with a given set of exemplars for classifying the new data. During the learning process, a set of generalized exemplars or hyperrectangles consisting of a set of examples are constructed. The dimension of a hyperrectangle is specified by a value range in case of numerical attributes. In case of nominal attributes, it is specified by the enumeration of values. Nnge is an effective classifier and was first implemented using Weka toolkit.

F. **OneR**: This is a simple classifier based on one rule (here R stands for rule), which is selected as the one that produces the slightest error for each predictor in the data. The rules are human understandable and output a less accurate classification in comparison with other learners.

G. **J48**: This is a classification algorithm that outputs a DT. This DT can be used as a classifier, and therefore, the J48 is termed as a statistical classifier. This algorithm was developed by the Weka project team.

H. **Shallow Neural Network (SNN)**: The SNN algorithms are based on neural networks where a set of processing elements or neurons are organized in two or more communicating layers. SNN consists of a limited number of neurons and layers, and is mostly used for classification in cybersecurity domain.

I. **K-Nearest Neighbor (K-NN)**: K-NN is also used for classification. It works well for multiclass problems. But the algorithm is computationally extensive in both training and testing phases since to classify a test sample, the sample is compared with all the training samples available.

7.6.1.2 Unsupervised SL Algorithm

The dataset provided here is not labeled, and the learning involves finding a pattern in the dataset and grouping the data by self-organization. The classifiers commonly used in this category are as follows:

A. **K-means Technique**: This is an unsupervised classifier that makes inference from unlabeled input vector. Here, k represents the number of centroids or a real or imaginary location that represents the center or mean of all instances of a cluster. Every data point is allocated to the closest cluster keeping the centroids to minimum. These centroids are randomly selected at the start of the algorithm. With each iteration, the position of the centroids is optimized until there is no appreciable change in their values.

7.6.2 DEEP LEARNING CLASSIFIERS

The deep learning algorithms are based on large deep neural networks (DNNs). The neural networks are organized in many layers and are capable of autonomous representation learning.

7.6.2.1 Supervised Deep Learning Classifiers

The common classifiers belonging to this category are listed as follows:

A. **Fully Connected Feed Forward Deep Neural Network (FNN)**: In this type of DNN, every neuron is connected to all the neurons in the previous layers. FNN does not make any assumption on input data and provides a flexible, general-purpose solution for classification. FNN is computationally extensive.

B. **Convolution Feed Forward Deep Neural Network (CNN)**: In this type of DNN, a neuron of a particular layer receives input from a subset of neurons from its previous layer as opposed to all in FNN. CNN works well with spatial data, but the accuracy is hampered when non-spatial data is used. CNN is less computationally extensive than FNN.

C. **Recurrent Deep Neural Network (RNN)**: This is another type of DNN where the neurons of the current layer can send their output to the previous layer. This network has a complex design and is harder to train than other types of DNN described above. It performs as excellent sequence generator.

7.6.2.2 Unsupervised Deep Learning Classifiers

The two most important classifiers in this category are as follows:

A. **Deep Belief Network (DBN)**: It is an excellent feature extractor. It is trained using unlabeled dataset. DBN is a model composed of unrestricted Boltzmann machine, which is a class of neural network with no output layer.

B. **Stacked Auto Encoders (SAE)**: SAE is a composition of multiple auto encoders, which is a class of neural networks with equal number of input and output neurons. Like DBN, SAE is also a good feature extractor and is mainly used in pretraining task. It performs well with a small set of training data.

7.7 IMPORTANT PARAMETERS TO MEASURE THE EFFECTIVENESS OF CLASSIFICATION

An IDS has been described as a classification problem in ML terms. In all the algorithms, an ML classifier is selected, and then the classifier is trained by extracting the relevant features from the training dataset. The effectiveness of the learning is quantitatively represented using the following parameters.

7.7.1 CONFUSION MATRIX

A simple true and false-based matrix model represents the effectiveness of the classification task. This model assumes a binary classification and assumes that the classifier has only a discrete response of either the condition is met with or it is not. In this case, it represents whether the data is an attack data or not. There are four outcomes of a confusion matrix, as shown in Figure 7.4, which are described with respect to the attack scenario.

7.7.1.1 True Positive (TP)

An event is classified as an attack, and indeed, the network has been attacked.

7.7.1.2 True Negative (TN)

An event is classified as a normal case, and it is a normal network traffic condition in reality.

7.7.1.3 False Positive (FP)

An event is classified as an attack event when in reality it is a normal network traffic scenario.

7.7.1.4 False Negative (FN)

The traffic is classified as a normal one when it is actually an attack event.

Based on the above four prediction results, the performance metrics of a classifier are defined as follows:

A. **Precision (P)**: This gives the percentage of results that are valid. A low precision indicates that the classifier has wrongly classified high number of events as positive. This is measured as

$$P = TP/(TP + FP)\,100 \tag{7.1}$$

		Class Prediction	
		Data Classified as Attack	Data Classified as Normal
Input Traffic	Attack Traffic	True Positive	False Negative
	Normal Traffic	False Positive	True Negative

FIGURE 7.4 Confusion matrix representation.

B. **Recall (R)**: Recall refers to the percentage of the total relevant results that are correctly classified by the algorithm. A low recall score indicates that the classifier has been unable to detect many of the conditions as an attack event. It is measured by

$$R = TP/(TP + FN)100 \qquad (7.2)$$

C. **F1 Score**: This score reflects the balance between precision and recall, and is measured as

$$F1 = 2\,(\text{Precision recall})/(\text{Precision} + \text{Recall}) \qquad (7.3)$$

D. **Accuracy (A)**: This refers to the percentage of samples correctly classified according to the total number of samples tested

$$A = (TP + TN)/(TP + FP + TN + FN)100 \qquad (7.4)$$

E. **False Alarm Rate**: This refers to the percentage of time the system has wrongfully classified an event as true.

$$\text{False alarm rate} = FP/(TN + FP)100 \qquad (7.5)$$

F. **Undetected Rate**: This refers to the percentage of events that was not classified as an anomaly.

$$\text{Undetected rate} = FN/(TP + FN)100 \qquad (7.6)$$

G. **Mathew's Correlation Coefficient (MCC)**: This computes the correlation coefficient of the predicted class and the true class by considering them as binary variables. The higher the correlation, the better the prediction value.

$$MCC = (TP \times TN - FP \times FN)$$
$$\div \sqrt{((TP + FP) \times (TP + FN) \times (TN + FP) \times (TN + FN))} \qquad (7.7)$$

7.8   TREATMENT OF ML CLASSIFICATION FOR DIFFERENT ATTACK SCENARIOS

This section briefly explains how the anomalies in the network traffic pattern for the attacks targeted to compromise different security goals of the network can be detected by the ML classifiers to detect an intrusion.

7.8.1   INTEGRITY

Most of the attacks in SCADA systems are targeted to compromise the integrity of system. Data and command injection attacks are very common among them. Since essentially, the function of IoT in ICS is to remotely procure data and send commands to and from the PLC for a smooth operation, changing or deleting the values passed can disrupt the entire process. ML classifiers are trained to learn the normal command parameters or the acceptable changes in the data value pattern. So a sudden change in the value of a sensor data or a command that is quite different from the usual system behavior can be classified as an anomaly. Also, the IDS can be designed to locate the source of the anomaly and that source can be blocked out to protect the system.

7.8.2 Availability

The most common way to compromise the availability of a system is the DoS attack. Common network analyzers can detect a DoS attack by maintaining a log of the traffic flow. But after that, human intervention is needed to analyze the log. But a properly trained ML classifier can, on the other hand, detect abnormal traffic flow or which node is sending unusually large traffic. Further, it can detect an address with an unfamiliar IP address or a network congestion when the PLC or HMI is overwhelmed with HTTP requests or if either or both HMI and PLC are unavailable.

7.8.3 Confidentiality

In eavesdropping, the normal flow of traffic is not hampered; therefore, the ML-based IDS finds it difficult to detect this attack. But as soon as the attacker tries to snoop or seize the packets to change the codes or data parameters passed, the attack can be identified using the spatial and temporal characteristics of the network.

7.8.4 Authentication

Authentication implies security control, so an effective cryptographic method or key management system can better deal with it. This category falls under prevention since attackers try to bypass the security to gain access to the system. ML algorithms cannot provide security to the system.

7.8.5 Authorization

Authorization activities can be detected by training the normal usage patterns of the registered users. Whenever a user tries to access part of the system or executes commands he/she does not have privilege to, that behavior can be classified as an anomaly. Users occasionally accessing the system might go undetected initially. But with the passage of time, if the learning parameters are set in a manner that the detection process is sensitive, the occasional unauthorized access can be detected as an anomaly.

7.8.6 Pros and Cons of Using ML Algorithms to Detect Network Intrusions

This subsection briefs the biggest advantage that ML algorithms offer when used to detect intrusions, and also lists the important challenges regarding their applicability in the real-world scenario.

7.8.6.1 Advantages of Using ML for IDS

The learning of the ML algorithms evolves with time; therefore, it can easily detect any deviation from the normal traffic even without knowing the kind of attack. Therefore, it might be helpful to detect unknown or improvised attacks to the system. This provides an advantage over rule-based detection system, which needs to know the type or attack scenario beforehand.

7.8.6.2 Challenges of Using ML for IDS

Despite the biggest advantage of ML that can detect an attack even without explicitly knowing its type, there are several challenges to implement the classifiers in order to detect an attack. The challenges are as follows:

A. Attacks that do not bring about any change to the network traffic pattern will go undetected (e.g., eavesdropping attack);
B. Adjusting sensitivity of the learning parameter needs to be well thought out. Since if the sensitivity is too high, there might be an overfitting of data, which might lead to an increase in the number of FPs. Again, decreasing the sensitivity will increase the number of FNs. Therefore, there is a need to strike a balance between the two keeping in mind the type of industrial system and the criticality of the application;
C. In the real world, the attack data comprises less than 1% of the normal data. The challenge of the classification problem increases with such a disbalanced dataset.

7.9   RELATED WORK

This section discusses in short several prominent works done so far in devising an ML-based IDS for IIoT. The heading of each subsection here indicates the name of the algorithm/paper as proposed by the respective authors. A comparative account of the algorithms discussed in this section is provided in Table 7.3.

7.9.1   AN EVALUATION OF MACHINE LEARNING METHODS TO DETECT MALICIOUS SCADA COMMUNICATIONS

In this paper [7], the authors have designed an IDS for a gas pipeline system. The system measures the pressure of the gas and exchanges data accordingly. Six different classifiers used here are J48, NB, RF, OneR, SVM, and Nnge. The labeled data used included injection attacks, both data and commands, apart from the normal data traffic. Some of the features selected are pipeline pressure, length of the command and response data, the control mode of RTU unit, and binary code to indicate if the function code is invalid. Each data packet contains command or response data, PLC address, and raw sensor data. Data and command injection attacks have been carried out. The data has been distorted in seven ways that include sending invalid data or injecting multiple data or data that eventually manipulates the control loop of PLC. Command injection attack has been carried out in four ways by adding multiple

TABLE 7.3

A Comparative Account of the Different Algorithms Discussed

Paper	Related Industry	Attack/Anomaly	Classifier Used
[7]	Gas pipeline system	Command and data injection	J48, NB, RF, OneR, SVM, and Nnge.
[8]	Network data of ICS	DoS, message spoofing.	NB, SVM, and RF
[9]	Water storage system	Command and data injection of PLC	ANN
[10]	Smart Grid	Data injection attack to manipulate meter reading	SVM
[11]	Gas pipeline system	MITM attack that interrupted traffic and intercepted and modified data.	J48
[12]	Conveyor belt control system	MITM attack using ARP poisoning	k-nearest neighbor
[13]	Dairy industry	Anomaly in data traffic using temporal features	Kalman filter
[14]	Smart grid	Multiple false data injection attack	A two-tiered ELM capable of detection and recovery of attack data
[15]	Smart grid	False data injection attack	DNN and DWT that look for deviation of spatial and temporal correlation to detect an attack.
[16]	Smart grid	False data injection attack	Conditional deep belief network
[17]	Tennessee Eastman process	Injection attacks	SVMs employed in two layers for detection and classification of attacks, respectively.
[18]	Smart grid	False data injection attack	CDBN with first layer being CGBRBM and other hidden layers based on RBM.

destination addresses, range of the functions, control parameters passed, or changing the set point of the PLC. The command injection training dataset contains 0.91% attacks, whereas the training data for data injection attack is 13.24% of the normal traffic. The data injection attack, however, is quite higher than the usual proportion of attack data. Each of the six classifiers is used, and the results are compared. Nnge and RF with a recall value of 0.75 and higher were the best classifiers for the data injection attack. The command injection attacks are simple to identify, and all the classifiers have shown good results. The work, however, has not considered other aspects of intrusion attacks like manipulation of time and protocol. Nevertheless, it gives a fair idea about the behavior of different classifiers for detecting an anomaly in the system.

7.9.2   SCADA Networks Anomaly-Based Intrusion Detection System

In this paper [8], Almehmadi has used NB, SVM, and RF for the classification of anomaly on a network dataset of an ICS captured using ETAP-2105. The classifiers were trained for 30 days on both normal traffic and compromised traffic using cross-validation. The data is divided into five categories, namely, packet flags, timespan of communication and traffic frequency, source and destination port, IP address, and protocols used,. The data from the two sets, normal and attacked, are labeled accordingly. In total, twenty features are applied, and it is observed that RF performed best in terms of accuracy in classification and NB performed the least in terms of accuracy.

7.9.3   Effect of Imbalanced Dataset on Security of
Industrial IoT Using Machine Learning

Zolanvari et al. in this paper [9] have attacked the PLC of a water storage system using command injection, and the data is captured using Wireshark and Argus. Twenty-three features have been extracted, including source and destination port, source and destination packet count, rate of packet flow, and average duration of traffic flow. An artificial neural network (ANN) has been trained with the imbalanced dataset with 10%, 1%, 0.7%, 0.3%, and 0.1% of attack data. Normally, a balanced data set helps to classify data effectively but it has been observed that in the real-world proportion of attack, traffic is significantly lower than normal one. It has been observed using Mathew's correlation coefficient (MCC) that for obvious reason, the better the proportion of attack data, the better the value. The attack traffic as low as 0.1% has shown less chance of being classified as an anomaly. This paper has tried to quantitatively point out how effective classification algorithm is used in the real world.

7.9.4   Detecting Stealthy False Data Injection Using
Machine Learning in Smart Grids

In this paper [10], Esfalifalak et al. have dealt with data injection attacks that aim to manipulate the meter reading or bring down parts or even the entire power grid system. Two types of algorithms have been proposed: One detects the anomaly and the other is a classifier. Principle component analysis (PCA) has been used to segregate the different noncorrelated data dimension and dimensions containing data of interest, thereby reducing the total data dimensional space. Next, statistical-based anomaly detection is used where a threshold is defined using historical data. In the second detection method, an SVM is used to detect the attack data. The numerical result is determined using MATPOWER considering a stochastic load in a particular range. From each transmission line, the measurements are taken. There are a total of 34 features used for measurements. These features are passed as input to the algorithms designed. Based on the F1 scores, it is observed that anomaly detection produces better results when the number of training data is less. As the number of training data increases, SVM shows a better performance.

7.9.5   A Hybrid Model for Anomaly-Based Intrusion Detection in SCADA Networks

In this paper [11], an anomaly detection system is developed using the test data that was obtained from a gas pipeline system. The dataset contained the data and control messages passed between PLC and sensors, and PLC and HMI. The protocol used was Modbus. The features consisted of different numeric data that included the set point of PLC, the addresses of the devices that communicated, sensor data reading, and binary data like command response. MITM attack that interrupted the flow of traffic, and intercepted and modified data, was carried out. The training set contained around 22% of attack traffic, which way above the real-life attack data percentage. Using the Weka tool, a J48 classifier was trained. Then, the model was developed using BayesNet. The model detects an anomaly with 100% accuracy.

7.9.6   Detection of Man-in-the-Middle Attacks on Industrial Control Networks

In this paper [12], a conveyor belt control system was built that contained PLC attached to sensors for estimating the position of the product on the conveyor, actuators to switch the system on or off, and HMI. The protocol used here was Modbus over TCP. Data from sensor and actuators were collected during normal operation, which served as the regular data traffic. Thirty-two features were extracted from each dataset. Then, K-NN was applied to the data with k values 3, 5, and 7 and distance measurement using Euclidian, Kernel Euclidian, and Bregman divergence. As per the results, Bregman divergence gave the best result when the three nearest neighbors were considered. The MITM attack was carried out when the conveyor belt was in operation. It was observed that the transmission time of data while the MITM was carried out was significantly higher than that generated during the normal traffic. Therefore, that behavior was detected as an anomaly by the IDS.

7.9.7   Cognitive Secure Shield – A Machine Learning Enabled Threat Shield for Resource-Constrained IoT Devices

Jaya Shankar et al. in their proposed work [13] have developed an IDS taking into consideration the preservation of battery life of the IoT devices. A dairy IoT IDS is designed using ATmega microcontroller, some sensors, accelerators and Bluetooth low energy (BLE). The normal traffic is said to be different from the attack data in two ways: First, the average frequency of waveform is generated by the data traffic and next is the time of communication between devices, which is observed to be significantly higher or lower than the average time. Hence, an attack data with a different frequency can be classified as an anomaly. Here, Kalman filter has been applied. Whenever a traffic is identified as an anomaly, the device disconnects itself and switches to sleep mode. The code is lightweight taking up only 9KB of space. Hence, it can be used for resource-constrained devices.

7.9.8   DETECTION OF FALSE DATA INJECTION ATTACKS IN SMART GRID USING ELM-BASED OCON FRAMEWORK

This paper [14] proposes a two-tiered extreme learning machine (ELM)-based multiple false data injection of smart grid system. IEEE 14 is used while simulating the model where there are 14 subnetworks corresponding to one of each of the total 14 bus nodes. Each layer employs a single-layered feed forward network to classify normal data from injected data. The data analysis of the first layer is sent to the next higher layer. Then, this layer analyzes the data and identifies which bus nodes are attacked. The FDI detection by the subnetworks usually has some percentage of inaccuracy; therefore, the global layer first compares the anomaly reported to a predefined threshold. If the anomaly reported is above that level, the bus node is considered to be attacked. A recovery strategy is designed by using the spatial correlation of the data as learnt from the unattacked history data by the ELM. The non-anomalous data is therefore used to predict the actual data that has been tampered with.

7.9.9   ONLINE FALSE DATA INJECTION ATTACK DETECTION WITH WAVELET TRANSFORM AND DEEP NEURAL NETWORKS

False data injection is often undetected in smart grids. Apart from the circuit laws of Kirchhoff, conventional AC state estimation methods are unable to detect that intrusion. So in this paper [15], the algorithm detects the deviation from the temporal and spatial correlation of system-state data using DNN and discrete wavelet transform (DWT). In the present-day FDI attack, usually attackers make sure that the spatial correlation is maintained. But a temporal correlation of state data exists in the consecutive time slots. The data manipulations usually do not satisfy this temporal correlation. Therefore, considering both spatial and temporal correlations, data deviation can detect inconsistencies in the system. Utilizing DWT, first the system estimates the states at a certain time interval. Then, this data is fed into another detector where a feature extractor computes the correlations of the spatial data. Next, the temporal data correlation is analyzed. The features extracted by using DWT are passed on to the DNN, which detects an FDI attack based on the deviation in the data. The DNN is used here to form a recurrent neural network that learns from the temporal and spatial state data features extracted by the DWT. A simulation is carried out in IEEE 118-bus system with a time interval of 33.3 ms and 5 history data. The system has been able to detect the FDI with 90% accuracy.

7.9.10   A DEEP LEARNING-BASED CYBER-PHYSICAL TO MITIGATE FALSE DATA INJECTION ATTACK IN SMART GRIDS

The system [16] proposes a two-tiered framework composed of CDBN (conditional deep belief network) that contains distributed agents receiving real-time data from the different phasor measurement unit (PMU). The correlation between real-time data is studied, and based on that, the data is classified. It is presumed by the algorithm that the attacker is aware of the topology of the ICS, cannot manipulate Phaser Concentrator data, and can manipulate up to a certain number of PMU.

7.9.11 MACHINE LEARNING-BASED DEFENSE AGAINST PROCESS-AWARE ATTACKS ON INDUSTRIAL CONTROL SYSTEMS

Keliris et al. have designed a process-aware IDS based on supervised learning in their work [17]. The system considers the operation of the ICS for the real-time detection of attack. Three types of attacks are considered separately, namely, the sensor, controller, and actuator attack. Sensor attack targets changing the data value or the sensor reading at an exponential rate. The cost of operation and the production rate also change at the same time. A possible consequence of a controller attack can be decreasing the reactor pressure, thereby increasing the cost of operation. In actuator attacks, the input values of actuators are changed to manipulate the system behavior. The process-aware attack detection includes a trained two-layered SVM.

A hybrid experimental setup has been adopted that linked a Simulink simulation software control loop to a PLC to replicate the environment of Tennessee Eastman process. The controller gain values have been changed and sent to the PLC over FTP. This attack data is then used to train the ML-based IDS. At first, the SVM detects data as an attack. Next separate SVMs are employed to find the type and category of the attack. According to researchers, SVMs have provided a better accuracy to detect the attack so they were chosen as the classifier. By changing the gain variables of the ladder logic, two attacks were launched at the same time: One decreases the reactor pressure and consequently increases the operation cost, and the other one causes the reactor pressure to slowly increase and finally reach the upper limit to shut down the process. Both attacks were successfully detected.

7.9.12 REAL-TIME DETECTION OF FALSE DATA INJECTION ATTACKS IN SMART GRIDS: A DEEP LEARNING-BASED INTELLIGENT MECHANISM

He et al. [18] have considered false data injection attacks that are aimed at stealing electricity from smart grid. A CDBN has been employed to extract temporal features to classify a data as attack data based on time parameters. The first layer uses a Conditional Gaussian Bernoulli-Restricted Boltzmann Machine (CGBRBM) and Restricted Boltzmann Machine (RBM) for other hidden layers. This paper considers DC state estimation model and considers an attack where the attacker knows the network topology and the basic mathematical equation used so as to incorporate the attack vector without getting detected. The CDBN is hence employed to extract the high-dimensional temporal data features to recognize the attack. The simulation has been carried out in IEEE-118 bus systems. Tests were carried out using five hidden layers with a timeframe window size of 3–6, and the highest accuracy was observed at the timeframe window size of 5. The attack data was more than 50% of the testing data, and the detection accuracy was reported to be 95.89. The scalability of the system was tested against IEEE-300 bus power test system. The detection accuracy has been plotted to be more than 95%.

7.9.13 PREVENTION OF FALSE DATA INJECTIONS IN SMART INFRASTRUCTURES

In this paper [19], Vasiliy et al. have proposed a generic algorithm that can detect false data injection attacks in various IIoT applications such as smart grids, healthcare, and smart cities with around 97% accuracy. The spatial correlation between data studied over temporal sequences is recognized in the first phase. The second phase uses a modified k-NN algorithm using Dempster–Shafer theory, which can distinguish between missing and conflicting information. Based on energy consumption data collected from a smart home, 1145 attacks were carried out. Twenty-two attacks went undetected, and about 36 FP attacks were reported. The injections carried out at several sensors at the same time showed a higher detection rate. When more than 50% sensors are attacked at the same time, the algorithm does not guarantee good results, since it is difficult to assess the general spatial–temporal correlation based on neighbor sensor behavior of the system.

7.9.14 HYBRID INTRUSION DETECTION FOR EDGE-BASED IIoT RELYING ON MACHINE LEARNING-AIDED DETECTION

Yao et al. in the paper [20] have implemented LightGBM, a new ML algorithm and a deep learning algorithm to build a two-dimensional IDS. The LightGBM is a modified DT algorithm that splits the leaf node rather than horizontally or depth-wise splitting. This facilitates parallel computation, thus reducing the time for training and communication. Apart from that, feature selection algorithm is also included in LightGBM; therefore, there is no additional necessity for doing feature engineering. The performance of the LightGBM is compared against those of traditional ML algorithms such as NB, DT, and RF on the same traffic dataset for intrusion. The detection accuracy is highest for LightGBM since it has better computational efficiency. The LightGBM algorithm is also lightweight in nature, so it can be implemented in resource-constrained edge devices. The authors have employed LightGBM algorithm in edge devices. On the master edge nodes, a second IDS is implemented based on deep learning algorithm. The edge devices select the advanced traffic features from the raw data and send those selected features over to the edge router. It is assumed that the master edge devices have sufficient resources to run the deep learning algorithm. This reduces the bandwidth consumption of the network to send the data over to the edge router. Further, this extra detection layer also effectively detects an intrusion in the system. This improves the detection efficiency since the edge nodes deliver the relevant advanced traffic features of the data to the master nodes.

7.10 EXPERIMENT ON MAN-IN-THE-MIDDLE ATTACK

Real-life datasets from industries are difficult to procure. So an experimental environment of a pharmaceutical plant has been set up. Due to the limitation in resources, a temperature sensor has been used to detect the temperature of a process, and a buzzer is used which raises an alarm when the temperature level rises above a certain level (here 35°C). The hardware components used in the experiment are listed in Figure 7.5. Using network analyzer tool Wireshark [21], the activity of the communication network over which the sensor node is sending the reading is captured. This is the normal data traffic.

Next, using the tool Ettercap [22], the MITM attack is launched using Address Resolution Protocol (ARP) poisoning technique. In ARP poisoning, the attacker can listen to traffic both ways. Since there is no authentication provided, the attacker can connect with the victim connected to the same LAN with the MAC address of its machine and IP address of another legitimate machine in communication with the victim. Again using Wireshark, the network information is logged. This is the attack data that is used in the training set. None of the packets have been dropped in the attack. The prominent features from packet headers are selected using cross-validation methods. Some of the features are given in Table 7.4.

Of the 30,000 total traffic samples selected, a ratio of 4:1 has been used to divide it into training set and testing set. Of the 30,000 total traffic samples, 29,700 are the normal traffic and 300 are the attack traffic data. This data is inputted into five classifiers, namely, SVM, NB, K-NN, RF, and DT using the sci-kit learn library of Python.

Hardware Used	Description
ESP 8266-12E	WIFI transceiver wireless Module
LM 35	Temperature Sensor
Buzzer	The one used raising an alarm
HP 15q Core i3 7th Gen Laptop	Receives the sensor reading from the WIFI Module, analyse data and send command to the buzzer module.

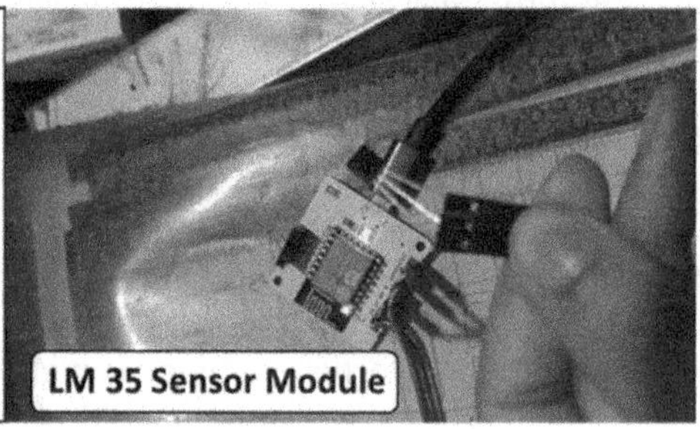

LM 35 Sensor Module

FIGURE 7.5   Hardware components used in the experiment.

TABLE 7.4
Some of the Features of the Traffic for Classification

Feature	Data Type	Description
IP identification number	Numeric	Uniquely identifies the IP packet
TCP sequence number	Numeric	32-bit number used by both clients of a TCP session. Keeps track of the amount of data interchanged.
TCP checksum	Numeric	Changes when the payload changes. If the attacker changes the payload, even if that change is difficult to detect, the checksum changes.
IP header length	Numeric	The length of the header of IP packet
IP total length	Numeric	Total packet length
TCP header length	Numeric	The length of header of TCP packet
Source port	Numeric	Source port number
Destination port	Numeric	Destination port number

7.11 RESULTS AND DISCUSSIONS

According to the parameters discussed in Section 7.7, a comparative account of the classifiers are charted according to their robustness in behavior to detect an anomaly.

A. **Accuracy in Classification**: Accuracy of all the classifiers used is measured in accordance with Equation 7.4. The accuracy results are charted in Figure 7.6 for the five classifiers used. Since the attack data percentage is low, all the classifiers show a good measure of accuracy. The RF is the most accurate classifier, and NB the least.

B. **False Alarm Rate**: Most of the classifiers apart from NB have shown a very low percentage of traffic misclassification when the outputs of the classifiers are analyzed using Equation 7.5 and the results of each are plotted in a bar chart as shown in Figure 7.7.

C. **Undetected Rate**: The undetected rate is measured according to Equation 7.6, and the plotted results for each classifier are displayed in Figure 7.8. SVM has shown the worst performance and has classified more than 30% of the attack data as normal data. However, NB and RF have shown the best performance.

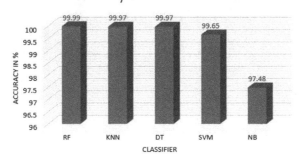

FIGURE 7.6 Comparison of accuracy of the classifiers.

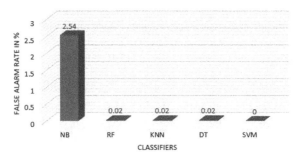

FIGURE 7.7 Comparison of false alarm rate of the classifiers.

Undetected Rate in Classifiers

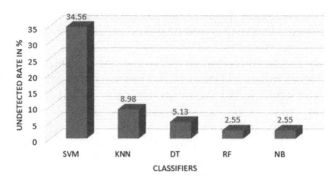

FIGURE 7.8   Comparison of undetected rate of the classifiers.

Recall value in Classification

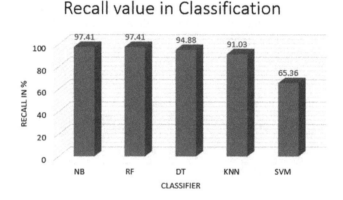

FIGURE 7.9   Comparison of recall value of the classifiers.

D. **Recall Value**: The recall values of the five classifiers as measured by Equation 7.2 are represented in the bar chart in Figure 7.9. A low true-positive rate of SVM has been the reason for a poor recall value during classification.

E. **Mathew's Correlation Coefficient (MCC)**: Among all the metrics selected, this parameter as represented by Equation 7.7 in Section 7.7 is the best so far to measure the effectiveness of classification. The phi coefficient in statistics has been renamed to MCC in ML context.

MCC gives equal importance to both the classes in consideration. In case of network attack as stated in challenges of using ML in Section 7.8, it has been mentioned that the real-world dataset is highly imbalanced. Therefore, it is very important to give stress to both the classification and misclassification results. Especially, the ratio of misclassification of an attack as normal traffic is an important parameter to determine whether the classifier or the algorithm can be implemented for the IDS. The MCC of

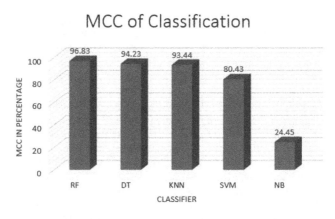

FIGURE 7.10 Comparison of MCC of the classifiers.

the classifiers is calculated according to Equation 7.7, and on plotting against each other in Figure 7.10, it is found that RF is the best performer and NB the worst.

7.12 CONCLUSION

The rate of increase of cyber-attacks has been alarmingly high, so an efficient cyber-security system is essential for industries. Before the incorporation of IoT in ICS, the system is used to work in a centralized manner inside a private network. But the present-day business competition demands a geographically distributed business operation for better efficiency of the business processes. Due to little advancement of the infrastructure of network components in traditional business systems and inefficiency of the communication protocols of IoT, it is extremely important to have an efficient IDS for industries.

This paper has provided a study on the various security goals specific to ICS and the associated vulnerabilities and why ML-based techniques provide a better detection system. Different filters have been discussed, and some of the prominent works on the SCADA system have been summarized. Lastly, a small experiment is conducted to show the effectiveness of some of the most commonly used classifiers when they need to classify an imbalanced dataset with an attack traffic of 1% keeping in mind the real-world attack data scenario. The future research area will be possibly focused on how to design a ML-based IDS that reduces the number of FNs close to 0 so as to classify almost all the attack scenarios properly.

REFERENCE

1. https://www.symantec.com/content/dam/symantec/docs/reports/istr-24-2019-en.pdf last accessed on 21.12.2019.
2. Nazir S, Patel S, Patel D. Assessing and augmenting SCADA cyber security: A survey of techniques. *Computers & Security*. 2017 Sep 1; 70: 436–54.
3. https://gcn.com/articles/2016/06/27/cyberattack-energy-transportation.aspx last accessed on 23.12.2019.

4. https://www.google.com/amp/s/www.computerweekly.com/news/252460353/Cyber-attacks-targeting-industrial-control-systems-on-the-rise%3famp=1 last accessed on 23.12.2019.
5. https://www.elprocus.com/scada-systems-work/ last accessed on 27.12.2019.
6. https://www.csoonline.com/article/2134000/access-control-spear-phishing-poses-threat-to-industrial-control-systems.html last accessed on 27.12.2019.
7. Beaver JM, Borges-Hink RC, Buckner MA. An evaluation of machine learning methods to detect malicious SCADA communications. In *2013 12th International Conference on Machine Learning and Applications* 2013 Dec 4 (Vol. 2, pp. 54–59). IEEE.
8. Almehmadi A. SCADA networks anomaly-based intrusion detection system. In *Proceedings of the 11th International Conference on Security of Information and Networks* 2018 Sep 10 (pp. 1–4).
9. Zolanvari M, Teixeira MA, Jain R. Effect of imbalanced datasets on security of industrial IoT using machine learning. In *2018 IEEE International Conference on Intelligence and Security Informatics (ISI)* 2018 Nov 9 (pp. 112–117). IEEE.
10. Esmalifalak M, Liu L, Nguyen N, Zheng R, Han Z. Detecting stealthy false data injection using machine learning in smart grid. *IEEE Systems Journal.* 2014 Aug 20; 11(3): 1644–52.
11. Ullah I, Mahmoud QH. A hybrid model for anomaly-based intrusion detection in SCADA networks. In *2017 IEEE International Conference on Big Data (Big Data)* 2017 Dec 11 (pp. 2160–2167). IEEE.
12. Eigner O, Kreimel P, Tavolato P. Detection of man-in-the-middle attacks on industrial control networks. In *2016 International Conference on Software Security and Assurance (ICSSA)* 2016 Aug 24 (pp. 64–69). IEEE.
13. Vuppalapati JS, Kedari S, Ilapakurti A, Vuppalapati C, Chauhan C, Mamidi V, Rautji S. Cognitive secure shield-a machine learning enabled threat shield for resource constrained IoT devices. In *2018 17th IEEE International Conference on Machine Learning and Applications (ICMLA)* 2018 Dec 17 (pp. 1073–1080). IEEE.
14. Xue D, Jing X, Liu H. Detection of false data injection attacks in smart grid utilizing ELM-based OCON framework. *IEEE Access.* 2019 Mar 4; 7: 31762–73.
15 James JQ, Hou Y, Li VO. Online false data injection attack detection with wavelet transform and deep neural networks. *IEEE Transactions on Industrial Informatics.* 2018 Apr 10; 14(7): 3271–80.
16. Wei J, Mendis GJ. A deep learning-based cyber-physical strategy to mitigate false data injection attack in smart grids. In *2016 Joint Workshop on Cyber-Physical Security and Resilience in Smart Grids (CPSR-SG)* 2016 Apr 12 (pp. 1–6). IEEE.
17. Keliris A, Salehghaffari H, Cairl B, Krishnamurthy P, Maniatakos M, Khorrami F. Machine learning-based defense against process-aware attacks on industrial control systems. In *2016 IEEE International Test Conference (ITC)* 2016 Nov 15 (pp. 1–10). IEEE.
18. He Y, Mendis GJ, Wei J. Real-time detection of false data injection attacks in smart grid: A deep learning-based intelligent mechanism. *IEEE Transactions on Smart Grid.* 2017 May 11; 8(5): 2505–16.
19. Krundyshev V, Kalinin M. Prevention of false data injections in smart infrastructures. In *2019 IEEE International Black Sea Conference on Communications and Networking (BlackSeaCom)* 2019 Jun 3 (pp. 1–5). IEEE.
20. Yao H, Gao P, Zhang P, Wang J, Jiang C, Lu L. Hybrid intrusion detection system for edge-based IIoT relying on machine-learning-aided detection. *IEEE Network.* 2019 Oct 9; 33(5): 75–81.
21. https://www.wireshark.org/ last accessed on 19.11.2019.
22. https://www.ettercap-project.org/ last accessed on 19.11.2019.

8 Person Authentication Based on Biometric Traits Using Machine Learning Techniques

Gautam Kumar, Debbrota Paul Chowdhury,
Sambit Bakshi, and Pankaj Kumar Sa
National Institute of Technology

CONTENTS

8.1   INTRODUCTION

Internet of Things is the collection of interconnected devices connected to the Internet. To collect, store, and exchange data among each other, these devices are embedded with different types of actuators, wired/wireless sensors, and software along with other required electronic objects. Each device has an Internet Protocol (IP) address and is capable of collecting and transmitting information across a network without human assistance or interference. Some examples of IoT are autonomous vehicles, smart homes, wearables, connected healthcare systems, and tracking and monitoring systems. The architecture of IoT can be represented by the following four stages:

Networked Devices: Embedded sensors, actuators, and other necessary electronic devices collect information from the real physical world for further processing.

Data Acquisition System: The information collected from the previous stage is usually in the form of an analog signal. The role of the data acquisition system is to aggregate and convert this signal into digital form for further processing.

Edge Analytics: After digitizing the data, the edge analytic stage is responsible for preprocessing and enhanced analysis before feeding it to cloud analytics.

Cloud Analysis: This is the final and crucial stage of IoT architecture where in-depth analysis happens and feedback is generated. Based on revision and meeting the quality and requirements, data is forwarded to the cloud-based system or physical data center. Figure 8.1 represents the four-stage IoT architecture.

8.1.1   APPLICATION OF IoT

IoT has been implemented in several domains, from smart home and smart city to industrial automation, personal to social applications, agriculture to the healthcare system. IoT's ability can still be used for the benefit of society to develop new applications. Some of the important real-world applications of IoT are mentioned below.

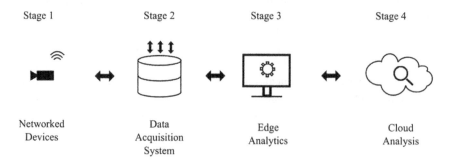

| Stage 1 | Stage 2 | Stage 3 | Stage 4 |

| Networked Devices | Data Acquisition System | Edge Analytics | Cloud Analysis |

FIGURE 8.1   Architecture of IoT showing all four stages.

Wearables: Wearable technology is probably the first IoT-deployed industry, and the demand for its products increased drastically. Now, smart watches are used to monitor sleeping/walking activity, measure heartbeats, and monitor blood pressure. Glucose monitoring devices are being used to measure the diabetic status of people. A tiny electrode that is responsible for collecting information is placed under the skin. Collected data from the sensor are sent to the monitoring device using radio frequency.

Smart Home: Sensors and actuators that are connected to household appliances, such as air conditioners and refrigerators, can monitor the environment in a house or office. These devices can also turn on/off light sensors such as electric bulbs/CFLs according to day/night. A temperature sensor can control air condition based on the current weather. This will save a greater amount of electric energy as well as money. An intelligent smoke detector sensor can be used to detect fire/smoke and generate an alarm. The system can also be used to send the alert message to the nearest hospital and fire brigade department. A smart camera can detect accidents and robbery and generate alarm as well as alert messages to the nearest hospital as well as police station for further action automatically without users' intervention. IoT-enabled water tab and washing machine can save a huge amount of water.

Nowadays, companies are collaborating with each other to make IoT-enabled products to provide better service to customers. Philips has worked on a project named "Home of the near future", where they designed intelligent physical objects used at home. A similar effort in IoT was made by Siemens, where they tried to make every device intelligent without any interconnection among them. The agent-based smart home simulation system was created at "The Multi-Agent Systems Lab" by the University of Massachusetts. Intel also released many home networking protocols (including UPnP), which are open-source implementations [1]. Although we have developed many intelligent devices to make home smart, there are some open challenges such as security, adoption to a new environment, and high cost of intelligence, which still need to be addressed [2].

Smart Cities: According to the UN report, by 2050, 66% of the world population will live in urban areas. In the survey, it is reported that Tokyo remains the world largest city (38 million inhabitants), followed by New Delhi (25 million), Shanghai (23 million), and Mexico and Mumbai (21 million each)[1]. With a drastic increase in the population, the consumption of resources and the distribution of energy become a challenge. To minimize consumption and provide better service to the people in the city, IoT emerged as a great solution to this problem. The use of IoT can optimize resources such as water supply management; traffic network – to avoid congestion; power grid; and parking space. Cleanliness is also one of the

---

[1] UN Report: World's population increasingly urban with more than half living in urban areas, July 10, 2014, New York, https://www.un.org/en/development/desa/news/population/world-urbanization-prospects-2014.html

important factors to make a city smart. The IoT technology can be used for the development of a smart waste management and sewage disposal system [3]. IoT-enabled meters can be used to monitor efficient electricity consumption and theft of electricity around the world [4]. Nowadays, IoT devices are used to monitor air and sound pollution and the purity of water supply. The IoT-enabled devices used to make a city smart are actually physical and can be vulnerable to break of a cyber-attack. Therefore, the data collected by these devices must be secured to assure the safety of people in the city where they live and work.

Industrial Automation: For any industry, the quality of the product and faster development play a crucial role in the return of the investment. A moderate-scale industry requires high manpower to maximize the development of products. Their production can be faster if the IoT technique is used for the automation of products, even from re-engineering to packaging. A plant/factory can be operated and maintained remotely; information exchange between IoT-enabled devices can maximize the productivity [5]. Real-time data acquired by sensors can be used to automate and quicken the manufacturing process at a low cost. However, using IoT in industry automation is very challenging. General challenges such as data and service security, trust, information privacy, and data integrity need to be assured.

Health Care: While we have achieved significant results in improving people's health through medicines and antidotes, accuracy is still needed in medical diagnostic reporting and disease detection at an early stage. Health care is another area where IoT can be used to track people (staff and patients) in real time; monitor the collection of test samples such as blood and urine; automatically gather information such as blood pressure, diabetes, and body temperature according to the scheduled date and time; and authenticate hospital staff to access a secure place and monitor patients based on their biometric characteristics and help them to avoid mistakes that are usually made, such as drug, dose, and time [6,7].

Agriculture: This is one of the sectors that can be largely influenced by the use of IoT [8]. In countries such as India, Bangladesh, and Brazil, most of the people stay in rural areas, and their profession is agriculture. The use of IoT-enabled devices not only decreases human effort but also increases productivity [9]. Enabling automation in agriculture by collecting information such as nutrients and humidity suggests the best time for irrigation; optimizing the use of fertilizers can save money, time, and effort of farmers.

The application of IoT is not limited to the domains discussed, but, in fact, it is growing exponentially in all other domains. It provides a reliable, fast, and efficient service to society by collecting real-time data of people. Therefore, privacy and security in IoT is one of the concerning subjects for researchers.

The rest of the chapter is organized as follows: Section 8.2 describes the types of security and attacks that can be made in IoT. The introduction of biometrics and its types along with the working model of a biometric system is presented in Section 8.3. Section 8.4 discusses the vulnerability in biometrics, and methods to secure biometric system components along with machine learning techniques to detect attacks on a biometric system. The ML algorithms used to develop a biometric system are presented in Section 8.5. An experiment using deep learning is performed in Section 8.6, followed by experimental outcomes and analysis of results. Finally, Section 8.7 concludes the study.

8.2 TYPES OF SECURITY AND ATTACKS IN IOT

IoT is nothing but a collection of devices that are connected with each other as well as the Internet. These devices collect each and every type of data, such as a person's name, age, address, financial transition status, credit/debit card details, health information, and biometric data, and store them in the device. They also share these details with other devices whenever needed. The interconnected and internetworking architecture of IoT makes it most vulnerable to attack. The following are the five common types of attacks that can be made in IoT [10]:

Man-in-the-Middle Attack: In this type of attack, an attacker invades communication between the sender and the receiver, and the invader acts as an original sender and sends a fake message to the receiver, while the receiver thinks that he/she is getting a message from the actual sender.

Botnet: A botnet is a network of devices integrated for remote control and malware delivery. This type of attack is used by hackers/criminals to steal personal data, banking information, and push emails [11].

Denial-of-Service (DoS): This type of attack generally happens when a service that usually works is unavailable. At the time of unavailability, devices through botnet are programmed to request the service [12].

Social Engineering: In this type of attack, the goal of an attacker is to get personal information such as email ID and bank account details from an individual. Attackers try to access the target system and install malicious software so that whenever authorized persons access the sensitive data, it can redirect these secure data to attackers.

Physical Attack: Due to the distributed nature of IoT, most of the devices are used outdoors, and attackers try to tamper with hardware components [10].

The need for security is not limited to accessing secure data over the Internet or computer system. There are many applications that we access in our daily life, such as banking, mobile devices, physical access of IoT-enabled devices, and computer resources, which need authentication. To certify IoT as a secured system, identification of a person who accesses IoT-enabled devices physically is essential. In order to enhance security in IoT, we generally use an identification card or smart card, password, or PIN. These conventional methods of authentication are still used by us, but there are certain limitations; for example, a smart card can be lost or stolen, and remembering complex passwords is difficult, while an easy password can be guessed by the hacker with few attempts. Biometric is one of the solutions to overcome these challenges and issues. Therefore, in the next section, we discuss biometrics, types of biometrics and its traits, and the working model of a biometric system in detail.

8.3 INTRODUCTION OF BIOMETRIC SECURITY

Compared to traditional methods of authentication, biometrics proved to be the most secure one. In simple words, biometrics is the process of identifying a person based on their biological traits. The most commonly used traits for authentication are fingerprint, face, iris, ear, gait, etc. The selection of these traits depends on the application and deployment environment. Biometrics is divided into two types of modality: (1) behavioral and (2) physiological. The traits under both types of modalities are shown in Figure 8.2. A detailed description of these modalities is given below.

8.3.1 BEHAVIORAL MODALITY

Behavioral biometrics is the field of study related to the analysis of behaviors in human activities that are unique in recognition and measurement. It includes analysis of the pattern of mouse use, gait pattern, and keystroke pattern; signature analysis; and analysis of the pattern of holding a mobile phone and pressing buttons [13]. Some of these traits that are mostly used for authentication are explained below.

Gait: The walking pattern of a human being is known as gait pattern. It has been proved that the gait pattern of each person is unique and it can be used to identify an individual [14]. Using IoT-enabled devices and sensors (wearable and non-wearable), it is easy to capture gait data and extract statistical features for further analysis [15]. Parameters and features that can be extracted by gait pattern using machine learning techniques are the velocity of a subject, placement distance between two successive steps, foot direction, the number of steps in a unit time, etc.

Signature: Signature matching is still used by some systems to identify people every day. However, it is easy to copy someone's style of signature. Many applications have been developed to recognize a person based on their signature. These systems measure the speed, shape, speed of strokes, acceleration, and pressure on the pad while signing [16]. Machine learning

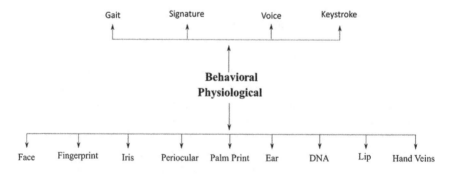

FIGURE 8.2 Biometric traits in physiological and behavioral modalities.

algorithms are widely used to improve the performance of the system. However, accuracy is still a challenging issue and needs to be improved.

Keystroke: In Ref. [13], it has been reported that a user can be identified based on their typing pattern. The speed at which the user presses a key, the time between two successive keypresses, finger placing pattern, etc., can be measured by machine learning techniques. A similar effort was made in Ref. [17], where a better result was obtained using an optimizer in deep learning.

Voice: IoT-enabled sensors can be used to measure the movement of lips, jaws, and tongue. Parameters such as the pitch and intensity of sound can be considered as a feature, and machine learning/deep learning models can be trained to recognize a person based on their voice [18].

8.3.2 PHYSIOLOGICAL MODALITY

Fingerprint: From academics to companies, the physiological biometric system is one of the most used applications. The reason behind this is that in any environment, it can be easily installed and used by the users. Cost-effective maintenance makes it suitable for small organizations. The texture pattern of the human fingers has been shown to have ridges and valleys (also known as "minutiae" points) and is unique for each person. These features can be extracted using image processing and machine learning techniques, which can be used to match a query with database samples [19].

Face: After fingerprint, face is another characteristic that is used to identify an individual. The benefit of this approach is that it does not require user participation and is ideal for an unconstrained environment. Features such as local binary pattern (LBP), principal component analysis (PCA), linear discriminant analysis (LDA), and phase intensive global pattern (PIGP) are some popular algorithms used to extract features from the face image. These features are used to train machine learning models using various ML techniques [20–23].

Iris: It is considered one of the most accurate systems among all biometric applications, but it requires a friendly environment to capture users' iris samples. It is believed that the colored structured pattern of the iris is special and stable throughout life. These characteristics make it a candidate to identify a person. Other features such as pits, stripes, and furrows are being used to differentiate one person from others. Daugman used 2D wavelet for feature encoding [24]. The experimental result claims recognition is done with good matching speed and excellent accuracy.

Periocular Region: The periphery of the ocular region is known as the periocular region. This system emerged when iris and face recognition failed to recognize a person. It was pioneered by Park et al. in 2009. The experimental result suggests that it can be used as an alternative to iris and face, especially when acquisition of high-resolution iris image is difficult, only partial faces are available, and identification in an unconstrained environment is needed [25,26].

Ear: Ear biometric has recently emerged as a feature for a person's authentication. It has been proved as one of the most consistent human anatomical characteristics that do not change with age. This biometric is mostly used by the police as well as forensic specialists and successfully used to solve many cases in the United States and Netherlands [27,28].

Palm Print: This biometric is mostly used by forensic specialists to recognize criminals based on the palm print available on a knife or gun. Like a fingerprint, it also contains a minutia-like structure, which is extracted as a feature and used to recognize criminals [29].

8.3.3 WORKING MODEL OF A BIOMETRIC SYSTEM

A biometric system consists of five modules as shown in Figure 8.3. The first module is biometric sample acquisition, where a sensor is used to capture the image of a subject. In the second module – extraction of the region of interest (ROI) and preprocessing – the ROI of a particular trait is cropped from the whole image, and preprocessing is done to remove noise and enhance image quality if required. The third module is feature extraction, where various image processing and machine learning techniques are used to extract important and unique patterns and for training and testing models with these features. This module is considered important because the accuracy of the biometric system largely depends on the uniqueness of extracted features from a particular trait. If the user is visiting for the first time, then the samples are stored in a database so that when he/she returns and claims to authenticate by giving query, then the system can identify. This process is called enrollment. If the user is already enrolled and gives a query to authenticate themselves, the system takes a query sample as input and matches it with the database template. If the algorithm finds a match with a score (S) less than the threshold (T), then it is identified as a match (genuine user), otherwise a non-match (imposter).

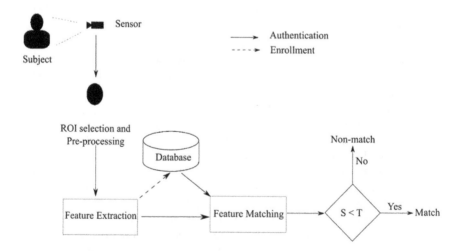

FIGURE 8.3 Working model of a biometric system.

8.4 VULNERABILITIES IN BIOMETRICS AND ML TECHNIQUES TO SECURE THE BIOMETRIC SYSTEM

Compared to traditional methods of authentications, the biometric system is the most secure one. However, attacks on modules such as biometric sample collection, feature extraction, database, and decision-making modules can make it vulnerable. This section highlights the possible attacks that can be made on a biometric module and methods to detect and secure biometric components using machine learning techniques.

8.4.1 VULNERABILITY IN BIOMETRICS

Four major modules that are most vulnerable to several types of attacks are explained below.

> **Attack on Data Collection**: These types of attacks can be made on sensors used to acquire data such as cameras and scanners. Fake fingerprint or face image can be used during the operation. It might be possible that an imposter can register themselves as a genuine person into the system.
> **Attack on Feature Extraction Technique**: By analyzing and exploring the weakness of feature extraction algorithms, an attackers can break into the system. They can also change the feature characteristics by introducing malware into the system. An attacker can set features that are preselected by them to be extracted by the algorithm from a genuine person.
> **Attack on the Biometric Database**: This is one of the crucial attacks that can be made by an attacker. Here, they can delete the target biometric sample from the dataset or can replace it with a fake sample made by themselves. It is also possible to duplicate an existing sample using reverse engineering.
> **Attack on Decision-Making Module**: Attacks on this unit can be made by introducing malware that can always produce a preselected matching score.

8.4.2 METHODS TO SECURE A BIOMETRIC SYSTEM

A typical biometric system can be insecure in terms of vulnerability and attacks on different components. Two major components that are more likely to be compromised are explained below, along with methods to secure it.

8.4.2.1 Security of a Biometric System Against Vulnerability of the System's Components

> **Database Security**: The interconnected devices in IoT continuously acquire data from the environment and store them in the devices themselves or in the database. These data are stored as features/templates, which are further used for identification/verification purposes. When a user authenticates themselves by giving a biometric impression, the given sample is matched with the database templates. This might be the time when an attack can be

made on the templates stored in the database, or during the communication process, a hacker can attack the channel to steal a template.

Four basic approaches that are used to protect a database from attacks are biometric cryptosystems, characteristic transformation, watermarking, and haptic–biometric template protection [30–32]. A secure cryptographic key is produced using password-based key generation systems or a biometric cryptographic key [33]. Biometrical features are reconstituted using a private key in transformation approaches, and a new version of the biometric template is saved in the database [34].

Data Acquisition Security: There are four types of attacks that are performed during data acquisition, which are denial-of-service, spoofing, false enrollment attacks, and latent print reactivation. To avoid attack in this phase, Noh et al. proposed a touchless sensor concept and collected all five fingerprints simultaneously [35]. As a result, it is reported that on average, a touchless sensor takes 57% of a whole rolled image and has 63 true minutiae, while a touch-based sensor takes 41% of a whole rolled image and has 40 true minutiae.

In Ref. [36], triangulation imaging principle-based 3D ear recognition was proposed. The suggested method was evaluated on 3000 samples collected from 500 subjects. The outcome of the experiment proves that the proposed method minimizes the cost and speeds up the identification process.

8.4.2.2 Security of a Biometric System Against Several Types of Attacks

The attacks on a biometric system can be classified as direct and indirect attacks [37]. Direct attacks are carried out by attacking the biometric identification devices themselves, while in indirect attacks, attackers try to break security by invading the communication channel or any software module. In direct attacks, a different identity can be created by the attacker without giving any biometric trait, while in spoofing, an attacker tries to enter into the biometric system by presenting an identity of someone else [38].

To secure the biometric system from spoofing attacks, it is necessary to distinguish between real biometric traits presented to the sensor and artificially created samples. For this purpose, in Ref. [39], an anti-spoofing technique was proposed for face recognition applications. The studied method is divided into three levels, i.e., data acquisition level (sensor), feature extraction level, and score generation level. At these levels, different machine learning techniques are used for different purposes.

A similar effort was made by Chingovska et al. in Ref. [40]. They developed an anti-spoofing framework named Expected Performance and Spoofability (EPS) using the Replay-Attack database. They used four baseline face verification systems based on Gaussian mixture model (GMM), local Gabor binary pattern histogram sequences (LGBPHS), Gabor jets (GJet), and intersession variability (ISV) modeling using discrete cosine transform (DCT) features. The performance of the system is evaluated by calculating the false acceptance rate (FAR), false rejection rate (FRR), half total error rate (HTER), and SFAR (the rate of wrongly accepted spoofing attacks). Compared to all, the performance of ISV is found to be good.

In Ref. [41], the vulnerability of face recognition-based biometric against spoofing attacks was investigated. A part-based face representation and GMM proposed by McCool and Marcel [42] was used for the face verification system. The studied method was evaluated on 1300 video clips, and the author investigated attacks attempting to collect 50 different identities. The experimental result shows the maximum vulnerability of the face database.

Choi et al. studied denial-of-service (DoS) attacks [43]. They found a vulnerability in Yoon and Kims [44], such as lack of anonymity, biometric recognition error, verification problem, and vulnerability to a DoS attack. Extending Yoon and Kims' work, they proposed fuzzy extraction-based biometric authentication in the wireless sensor network. They conducted a security analysis experiment to evaluate the performance of the proposed method and found the suggested approach is more secure compared to other authentication methods.

To cope with the replay attack, in Ref. [45], genetic and evolutionary biometric security (GEBS) was proposed by Shelton et al. They developed disposable feature extractors (FEs) using genetic and evolutionary computation and proposed two protocols based on the use of FEs and their template or feature vector (FV), which are used to authenticate a person. Experimental outcomes claim that the developed applications are successfully used to overcome the issue of a replay attack.

Most of the time, people think that templates stored in the database are secure, but in fact, those stored templates can be manipulated by the attacker. Therefore, in Ref. [46], Shehu et al. studied a method to detect tampering in the features stored in the database. SVM is trained using original as well as tempered templates of the fingerprint database. The experimental result shows that the suggested approach can detect alteration with 90% accuracy.

The attacks made on the various modules, especially at data acquisition devices, can be avoided by extracting discriminating features and using appropriate machine learning classification techniques to make a final decision. A classification technique is said to be good if it is able to classify genuine and imposter users and minimize false acceptance and rejection with sufficient accuracy. The most popular and commonly used classification techniques in machine learning are discussed below.

8.5 MACHINE LEARNING ALGORITHMS USED TO DEVELOP BIOMETRIC SYSTEMS IN IOT

The performance of a biometric system can be improved by using various machine learning techniques to extract and analyze dominant features. Several classification algorithms have been proposed and successfully used to classify genuine and imposter subjects from a large population. This section discusses the classification methods that exist in the literature, which are generally used to develop a biometric system.

k-Nearest Neighbors (KNN): KNN is one of the most fundamental but essential classification algorithms. It is also called a lazy method of learning, which stores the vectors in the training dataset, and until classification, all processing is delayed. The feature vector is extracted and compared with all stored samples by means of a distance function when the system is requested with a query sample. The algorithm returns class as a result along with the *k*-nearest neighbor of the queried sample. It is used in the supervised learning field and can be used to detect intrusion and in data mining applications and pattern recognition. Due to its ability to classify data efficiently, it is also used in biometrics.

In Ref. [47], an efficient method of personal identification using KNN was proposed. Extending the work of Gunetti et al. (where they compared query input to every sample of the database [48]), in this study, the use of KNN classifier is limited to the number of comparisons within a cluster. However, the calculated FAR and FRR are found to be as in [48], but authentication performance is improved by 66.7%. ANN is one of the powerful and frequently used techniques to train a model. However, with less amount of data, the system fails to classify a new test sample. Therefore, in Ref. [49], Azmi et al. trained KNN by extracting the Freeman chain code (FCC) as a feature. Performance result is obtained by evaluating it on the MCYT bimodal database using Euclidean distance as a similarity measure, and verification is performed using KNN. The experimental result claims that the proposed method produces better outcomes compared to other methods when tested on the same database with an FRR and FAR of 6.67% and 12.44%, respectively, with an AER of 9.85%.

Support Vector Machine (SVM): SVM is a widely used classifier for classification and regression problems. Its goal is to fit a hyperplane that can divide different classes. Due to its power and simplicity to train/test the ML model, it is widely used in biometric applications [50,51]. It has been successfully used to classify people based on their physiological and behavioral characteristics. In Ref. [52], people identification based on user interaction patterns with the mobile device was proposed. Touch data were collected, and a model was trained by learning the touch pattern of users during the enrollment phase. KNN, as well as SVM with RBF kernel, was used for classification. The experimental result shows that the system is able to accept/reject people based on the way he/she uses the device. The proposed method yields an EER of 0% and between 2% and 3% in an intra-session and intersession authentication environment, respectively.

Person identification based on the finger vein pattern was proposed by Wu et al. [53]. In their study, they used CCD and infrared LED cameras to capture samples. Images were preprocessed to remove noise, and features were extracted using linear discriminant analysis (LDA). To speed up the identification process, principal component analysis (PCA) was applied to reduce the feature dimension. SVM was used for classification. From the experiment, 98% classification accuracy was observed, and the system can identify people in 0.015 seconds.

Naïve Bayes (NB): The NB classifier assumes that the probability distribution of the feature vector of a subject is the same as if that user returns in future and claims for authentication by giving the same biometric trait that was given during enrollment. It also assumes the values of feature vectors extracted from a subject are different and unique to the features of other subjects.

To show the effectiveness of the NB classifier, in Ref. [54], a finger vein-based biometric identification system was developed. Database images were preprocessed, followed by the region-of-interest (ROI) selection; afterward, fast filter, Gabor filter, and freak descriptor were used to extract the feature that was trained using the naive Bayes classifier. For comparison of the results, discriminant analysis of extracted features was done. The experimental result shows the superiority of the NB classifier with an accuracy of 98.38% over discriminant analysis (94.46%). Although it was possible to extract discriminating patterns using feature extraction techniques, all these patterns were not useful and increased the dimension of features. Therefore, a feature subset selection approach was studied by Kumar et al. in Ref. [55]. They extracted features from hand images and selected a subset of the pattern from whole features, which are necessary to build an identification system. They used several classifiers such as KNN, SVM, decision tree, naive Bayes, and FFN to compare the performance of the proposed method.

Decision Trees (DT): For prediction and classification problems, decision tree is considered as an effective technique that helps the system to make decisions based on the set of rules. In the domain of biometrics, feature vector components are evaluated by decision nodes, while leaf represents the subject assigned to each biosignal [51].

In Ref. [56], a decision tree-based study done by Kumar et al., the authors classified genuine and imposter users by making a decision based on the fuzzy binary decision tree (FBDT) algorithm. To authenticate a user using the keystroke pattern, in Ref. [57], user identification based on a parallel decision tree was proposed. In this study, eight training subsets of each user were obtained by dividing the training set into four subsets, followed by a wavelet transform of each subject. The performance of the algorithm was evaluated by calculating the false rejection rate and false acceptance rate of 9.62% and 0.88%, respectively.

Random Forest (RF): Like other ML algorithms, random tree is used in classification and regression problems. It extracts data from data. As the name suggests, it is a collection of decision trees. During classification using the random forest, each decision tree predicts a class label, and based on the maximum vote, random tree makes a decision. Due to its great performance, it is one of the most used techniques to make predictions in biometric security.

An indoor location recognition system based on Wi-fi using the smart watch and the random forest was proposed in Ref. [58]. To position recognition problems, the authors used the Basic Service Set Identifier (BSSID) and the Received Signal Strength Indication (RSSI). Execution speed and accuracy of the system were improved by applying the filtering process and random forest. The experimental result shows the consistent performance irrespective of fewer data and location information. In Ref. [59], the random forest was used to make decisions based on the matching scores generated by multiple biometric devices. Three biometric devices based on face, fingerprint, and hand geometry were used for experimentation. Combining scores generated by these devices, the performance of random forest was investigated. The authors also conducted experiments on a different subset of the dataset.

Gaussian Mixture Model (GMM): The objective of GMM is to group the data points that belong to the same distribution. In the biometric system, a GMM is generated for each class, and for a given query image, the sample is compared with the GMM of all classes to generate likelihood as an output [60]. Based on the result, the sample is considered to belong to the class that has a maximum likelihood. The Gaussian mixture model is mostly used to develop a voice-based biometric system [61] and to develop a multi-model authentication system based on face and speech [62,63].

Convolutional Neural Network (CNN): Using machine learning techniques, we get significant success to secure system using biometric features. However, sometimes, the system fails to authenticate a person in an unconstrained environment. Traditional methods of authentication based on ML extract features manually and classify them using SVM, KNN, etc. These handpicked features are not sufficient to identify a subject; therefore, an automatic way of feature extraction is needed. This is why the application of deep learning in the field of biometrics comes into the picture. Deep learning (convolutional neural network) learns features from the given data automatically and uses the learned knowledge to make a decision on new data. CNN not only produces high accuracy but also speeds up the identification process, which is a great need for devices connected in the IoT environment.

To authenticate a person using an IoT-enabled device using iris as a biofeature, Liao et al. trained a U-shaped CNN architecture [71]. The performance of the method was evaluated on the CASIA and BATH databases, which consist of 20,000 and 24,156 image samples, respectively. The experimental results show that the proposed method can yield 98.55% and 99.71% accuracy on the BATH and CASIA iris databases, respectively. In Ref. [72], a finger vein recognition-based authentication system using deep learning was studied. They developed a framework named FVR-DLRP, which can protect original biometric data even if the attacker decodes the user's password.

A liveness and face recognition-based access control model was proposed by Chandraker et al. [73]. In Ref. [74], a digital door lock system was designed to strengthen the security in the IoT environment. The developed system can capture the image of a person and send it to the mobile device if an unauthorized user attempts to break the door lock. A similar effort toward the development of home security in IoT using face recognition was made in Ref. [75].

In Ref. [76], privacy management for live video analytics in IoT was studied. RTFace was build based on inter-frame tracking and OpenFace. Using RTFace, a privacy-aware architecture was proposed.

Table 8.1 represents the different machine learning techniques and biometric features that are generally used to develop a biometric system along with their performance results.

TABLE 8.1

ML Techniques Used to Develop a Biometric System

Author	ML Technique	Bio-Features	Database	Performance Result
Buriro et al. [64]	NB, SVM, KNN BN, DT, DT-RF	Behavioral pattern (sitting, standing, and walking)	10,200 samples from 85 subjects	Accuracy: 98.98% (RF – full feature) and 99.35% (reduced feature)
Jagadeesan et al. [65]	KNN	Keyboard and mouse operation	A pool of 20 users	Accuracy: 96.4% (application-based) and 82.2% (application-independent)
Shen et al. [66]	Nearest neighbor Neural network and SVM (one-class)	Mouse operation	90,000 samples 28 subjects	FAR (%) / FRR (%): Nearest neighbor 2.73 / 3.67; Neural network 0.89 / 2.15; SVM (one-class) 0.37 / 1.12
Bailey et al. [67]	BayesNet, LibSVM J48	Keyboard dynamics Mouse dynamics GUI	14,552 keystroke dynamics 465 mouse dynamics 85 window class names (GUI)	Accuracy (in %) using BayesNet — Keyboard: 97.05 ± 3.03, Mouse: 82.77 ± 2.96, GUI: 86.57 ± 2.69, Fusion: 99.39 ± 1.11. Accuracy (in %) using LibSVM — Keyboard: 96.86 ± 2.38, Mouse: 85.53 ± 4.26, GUI: 69.74 ± 5.00, Fusion: 96.66 ± 2.23. Accuracy (in %) using J48 — Keyboard: 85.68 ± 4.37, Mouse: 74.26 ± 5.55, GUI: 81.72 ± 5.45, Fusion: 86.64 ± 4.62
Hussain et al. [68]	GMM	Voice	15 clients 56 imposters	EER: 3.25%, at 256 GMM (best result)

(Continued)

TABLE 8.1 (Continued)
ML Techniques Used to Develop a Biometric System

Author	ML Technique	Bio-Features	Database	Performance Result			
				Method	Palm print	Iris	Bimodal system
Kumar et al. [56]	PNN SVM DT FBDT	Palm print, iris, and the fusion of palm print and iris	IITD palm print and iris databases, 150 subjects	PNN	FAR: 0.0051 FRR: 4.5421	FAR: 0.5453 FRR: 10.5765	FAR: 0.00345 FRR: 4.00000
				SVM	FAR: 0.0026 FRR: 5.8270	FAR: 0.6754 FRR: 9.333	FAR: 0.00752 FRR: 3.32000
				DT	FAR: 0.0046 FRR: 6.0250	FAR: 0.5302 FRR: 12.5051	FAR: 0.0345 FRR: 4.5202
				FBDT	FAR: 0.0050 FRR: 3.0270	FAR: 0.0250 FRR: 8.1081	FAR: 0 FRR: 2.0270
Parkhi et al. [69]	Deep learning	Face	LFW and YTD	Accuracy:		LFW YTD	98.95% 97.3%
Gadaleta et al. [70]	CNN SVM	Gait	50 subjects	Misclassification rate: 0.15%			

8.6 APPLICATION OF DEEP LEARNING IN BIOMETRICS

The state-of-the-art deep convolutional network is widely used to secure access using biometric traits. The most used bio-features are fingerprint, face, iris, and ear, and periocular region – when face and iris fail to recognize the subject. Many CNN architectures such as AlexNet [77], VGG16 [78], ResNet50 [79], and DenseNet [80] are available in the literature and used by researchers and practitioners for the development of biometric systems. The performance of these architectures differs based on the dataset used to evaluate the algorithm. In this section, we used a smaller version of ResNet50 to evaluate the performance of face and iris biometrics. We trained the CNN model with the Caltech face database and the UBIRIS.v1 iris database and compared its performance with traditional machine learning techniques.

8.6.1 EXPERIMENTAL STUDY

8.6.1.1 Data Preprocessing and Augmentation

The Caltech face database comes with the front face of the subject, but it also includes hair and shoulder, which do not contribute to the recognition performance. Therefore, it is necessary to remove those parts from the image. To extract only the facial region from the database image, we used the Viola–Jones algorithm [81]. The detected facial region was cropped and resized to 64×64. To make system illumination invariant, we converted RGB images into grayscale before extracting features from face and iris images. Augmentation is frequently used to reduce overfitting and enhance generalization of the model by applying random jitters and perturbations to data. It is proved to be an effective way to increase the result of classification tasks [82]. Generally, there are two types of data augmentation: The first one is to increase the training image size, and the second one is in-place (or on-the-fly) data augmentation. In the first type of data augmentation process, the algorithm returns the transformed image and writes it in the disk; therefore, it has both original and augmented images. Therefore, this process expands the database size but does not increase the generalization ability of the trained model. However, the second type of augmentation process applies a series of random transformations (such as resizing, shearing, and rotation) and uses the transformed images to train the CNN model. In this study, we used the on-the-fly data augmentation technique.

8.6.1.2 Model Training

We trained a CNN model using an architecture that is similar to ResNet50 but can accept low-resolution image as an input. The complete architecture of ResNet50 is available in [79], which accepts input images having a fixed resolution, i.e., 224×224. We used a subset of this architecture to train with images of smaller size, i.e., 64×64. As databases considered for the study have different image resolutions, we resized the images of both databases to the dimension 64×64 and augmented them. For data augmentation, we transformed the images by shifting the height and width by a factor of 0.2, zooming, setting the shear range to 0.15, and rotating the images by a factor of 20, followed by flipping the images horizontally. We modified the ResNet architecture by reducing its depth as suggested in Ref. [83].

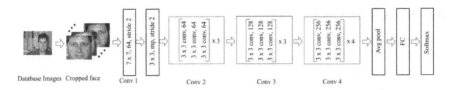

FIGURE 8.4 Architecture of CNN used to train the model.

The architecture is divided into four "Conv" layers. The first Conv layer (i.e., Conv 1) generates 64 filters. In the next two Conv layers, i.e., Conv 2 and Conv 3, we stack three sets of residual modules in both layers, and each set has three convolution operations. Finally in the last Conv layer, i.e., Conv 4, we stack four sets of residual models. The number of filters learned in Conv 2, Conv 3, and Conv 4 is 64, 128, and 256, respectively. The architecture of the studied model is shown in Figure 8.4. Finally, we have total parameters = 384,881, trainable parameters = 380,139, and non-trainable parameters = 4742.

Each database is divided into training set and validation set at a ratio of 80:20. The model is trained for 250 epochs with initial learning rate *INIT ˙LR* = 1*e* 1 and decay = *INIT ˙LR/EPOCHS*. Performance parameters such as recall, precision, F1-score, and accuracy are calculated. We also plot the precision–recall curve and ROC curve. For comparison of the results obtained by deep learning, we calculated 256 bin histogram features from each database image and classified them using traditional machine learning algorithms such as KNN, DT, RF, NB, SVM, and MLP. We used nested k-fold cross-validation for training ML model [84]. The performance of traditional ML algorithms, as well as state-of-the-art deep learning, was compared.

8.6.1.3 Experimental Results

8.6.1.3.1 Databases Used for Evaluation
As per the guidelines suggested in Ref. [85], we used the following two databases to evaluate the performance of the model:

Caltech: This database was collected by Markus Weber at the California Institute of Technology. It consists of 450 frontal face images from 27 subjects. Images in the database are stored in JPEG format.

UBIRIS.v1: It has a total of 1877 iris images captured from 241 subjects in a different session. It is one of the most widely used standard iris databases that is publicly available for experimentation. Several noise factors and illumination variations are added by acquiring image samples in the indoor/outdoor environment. The sample images of these databases are shown in Figure 8.5.

8.6.1.3.2 Evaluation Parameters
Precision: It measures the ratio of correctly predicted class from all positive classes. Mathematically, precision can be measured by Equation 8.1:

Caltech

UBIRIS.v1

FIGURE 8.5   Sample database images by which the model is trained.

$$\text{Precision} = \rightarrow \frac{\text{TP}}{\text{TP} + \text{FN}} \tag{8.1}$$

TP and TN represent true positive and true negative, respectively, while FP and FN represent false positive and false negative, respectively.

Recall: It represents the ratio of correctly predicted positive class to total actual positive class.

$$\text{Recall} = \rightarrow \frac{\text{TP}}{\text{TP} + \text{FN}} \tag{8.2}$$

F1-Score: There is a trade-off between precision and recall. Therefore, F1-score can be considered a function of precision and recall which measures a balance between these two parameters by calculating the harmonic mean. Therefore, F1-score can be calculated by Equation 8.3:

$$\text{F1} \Rightarrow \text{score} = 2 \times \frac{\text{Precision} \times \text{Recall}}{\text{Precision} + \text{Recall}} \tag{8.3}$$

Accuracy: It is the ratio of the correctly identified subject from the pool of subjects:

$$\text{Accuracy} = \rightarrow \frac{\text{TP} + \text{TN}}{\text{TP} + \text{FP} + \text{TN} + \text{FN}} \tag{8.4}$$

8.6.1.3.3   Results and Findings

The performance parameters were calculated using traditional ML algorithms as well as deep learning techniques. The parameter metrics discussed above were calculated for the face and iris databases, i.e., Caltech and UBIRIS.v1, respectively. The comparative results are shown in Table 8.2. For deep convolutional neural network

TABLE 8.2
Comparison of Results of Traditional ML Techniques and Deep Learning on the Face and Iris Databases

ML Technique	Bio-Features	Database	Performance Measures (in %) Using Caltech				Performance Measures (in %) Using UBIRIS.v1			
			Precision	Recall	F1-score	Accuracy	Precision	Recall	F1-score	Accuracy
KNN			44	44	44	44.10	92	92	91	92.00
DT			34	34	34	34.00	40	39	38	39.00
RF			53	57	55	53.80	87	87	86	87.00
NB	face	Caltech	43	42	42	43.00	63	49	49	63.00
SVM	iris	UBIRIS.v1	47	50	48	50.01	93	92	92	92.00
LR			37	42	39	42.00	83	82	82	82.00
MLP			43	47	45	47.20	63	63	62	63.00
CNN			97	98	97	96.90	98	97	97	97.30

experiments, we plot a learning curve, which represents the increase/decrease in the accuracy/loss of training and validation process with respect to epochs in Figure 8.6. The ROC curve for both (face and iris) databases is plotted in Figure 8.7. Figure 8.8 shows the precision–recall curve.

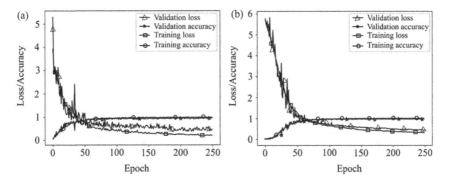

FIGURE 8.6   Learning curve obtained while training the network with the Caltech database (a) and the UBIRIS.v1 database (b).

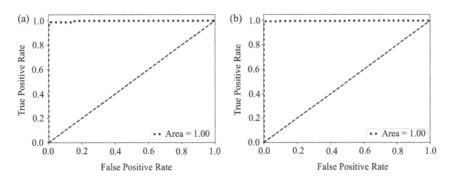

FIGURE 8.7   ROC curve obtained while training the network with the Caltech database (a) and the UBIRIS.v1 database (b).

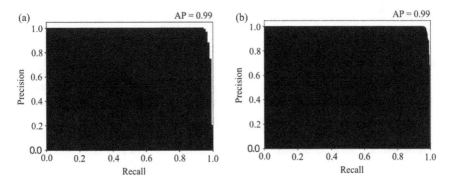

FIGURE 8.8   Precision–recall curve obtained while training the network with the Caltech database (a) and the UBIRIS.v1 database (b).

8.6.1.3.4   Analysis of Results

Figure 8.6 represents the learning behaviors of the CNN model. From plot character-istics, we can analyze that the model is neither overfitting nor underfitting. Training and validation loss continuously decreases with epochs, and after 230 epochs, it is almost constant, and there is a small gap between these two curves. A similar analy-sis can be made for training and validation accuracy. The gap between training and validation loss is less while the model is being trained with the UBIRIS.v1 dataset compared to the Caltech face database. Also, the model learning is smooth with the UBIRIS.v1 database, but we can see the fluctuations in validation loss while training the network with the Caltech face database.

Table 8.2 shows the comparison of the performance of deep learning with traditional ML algorithms. Histogram features are extracted and classified using conventional ML techniques. Among all traditional methods, we found that KNN and SVM per-form better for iris-based person recognition with an accuracy of 92%, while for face recognition, random forest yields good results comparatively. The decision tree shows worst performance for both, i.e., face and iris recognition. Deep learning automatically learns features from the raw input image and gives better performance compared to conventional ML algorithms and produces 96.90% and 97.30% accuracy for face and iris recognition, respectively. Another analysis can be done in terms of bio-features. From Table 8.2, it can be concluded that iris-based person identification outperforms the face identification with traditional ML algorithms as well as deep learning techniques.

The receiver operating characteristic (ROC) curve is shown in Figure 8.7. It repre-sents the relation between the true-positive rate and the false-positive rate at different thresholds. Using ROC plots, we measure the average area under the curve (AUC). If AUC = 1, it means the model is able to separate two classes, while the worst value is 0, which means the model cannot separate classes. In our experiment, we got AUC = 1 using CNN, which means the developed system has excellent separability.

Figure 8.8 represents the precision–recall curve. Like the ROC plot, a precision–recall curve is obtained by plotting precision and recall values at different values of thresholds. A classifier is said to be ideal if it does not make any prediction error. These types of classifiers have an average precision (AP) or area under the curve (AUC-AP) equal to 1 or close to 1. The studied model produces AP = 0.99, which means the developed system has minimum prediction error.

8.7   CONCLUSION

With improvements in of Internet and communication technologies, the demand for IoT has increased. Therefore, privacy and security of collection and storage of sensi-tive information becomes a challenge. There are many types of security attacks that can occur at a different level of IoT architecture. In this study, we discussed some of these attacks, such as man-in-the-middle, denial-of-service, botnet, and physical attack. To access a secured system, authentication and authorization of end users based on traditional means as well as the use of biometrics is explained. We pre-sented a working model of a biometric system, different biometric modalities, and the traits used to identify a person. Vulnerability in biometrics such as attack on data collection devices, feature extraction techniques, the biometric database itself, and decision-making module is highlighted. We discussed various methods of detection of attacks such as spoofing attack, replay attack, and denial-of-service on the biometric system using machine learning techniques. We also discussed methods to secure biometric components such as templates in the database and data acquisition devices themselves. Machine learning techniques used for training as well as classifying genuine and fake users such as k-nearest neighbors, support vector machine, naive Bayes classifier, decision trees, random forest, and Gaussian mixture model are explained in brief along with their applications available in the literature. Finally, we present a state-of-the-art deep learning technique and its performance to recognize a person based on their biometric traits. For this purpose, we trained a CNN model similar to but smaller (in depth) than ResNet with two standard and publicly available face and iris databases. Performance parameters were calculated, and the performance of the studied model was compared with traditional ML techniques. The experimental result clearly shows the excellency of deep learning in terms of precision, recall, F1-score, and accuracy for both the face and iris databases.

The usage of biometrics is growing along with growth in the demand of IoT. Therefore, we can conclude that techniques such as deep learning in biometrics can be used to improve the performance and security of IoT-enabled devices, secure data transfer, identify fake and genuine users, surveil a city and traffic in a smart way, and recognize a person and criminal activity easily.

REFERENCES

1. Li Jiang, Da-You Liu, and Bo Yang. Smart home research. In *Proceedings of International Conference on Machine Learning and Cybernetics (IEEE Cat. No. 04EX826)*, vol. 2, pp. 659–663. IEEE, 2004. DOI: 10.1109/ICMLC.2004.1382266.
2. Meensika Sripan, Xuanxia Lin, Ponchan Petchlorlean, and Mahasak Ketcham. Research and thinking of smart home technology. In *International Conference on Systems and Electronic Engineering (ICSEE 2012)*, pp. 61–63, 2012.
3. Andrea Zanella, Nicola Bui, Angelo Castellani, Lorenzo Vangelista, and Michele Zorzi. Internet of things for smart cities. *IEEE Internet of Things Journal*, 1(1): 22–32, 2014. DOI: 10.1109/JIOT.2014.2306328.
4. Miao Yun and Bu Yuxin. Research on the architecture and key technology of internet of things (IoT) applied on smart grid. In *International Conference on Advances in Energy Engineering*, pp. 69–72. IEEE, 2010. DOI: 10.1109/ICAEE.2010.5557611.
5. Hongyu Pei Breivold and Kristian Sandstro¨m. Internet of things for industrial automation–challenges and technical solutions. In *IEEE International Conference on Data Science and Data Intensive Systems*, pp. 532–539. IEEE, 2015. DOI: 10.1109/ DSDIS.2015.11.
6. Nicola Bui and Michele Zorzi. Health care applications: A solution based on the internet of things. In *Proceedings of the 4th International Symposium on Applied Sciences in Biomedical and Communication Technologies*, p. 131. ACM, 2011. DOI:10.1145/2093698.2093829.
7. Amir-Mohammad Rahmani, Nanda Kumar Thanigaivelan, Tuan Nguyen Gia, Jose Granados, Behailu Negash, Pasi Liljeberg, and Hannu Tenhunen. Smart e-health gateway: Bringing intelligence to internet-of-things based ubiquitous healthcare systems. In *12th Annual IEEE Consumer Communications and Networking Conference (CCNC)*, pp. 826–834. IEEE, 2015. DOI: 10.1109/CCNC.2015.7158084.

8. Yifan Bo and Haiyan Wang. The application of cloud computing and the internet of things in agriculture and forestry. In *International Joint Conference on Service Sciences*, pp. 168–172. IEEE, 2011. DOI: 10.1109/IJCSS.2011.40.

9. Ji-chun Zhao, Jun-feng Zhang, Yu Feng, and Jian-xin Guo. The study and application of the IoT technology in agriculture. In *3rd International Conference on Computer Science and Information Technology*, vol. 2, pp. 462–465. IEEE, 2010. DOI: 10.1109/ICCSIT.2010.5565120.

10. Mohamed Abomhara and Geir M. Køien. Cyber security and the internet of things: Vulnerabilities, threats, intruders and attacks. *Journal of Cyber Security and Mobility*, 4(1): 65–88, 2015. DOI: 10.13052/jcsm2245-1439.414.

11. Elisa Bertino and Nayeem Islam. Botnets and internet of things security. *Computer*, (2): 76–79, 2017. DOI: 10.1109/MC.2017.62.

12. Lulu Liang, Kai Zheng, Qiankun Sheng, and Xin Huang. A denial of service attack method for an IoT system. In *2016 8th International Conference on Information Technology in Medicine and Education (ITME)*, pp. 360–364. IEEE, 2016. DOI: 10.1109/ITME.2016.0087.

13. Mohammad S Obaidat, Soumya Prakash Rana, Tanmoy Maitra, Debasis Giri, and Subrata Dutta. Biometric security and internet of things (IoT). In *Biometric-Based Physical and Cybersecurity Systems*, pp. 477–509. Springer, 2019. DOI: 10.1007/978-3-319-98734-719.

14. David Cunado, Mark S Nixon, and John N Carter. Using gait as a biometric, via phase-weighted magnitude spectra. In *International Conference on Audio-and Video-Based Biometric Person Authentication*, pp. 93–102. Springer, 1997. DOI: 10.1007/BFb0015984.

15. Alvaro Muro-De-La-Herran, Begonya Garcia-Zapirain, and Amaia Mendez-Zorrilla. Gait analysis methods: An overview of wearable and non-wearable systems, highlighting clinical applications. *Sensors*, 14(2): 3362–3394, 2014. DOI: 10.3390/s140203362.

16. Andreas Fischer and Réjean Plamondon. Signature verification based on the kinematic theory of rapid human movements. *IEEE Transactions on Human-Machine Systems*, 47(2): 169–180, 2016. DOI: 10.1109/THMS.2016.2634922.

17. Yohan Muliono, Hanry Ham, and Dion Darmawan. Keystroke dynamic classification using machine learning for password authorization. *Procedia Computer Science*, 135: 564–569, 2018.

18. Andrew Boles and Paul Rad. Voice biometrics: Deep learning-based voiceprint authentication system. In *12th System of Systems Engineering Conference (SoSE)*, pp. 1–6, June 2017.

19. Sarat C Dass and Anil K Jain. Fingerprint-based recognition. *Technometrics*, 49(3): 262–276, 2007. DOI: 10.1198/004017007000000272.

20. Timo Ahonen, Abdenour Hadid, and Matti Pietikäinen. Face recognition with local binary patterns. In *European Conference on Computer Vision*, pp. 469–481. Springer, 2004. DOI: 10.1007/978-3-540-24670-136.

21. Guo-Can Feng, Pong Chi Yuen, and Dao-Qing Dai. Human face recognition using PCA on wavelet subband. *Journal of Electronic Imaging*, 9(2): 226–234, 2000. DOI: 10.1117/1.482742.

22. Juwei Lu, Konstantinos N Plataniotis, and Anastasios N Venetsanopoulos. Face recognition using LDA-based algorithms. *IEEE Transactions on Neural Networks*, 14(1): 195–200, 2003. DOI: 10.1109/TNN.2002.806647.

23. Sambit Bakshi, Pankaj K Sa, and Banshidhar Majhi. Phase intensive global pattern for periocular recognition. In *Annual IEEE India Conference (INDICON)*, pp. 1–5. IEEE, 2014. DOI: 10.1109/INDICON.2014.7030362.

24. John Daugman. How iris recognition works. In *The Essential Guide to Image Processing*, pp. 715–739. Elsevier, 2009. DOI: 10.1109/TCSVT.2003.818350.

25. Unsang Park, Arun Ross, and Anil K Jain. Periocular biometrics in the visible spectrum: A feasibility study. In *IEEE 3rd International Conference on Biometrics: Theory, Applications, and Systems*, pp. 1–6. IEEE, 2009. DOI: 10.1109/BTAS.2009.5339068.

26. Gautam Kumar, Sambit Bakshi, Pankaj Kumar Sa, and Banshidhar Majhi. Non-overlapped blockwise interpolated local binary pattern as periocular feature. *Multimedia Tools and Applications*, 2020. DOI: 10.1007/s11042-020-08708-w.

27. AJ Hoogstrate, Hazem Van den Heuvel, and E Huyben. Ear identification based on surveillance camera images. *Science & Justice*, 3(41): 167–172, 2001. DOI: 10.1016/S1355-0306(01)71885-0.

28. Alfred V Lannarelli. Ear identification. *Forensic Identification Series,* 1989.

29. Lin Zhang, Lida Li, Anqi Yang, Ying Shen, and Meng Yang. Towards contactless palm-print recognition: A novel device, a new benchmark, and a collaborative representation based identification approach. *Pattern Recognition*, 69: 199–212, 2017. DOI: 10.1016/j.patcog.2017.04.016.

30. Rohit Thanki and Komal Borisagar. Biometric watermarking technique based on CS theory and fast discrete curvelet transform for face and fingerprint protection. In *Advances in Signal Processing and Intelligent Recognition Systems*, pp. 133–144. Springer, 2016. DOI: 10.1007/978-3-31928658-712.

31. Rohit Thanki and Komal Borisagar. Sparse watermarking technique for improving security of biometric system. *Procedia Computer Science*, 70: 251–258, 2015. DOI: 10.1016/j.procs.2015.10.083.

32. P Poongodi and P Betty. A study on biometric template protection techniques. *International Journal of Engineering Trends and Technology (IJETT)*, 7(4): 202–204, 2014. DOI: 10.14445/22315381/IJETT-V7P244.

33. Ganiyev Salim Karimovich and Khudoykulov Zarif Turakulovich. Biometric cryptosystems: Open issues and challenges. In *International Conference on Information Science and Communications Technologies (ICISCT)*, pp. 1–3. IEEE, 2016. DOI: 10.1109/ICISCT.2016.7777408.

34. Bilgehan Arslan, Ezgi Yorulmaz, Burcin Akca, and Seref Sagiroglu. Security perspective of biometric recognition and machine learning techniques. In *15th IEEE International Conference on Machine Learning and Applications (ICMLA)*, pp. 492–497. IEEE, 2016. DOI: 10.1109/ICMLA.2016.0087.

35. Donghyun Noh, Heeseung Choi, and Jaihie Kim. Touchless sensor capturing five fingerprint images by one rotating camera. *Optical Engineering*, 50(11): 113202, 2011. DOI: 10.1117/1.3646327.

36. Yahui Liu, Guangming Lu, and David Zhang. An effective 3D ear acquisition system. *PloS One*, 10(6): e0129439, 2015. DOI: 10.1371/journal.pone.0129439.

37. Muhammad Rehman Zafar and Munam Ali Shah. Fingerprint authentication and security risks in smart devices. In *22nd International Conference on Automation and Computing (ICAC)*, pp. 548–553. IEEE, 2016. DOI:10.1109/IConAC.2016.7604977.

38. Christina-Angeliki Toli and Bart Preneel. Provoking security: Spoofing attacks against crypto-biometric systems. In *World Congress on Internet Security (WorldCIS)*, pp. 67–72. IEEE, 2015. DOI: 10.1109/WorldCIS.2015.7359416.

39. Javier Galbally, Sébastien Marcel, and Julian Fierrez. Biometric antispoofing methods: A survey in face recognition. *IEEE Access*, 2: 1530–1552, 2014. DOI: 10.1109/ACCESS.2014.2381273.

40. Ivana Chingovska, Andre Rabello Dos Anjos, and Sebastien Marcel. Biometrics evaluation under spoofing attacks. *IEEE Transactions on Information Forensics and Security,* 9(12): 2264–2276, 2014. DOI: 10.1109/TIFS.2014.2349158.

41. Abdenour Hadid. Face biometrics under spoofing attacks: Vulnerabilities, countermeasures, open issues, and research directions. In *Proceedings of the IEEE Conference on Computer Vision and Pattern Recognition Workshops*, pp. 113–118, 2014. DOI: 10.1109/CVPRW.2014.22.

42. Christopher McCool and Se´bastien Marcel. Parts-based face verification using local frequency bands. In *International Conference on Biometrics*, pp. 259–268. Springer, 2009. DOI: 10.1007/978-3-642-01793-327.

43. Younsung Choi, Youngsook Lee, and Dongho Won. Security improvement on biometric based authentication scheme for wireless sensor networks using fuzzy extraction. *International Journal of Distributed Sensor Networks*, 12(1):8572410, 2016. DOI: 10.1155/2016/8572410.

44. Eun-Jun Yoon and Cheonshik Kim. Advanced biometric-based user authentication scheme for wireless sensor networks. *Sensor Letters*, 11(9): 1836–1843, 2013. DOI: 10.1166/sl.2013.3014.

45. Joseph Shelton, Kelvin Bryant, Sheldon Abrams, Lasanio Small, Joshua Adams, Derrick Leflore, Aniesha Alford, Karl Ricanek, and Gerry Dozier. Genetic & evolutionary biometric security: Disposable feature extractors for mitigating biometric replay attacks. *Procedia Computer Science*, 8: 351–360, 2012. DOI: 10.1016/j.procs.2012.01.072.

46. Yahaya Isah Shehu, Anne James, and Vasile Palade. Detecting an alteration in biometric fingerprint databases. In *Proceedings of the 2nd International Conference on Digital Signal Processing*, pp. 6–11. ACM, 2018. DOI: 10.1145/3193025.3193029.

47. Jiankun Hu, Don Gingrich, and Andy Sentosa. A k-nearest neighbor approach for user authentication through biometric keystroke dynamics. In *IEEE International Conference on Communications*, pp. 1556–1560, May 2008.

48. Daniele Gunetti and Claudia Picardi. Keystroke analysis of free text. *ACM Transactions on Information and System Security (TISSEC)*, 8(3):312–347, 2005. DOI: 10.1145/1085126.1085129.

49. Aini Najwa Azmi, Dewi Nasien, and Fakhrul Syakirin Omar. Biometric signature verification system based on freeman chain code and k-nearest neighbor. *Multimedia Tools and Applications*, 76(14): 15341–15355, 2017. DOI: 10.1007/s11042-016-3831-2.

50. Sun Yuan Kung, Man-Wai Mak, and Shang-Hung Lin. *Biometric Authentication: A Machine Learning Approach*. Prentice Hall Professional Technical Reference New York, 2005.

51. Jorge Blasco, Thomas M Chen, Juan Tapiador, and Pedro Peris-Lopez. A survey of wearable biometric recognition systems. *ACM Computing Surveys (CSUR)*, 49(3): 43, 2016. DOI: 10.1145/2968215.

52. Mario Frank, Ralf Biedert, Eugene Ma, Ivan Martinovic, and Dawn Xiaodong Song. Touchalytics: On the applicability of touchscreen input as a behavioral biometric for continuous authentication. *IEEE Transactions on Information Forensics and Security*, 8: 136–148, 2013. DOI: 10.1109/TIFS.2012.2225048.

53. Jian-Da Wu and Chiung-Tsiung Liu. Finger-vein pattern identification using SVM and neural network technique. *Expert Systems with Applications*, 38(11): 14284–14289, 2011.

54. Insha Qayoom and Sameena Naaz. Discriminant analysis and naïve Bayes classifier-based biometric identification using finger veins. *International Journal of Computer Vision and Image Processing (IJCVIP)*, 9(4): 15–27, 2019. DOI: 10.4018/IJCVIP.2019100102.

55. Ajay Kumar and David Zhang. Biometric recognition using feature selection and combination. In *International Conference on Audio-and Video-Based Biometric Person Authentication*, pp. 813–822. Springer, 2005. DOI: 10.1007/1152792385.

56. Amioy Kumar, Madasu Hanmandlu, and H. M Gupta. Fuzzy binary decision tree for biometric based personal authentication. *Neurocomputing*, 99: 87–97, 2013. DOI: 10.1016/j.neucom.2012.06.016.

57. Yong Sheng, Vir V Phoha, and Steven M Rovnyak. A parallel decision tree-based method for user authentication based on keystroke patterns. *IEEE Transactions on Systems, Man, and Cybernetics, Part B (Cybernetics)*, 35(4): 826–833, 2005. DOI: 10.1109/TSMCB.2005.846648.

58. Sunmin Lee, Jinah Kim, and Nammee Moon. Random forest and WiFi fingerprint-based indoor location recognition system using smart watch. *Human-Centric Computing and Information Sciences*, 9(1): 6, 2019. DOI: 10.1186/s13673-019-0168-7.

59. Yan Ma, Bojan Cukic, and Harshinder Singh. A classification approach to multi-biometric score fusion. In *International Conference on Audio-and Video-Based Biometric Person Authentication*, pp. 484–493. Springer, 2005. DOI: 10.1007/1152792350.

60. Douglas Reynolds. Gaussian mixture models. *Encyclopedia of Biometrics*: 827–832, 2015. DOI: 10.1007/978-0-38773003-5196.

61. Douglas A Reynolds, Thomas F Quatieri, and Robert B Dunn. Speaker verification using adapted Gaussian mixture models. *Digital Signal Processing*, 10(1–3): 19–41, 2000. DOI: 10.1006/dspr.1999.0361.

62. Mohamed Soltane, Noureddine Doghmane, and Noureddine Guersi. Face and speech based multi-modal biometric authentication. *International Journal of Advanced Science and Technology*, 21(6): 41–56, 2010.

63. Girija Chetty and Michael Wagner. Multi-level liveness verification for face- voice biometric authentication. In *Biometrics Symposium: Special Session on Research at the Biometric Consortium Conference*, pp. 1–6. IEEE, 2006. DOI: 10.1109/BCC.2006.4341615.

64. Attaullah Buriro, Bruno Crispo, and Mauro Conti. AnswerAuth: A bimodal behavioral biometric-based user authentication scheme for smartphones. *Journal of Information Security and Applications*, 44: 89–103, 2019.

65. Harini Jagadeesan and Michael S Hsiao. A novel approach to design of user re-authentication systems. In *IEEE 3rd International Conference on Biometrics: Theory, Applications, and Systems*, pp. 1–6. IEEE, 2009. DOI: 10.1109/BTAS.2009.5339075.

66. Chao Shen, Zhongmin Cai, and Xiaohong Guan. Continuous authentication for mouse dynamics: A pattern-growth approach. In *IEEE/IFIP International Conference on Dependable Systems and Networks (DSN 2012)*, pp. 1–12. IEEE, 2012. DOI: 10.1109/DSN.2012.6263955.

67. Kyle O Bailey, James S Okolica, and Gilbert L Peterson. User identification and authentication using multi-modal behavioral biometrics. *Computers & Security*, 43: 77–89, 2014.

68. H Hussain, SH Salleh, CM Ting, AK Ariff, I Kamarulafizam, and RA Suraya. Speaker verification using Gaussian mixture model (GMM). In *5th Kuala Lumpur International Conference on Biomedical Engineering 2011*, pp. 560–564. Springer, 2011. DOI: 10.1007/978-3-642 21729-6140.

69. Omkar M Parkhi, Andrea Vedaldi, Andrew Zisserman. Deep face recognition. In *bmvc*, vol. 1, p. 6, 2015. DOI: 10.5244/C.29.41.

70. Matteo Gadaleta and Michele Rossi. IDNet: Smartphone-based gait recognition with convolutional neural networks. *Pattern Recognition*, 74: 25–37, 2018.

71. Yi-Pin Liao and Chih-Ming Hsiao. A novel multi-server remote user authentication scheme using self-certified public keys for mobile clients. *Future Generation Computer Systems*, 29(3): 886–900, 2013. DOI: 10.1016/j.future.2012.03.017.

72. Yi Liu, Jie Ling, Zhusong Liu, Jian Shen, and Chongzhi Gao. Finger vein secure biometric template generation based on deep learning. *Soft Computing*, 22(7): 2257–2265, 2018. DOI: 10.1007/s00500-017-2487-9.

73. Manmohan Chandraker, Xiang Yu, Eric Lau, and Wong Elsa. Login access control for secure/private data, 2019. US Patent App. 10/289,825.

74. Ilkyu Ha. Security and usability improvement on a digital door lock system based on internet of things. *International Journal of Security and Its Applications*, 9(8): 45–54, 2015. DOI: 10.14257/ijsia.2015.9.8.05.

75. Mrutyunjaya. Sahani, Chiranjiv Nanda, Abhijeet Kumar Sahu, and Biswajeet Pattnaik. Web-based online embedded door access control and home security system based on face recognition. In *International Conference on Circuits, Power and Computing Technologies [ICCPCT-2015]*, pp. 1–6, 2015.

76. Junjue Wang, Brandon Amos, Anupam Das, Padmanabhan Pillai, Norman Sadeh, and Mahadev Satyanarayanan. A scalable and privacy-aware IoT service for live video analytics. In *Proceedings of the 8th ACM on Multimedia Systems Conference*, pp. 38–49. ACM, 2017. DOI: 10.1145/3083187.3083192.

77. Alex Krizhevsky, Ilya Sutskever, and Geoffrey E Hinton. Imagenet classification with deep convolutional neural networks. In *Advances in Neural Information Processing Systems*, pp. 1097–1105, 2012. DOI: 10.1145/3065386.

78. Karen Simonyan and Andrew Zisserman. Very deep convolutional networks for large scale image recognition. *arXiv preprint arXiv:1409.1556,* 2014.

79. Kaiming He, Xiangyu Zhang, Shaoqing Ren, and Jian Sun. Deep residual learning for image recognition. In *Proceedings of the IEEE Conference on Computer Vision and Pattern Recognition*, pp. 770–778, 2016. DOI: 10.1109/CVPR.2016.90.

80. Gao Huang, Zhuang Liu, Laurens Van Der Maaten, and Kilian Q Weinberger. Densely connected convolutional networks. In *Proceedings of the IEEE Conference on Computer Vision and Pattern Recognition*, pp. 4700–4708, 2017. DOI: 10.1109/CVPR.2017.243.

81. Paul Viola, Michael Jones. Robust real-time object detection. *International Journal of Computer Vision*, 4(34–47): 4, 2001. DOI: 10.1023/B:VISI.0000013087.49260.fb.

82. Luis Perez and Jason Wang. The effectiveness of data augmentation in image classification using deep learning. *arXiv preprint arXiv:1712.04621,* 2017.

83. Adrian Rosebrock. Deep learning for computer vision with python: ImageNet Bundle. *Deep Learning for Computer Vision with Python*. PyImageSearch, 2017.

84. Gavin C Cawley and Nicola LC Talbot. On over-fitting in model selection and subsequent selection bias in performance evaluation. *Journal of Machine Learning Research*, 11: 2079–2107, Jul 2010.

85. Zahid Akhtar, Gautam Kumar, Sambit Bakshi, and Hugo Proenca. Experiments with ocular biometric datasets: A practitioner's guideline. *IT Professional*, 20(3): 50–63, May 2018.

9 Data Security and Privacy through Text Steganography Approaches

Suvamoy Changder
NIT Durgapur

Ratnakirti Roy
BCREC

Narayan C. Debnath
EIU

CONTENTS

9.1 INTRODUCTION

9.1.1 Steganography as a Concept

Steganography or "covered writing" is defined as the art of hiding information within innocuous objects such that its very existence cannot be detected (Cachin 1998). In a typical steganography system, the secret message is encoded in a way such that an external adversary fails to detect its very existence (Chandramouli et al. 2004) and neither can draw suspicion about the transmission of secret data (Johnson and Jajodia 1998). The aim of steganography is not to prevent anyone from knowing that any two parties are communicating, but to keep the adversary from suspecting that there is secret communication. A steganography method fails if an external adversary successfully figures out the existence of secret communication (Artz 2001). A typical steganography system hides a secret message into an innocuous medium called the cover medium to generate the steganogram or the stego-message. The stego-system may have a key-based encryption–decryption system. In such cases, the security of the system is governed by the Kirchhoff's principle that suggests that the security of a key-based system is dependent on the key and only on the key alone.

9.1.2 History of Steganography

The history of steganography dates back to ancient Greece where Herodotus explains how ancient Spartans were warned about the Persian attack by Demeratus by secretly conveying the message using a writing tablet (Herodotus 440 BC). There were many such techniques that were practiced in the ancient times. Writing messages on the back of the wax writing tablets, tattooing the scalp of slaves, and writing on the rabbits' belly are few of the most well-known ones. Data hiding techniques have been found to be useful by top diplomats and administrative heads, too. There are examples where Margaret Thatcher, the former British Prime Minister, placed counted white-spaces in the documents pertaining to each cabinet minister so that the owner of the document can be traced. Similarly, invisible ink has been in use since a long time and has both fun and spurious uses. Microdots and microfilm, another form of secret communication, came to the light after the advent of photography (Arnold et al. 2003; Johnson et al. 2001; Kahn 1996; Wayner 2002).

9.1.3 General Model for Steganography

Steganography in the modern context is expressed through the classical prisoner's problem (Simmons 1985). The problem states that two inmates Alice and Bob wish to hatch an escape plan. However, everything they communicate is invigilated by the warden, Wendy, who will put them into solitary confinement if he smells secret communication.

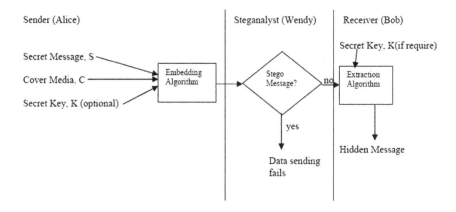

FIGURE 9.1   General model for steganography.

In the general model for steganography (Figure 9.1), communication between Alice and Bob is governed by the principles of steganography. This situation gives rise to two possibilities regarding the role of Wendy. Wendy can be either passive (passive warden) where he does not deliberately noise the communication channel between Alice and Bob, or Wendy can deliberately noise the communication channel whether or not any secret communication is taking place. This is the active warden concept and gives rise to the concept of highly robust steganography or watermarking.

Pure steganography is a covert communication system where no a priori exchange of secret information takes place. The embedding process is a mapping HIDE: $C \times S \rightarrow Ć$, where C is the cover, S is the message, and Ć is the generated cover file. The extraction is a mapping EXTRACT: $Ć \rightarrow S$, extracting S from Ć. So, we can define a pure steganography system as follows:

The tuple H = <C, S, Ć, HIDE, EXTRACT>, where C is possible covers, S the secret messages with |C| >= |S|, HIDE: $C \times S \rightarrow Ć$ the embedding function, and EXTRACT: $Ć \rightarrow S$ the extraction function such that EXTRACT (HIDE(C, S)) = S.

A secret key steganography system can be defined as follows:

The tuple H = <C, S, K, Ć, HIDE, EXTRACT>, where C is the cover file, S the secret message with |C| >=|S|, K the set of secret keys, HIDE: $C \times S \times K \rightarrow Ć$, and EXTRACT: $Ć \times K \rightarrow S$ such that EXTRACT (HIDE(C, S, K), K)) = S.

Similarly, public key steganography systems require the use of two keys: private and public keys, and one of them is stored in a public database. Whereas the public key is used in the embedding process, the secret key is used to reconstruct the secret message.

9.2   TYPES OF STEGANOGRAPHIC METHODS AND CARRIER MEDIUMS

Technical steganography uses scientific methods to hide a message, such as the use of invisible ink, microdots, and other size reduction methods. Other types of steganography include linguistic steganography that employs non-conventional approaches to hide the data in the carrier, jargon codes, and concealment ciphers.

The linguistic steganography is further classified into semagrams and open codes. Semagrams hide data by using signs or symbols. These can be innocent-looking objects of daily life (visual semagrams) or subtle modification of the appearance of the texts (text semagrams). On the other hand, open codes uses jargons (Warchalking 2003) or techniques such as the concealment ciphers. A concealment cipher openly hides data in visual text. The extraction is only known to the person who knows the data was concealed.

Steganography techniques that employ text documents as the choice of cover are arguably the most difficult ones in practice (Brassil et al. 1995). This is because of the fact the text documents are in most cases structurally similar to how they appear visually. Situations like this are not the same for other multimedia file types such as images and audio. They have different structural representations that they appear perceptibly. As a result, the amount of redundancy in these types of files is way higher than that of the text files. Therefore, unlike text documents, it is possible to introduce changes to their structure without affecting their external output. However, text documents are ubiquitous, consume less memory and bandwidth, and are sent easily over a network. This makes text documents an excellent contender as a steganography cover medium.

9.2.1 Understanding Text Steganography

Considering the different data hiding approaches chosen for text steganography, the following classification exists:

- **Format-Based Text Steganography** (Shirali-Shahreza and Shirali-Shahreza 2008a,b; Shirali-Shahreza and Shirali-Shahreza 2010; Alattar and Alattar 2004; Low et al. 1995; Huang and Yan 2001; Kim et al. 2003; Khairullah 2009): Data are being hidden by changing the format of some visible features of the cover file. The features that are considered include the font size, font color, the structure of the characters, angle of a line, distances between the words and open space. With an unnoticeable change made to the cover file, data is hidden by these methods.
- **Linguistic Steganography** (Kim et al. 2003; Khairullah 2009; Ali 2010; Rabah 2004; Roy and Manasmitha 2011; Wayner 1992; Wayner 1995; Atallah et al. 2001; Atallah et al. 2002; Taskiran et al. 2006; Topkara, Riccardi et al. 2006; Topkara et al. 2005; Topkara et al. 2006a,b; Grothoff et al. 2005a,b; Stutsman et al. 2006; Chapman and Davida 1997; Chapman and Davida 2002; Chapman et al. 2001; Bergmair 2004; Bolshakov 2004; Calvo and Bolshakov 2004; Bolshakov and Gelbukh 2004; Murphy 2001; Murphy and Vogel 2007; Murphy and Vogel 2007b): In linguistic steganography, a function makes changes to a file A in a way that it accrues the statistical features of another file B. Such functions are called mimic functions. Mathematically, if $P(t, A)$ is the probability that a substring t occurs in A, then a mimic function f recodes A so that $P(t, f(A))$ approximates $P(t, B)$ such that $\forall t \exists n \big(\operatorname{len}(t) < \operatorname{len}(n) \big)$.
- **Others Such as through HTML Tag, CHAT, and SMS**: In these techniques, the format flexibility of the cover files such as the HTML tags is exploited to convey secret messages.

- **Text Steganography through Specific Languages**: Steganography practices can be implemented using the characteristics of languages such as Arabic, Persian, Thai, Chinese, and even Indian languages such as Hindi and Bengali. These techniques exploit the redundancy in the language scripts and also the structural details of the alphabets. For example, in Persian and Arabic, each letter can have four different shapes regarding its position in the word. This feature of Persian and Arabic languages and the way in which the documents are saved in the Unicode Standard are used to hide information in Unicode text documents.

9.3   TEXT STEGANOGRAPHY THROUGH INDIAN LANGUAGES

In order to provide security to data over the public channel, steganography is one of the major and interesting areas such as cryptography and coding. Due to the absence of large degree of redundancy in the internal structure of a text file, compared to other steganography such as on images and other multimedia files, text steganography may be considered the most difficult one. In the space of linguistic steganography, to overcome the problem of forming semantically incorrect sentences, the choice of Indian languages can provide both syntactically and semantically correct cover file, due to the flexible grammar and phrase structure of the languages. Furthermore, by using the very uncommon properties of Indian languages, new algorithms can be developed to ensure more protection to the data.

Text steganography techniques through Indian languages can be classified into various implementations as follows.

9.3.1   TEXT STEGANOGRAPHY BY WORD SHIFTING METHOD

The word shifting technique conceals the data by shifting a word horizontally, i.e., by altering the distance between the words. In general, the algorithm can be developed by considering a secret message consisting of a sequence of binary digits and a text file consisting of texts of an Indian language. The algorithm will create the cover file by changing the structure of the text file as follows:

    a. Single gap between words represents a "0" (zero).
    b. More than one gap represents a "1" (one).

Pictorially, the hiding process can be represented as in Figure 9.2 and 9.3.
The Hiding Algorithm:

HIDE_WORDSHIFTING ( )

```
1. Input Secret Message S and a text file T, containing Indian
Texts, as Cover Media.
2. Compress the message and convert it to its equivalent
binary CONVSECB.
3. Find the length L of CONVSECB.
4. Add binary equivalent of L as the header of CONVSECB.
5. For each bit of CONVSECB.
```

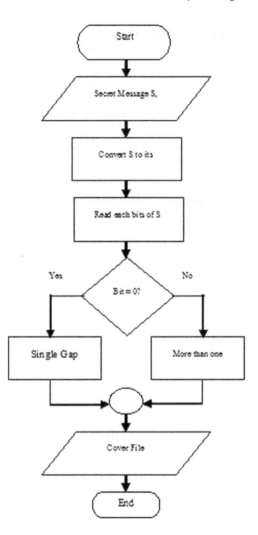

FIGURE 9.2 Flowchart, representing the data hiding process of the word shifting method.

6. Read the Text file T and create the cover file *COVER* as:
7. If *CONVSECB* bit is 0 and the distance between two
consecutive words is one then no change; else change the
distance as a single gap.
8. If *CONVSECB* bit is 1 and the distance between two
consecutive words is more than one then no change; else change
the distance by introducing more gap.
9. End.

To explain the algorithm, let us consider the secret message "*BOMB*" to be sent using
the cover media shown in Figure 9.4. To do this, the algorithm converts the secret
message to its equivalent binary which is "1000010100111110011011000010".

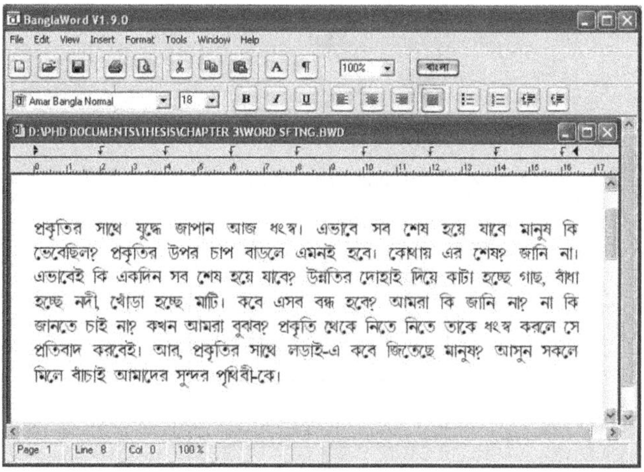

FIGURE 9.3   Original text file used for data hiding by HIDE_WORDSHIFTING.

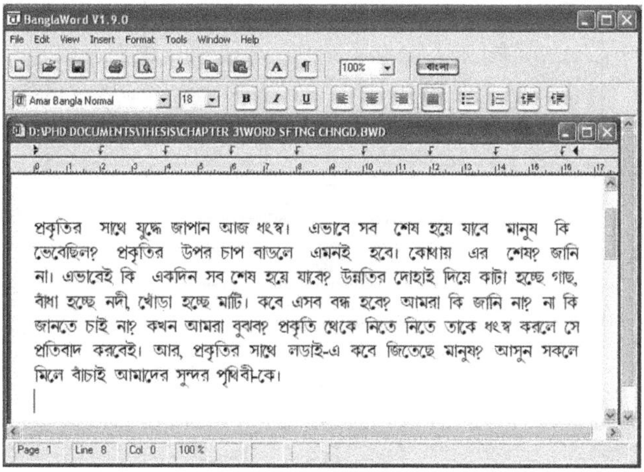

FIGURE 9.4   Cover file, generated by HIDE_WORDSHIFTING.

The algorithm starts to read the incoming bits of the secret message and makes the necessary changes to the cover file. The generated cover file is shown in Figure 9.5. Considering the distances between the words, the algorithm hides the data with the following function:

HIDE_WORDSHIFTING$_{DIST}$ $(S, T) \rightarrow COVER.$

At the receiving side, to extract the information, the algorithm will read the cover file generated by the hiding algorithm and place the data bit as per the space introduced between two consecutive words. Figure 9.5 represents the pictorial diagram of the process. The function for extracting the message is the reverse mapping of

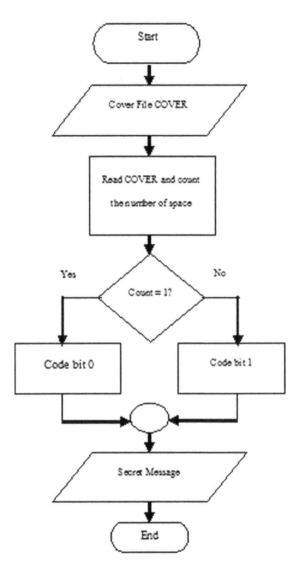

FIGURE 9.5 Extraction flowchart of the word shifting method.

the algorithm HIDE_WORDSHIFTING ( ). During the execution of the algorithm, if the distance between two consecutive words is found as one, the code bit zero is placed as the data bit of the output message string. Otherwise, if more than one space is found in between two consecutive words, then code bit one is placed. By finding the length of the message (added as header), and after decompressing (if necessary), the original message is extracted by the pre-intended recipient from the cover file that is received.

The function **EXTRACT_WORDSHIFTING**$_{DIST}$ $(COVER) \rightarrow S$ extracts the original message. The following pseudocode represents the extraction algorithm.

EXTRACT_WORDSHIFTING ( )

Input: Cover file *COVER*
Algorithm:
1. Read the *COVER* and count the number of spaces between each two consecutive words.
2. Place the code bits as:
3. If the number of spaces is 1, code bit = 0
4. If the number of spaces is greater than 1, code bit = 1.
5. Find L = Length(message).
6. Decompress L number of bits to extract the secret message.
7. End.

9.3.2 TEXT STEGANOGRAPHY BY LINE SHIFTING

What conclusion can be drawn if the followers of an opponent political party start a waited agitation or movement against the government after receiving the document, shown in Figure 9.6, from their leader? The answer may be that the text of the document is saying to start the movement. NO! The conclusion is incorrect as

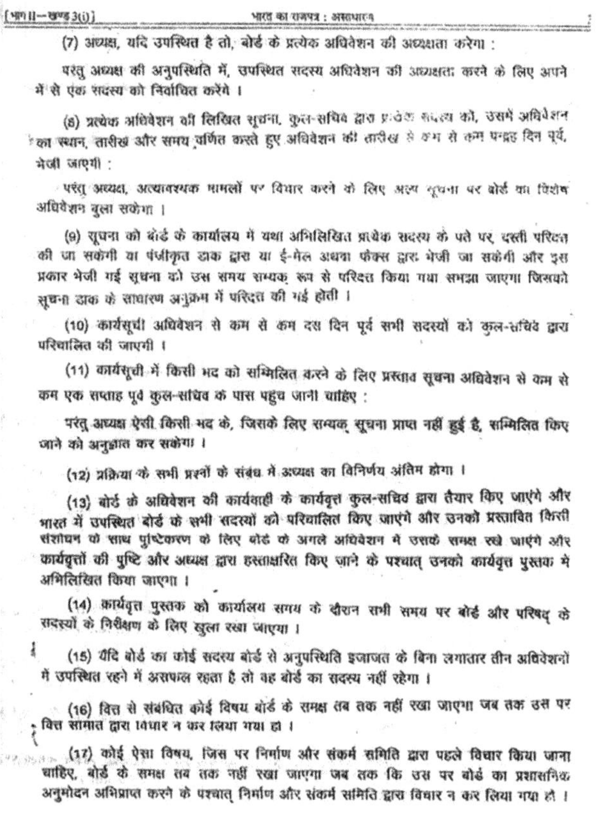

[भाग II—खण्ड 3(i)]   भारत का राजपत्र : असाधारण

(7) अध्यक्ष, यदि उपस्थित है तो, बोर्ड के प्रत्येक अधिवेशन की अध्यक्षता करेगा :

परन्तु अध्यक्ष की अनुपस्थिति में, उपस्थित सदस्य अधिवेशन की अध्यक्षता करने के लिए अपने में से एक सदस्य को निर्वाचित करेंगे ।

(8) प्रत्येक अधिवेशन की लिखित सूचना, कुल-सचिव द्वारा प्रत्येक सदस्य को, उसमें अधिवेशन का स्थान, तारीख और समय वर्णित करते हुए अधिवेशन की तारीख से कम से कम पन्द्रह दिन पूर्व, भेजी जाएगी :

परन्तु अध्यक्ष, अत्यावश्यक मामलों पर विचार करने के लिए अल्प सूचना पर बोर्ड का विशेष अधिवेशन बुला सकेगा ।

(9) सूचना बोर्ड के कार्यालय में यथा अभिलिखित प्रत्येक सदस्य के पते पर, दस्ती परिदत्त की जा सकेगी या पंजीकृत डाक द्वारा या ई-मेल अथवा फैक्स द्वारा भेजी जा सकेगी और इस प्रकार भेजी गई सूचना को उस समय सम्यक् रूप से परिदत्त किया गया समझा जाएगा जिसमें सूचना डाक के साधारण अनुक्रम में परिदत्त की गई होती ।

(10) कार्यसूची अधिवेशन से कम से कम दस दिन पूर्व सभी सदस्यों को कुल-सचिव द्वारा परिचालित की जाएगी ।

(11) कार्यसूची में किसी मद को सम्मिलित करने के लिए प्रस्ताव सूचना अधिवेशन से कम से कम एक सप्ताह पूर्व कुल-सचिव के पास पहुंच जानी चाहिए :

परन्तु अध्यक्ष ऐसी किसी मद के, जिसके लिए सम्यक् सूचना प्राप्त नहीं हुई है, सम्मिलित किए जाने को अनुज्ञात कर सकेगा ।

(12) प्रक्रिया के सभी प्रश्नों के संबंध में अध्यक्ष का विनिर्णय अंतिम होगा ।

(13) बोर्ड के अधिवेशन की कार्यवाही के कार्यवृत्त कुल-सचिव द्वारा तैयार किए जाएंगे और भारत में उपस्थित बोर्ड के सभी सदस्यों को परिचालित किए जाएंगे और उनको प्रस्तावित किसी संशोधन के साथ पुनरीक्षण के लिए बोर्ड के अगले अधिवेशन में उसके समक्ष रखे जाएंगे और कार्यवृत्तों की पुष्टि और अध्यक्ष द्वारा हस्ताक्षरित किए जाने के पश्चात् उनको कार्यवृत्त पुस्तक में अभिलिखित किया जाएगा ।

(14) कार्यवृत्त पुस्तक को कार्यालय समय के दौरान सभी समय पर बोर्ड और परिषद् के सदस्यों के निरीक्षण के लिए खुला रखा जाएगा ।

(15) यदि बोर्ड का कोई सदस्य बोर्ड से अनुपस्थिति इजाजत के बिना लगातार तीन अधिवेशनों में उपस्थित रहने में असफल रहता है तो वह बोर्ड का सदस्य नहीं रहेगा ।

(16) वित्त से संबंधित कोई विषय बोर्ड के समक्ष तब तक नहीं रखा जाएगा जब तक उस पर वित्त सामित द्वारा विचार न कर लिया गया हो ।

(17) कोई ऐसा विषय, जिस पर निर्माण और संकर्म समिति द्वारा पहले विचार किया जाना चाहिए, बोर्ड के समक्ष तब तक नहीं रखा जाएगा जब तक कि उस पर बोर्ड का प्रशासनिक अनुमोदन अभिप्राप्त करने के पश्चात् निर्माण और संकर्म समिति द्वारा विचार न कर लिया गया हो ।

FIGURE 9.6 Text file containing a secret message using the line shifting method.

the document is a page, written in an Indian language (Hindi), from The Gazette of India, a general circular from Government of India.

Then the conclusion may be that it is just a coincident as there is no relationship between the movement and the document. But again, NO! This document is containing the message *"DO"*, which is hidden within the document that has been done by using the line shifting method (Figure 9.7).

To start with this steganography method, let us have a look at Figure 9.8, which shows the extraction algorithm of the word shifting method. The algorithm extracts the information from the cover media created by the hiding algorithm of the word shifting method.

The content of the text is the same as the data extraction algorithm used by the word shifting method. But this text is containing the information "A" along with the

Algorithm EXTRACT_WORD SHIFTING ( )

1. Input the cover file COVER
2. Read the COVER and count the number of space between each two consecutive words.
3. Place the code bits as:
4. If number of space is 1, code bit = 0
5. If number of space is greater than 1, code bit = 1.
6. Find the length of the message L.
7. Decompress L number of bits and extract the secret message from the cover file.
8. End.

FIGURE 9.7 Example of the line shifting method.

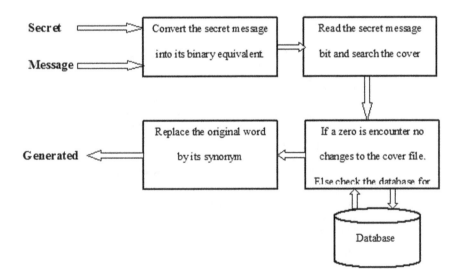

FIGURE 9.8 Hiding process flowchart of synonym replacement method.

method to extract the data from the cover media. If the text is looked properly, it can be noticed that line 1 and line 7 of the algorithm are not proper as these should be. These two lines are displaced vertically with an angle 1°. And this is the key idea of data hiding by the line shifting method.

9.3.3 TEXT STEGANOGRAPHY USING SYNONYM REPLACEMENT

In synonym replacement steganography, a word is replaced by its synonym. The replacement word selected out of the pool of synonyms depends on the secret bits. As an example, let the following sentence be considered: "Joseph is a good boy". If good is represented by 00, then considering the input bits in sequence (01/10/11), the word good can be replaced by its synonyms such as nice, kind, and pleasant, respectively, for bit hiding. The simplified concept, shown in Figure 9.8, for steganography by synonym replacement is that to hide a zero there will be no change in the cover file and to hide a one the original word will be replaced by its synonym.

The database is created by maintaining a dictionary of words and their corresponding synonym.

The following pseudocode represents the algorithm for steganography by synonym replacement method:

HIDE_SYNONYM ( )

```
1. Input Secret Message S and a text file T, containing Indian
Texts, as Cover Media.
2. Compress the message and convert it to its equivalent
binary CONVSECB.
3. Find the length L of CONVSECB.
4. Add binary equivalent of L as the header of CONVSECB.
5. For each bit of CONVSECB.
6. Search the text file T for the words having synonym
7. If CONVSECB bit is 0 the no changes to be made to the cover
file, i.e., the original word will be in the cover file.
8. If CONVSECB bit is 1, search the database for its synonym
and replace the original word of the cover file with its
synonym.
9. End.
```

The algorithm HIDE_SYNONYM ( ) therefore converts the considered text file into a new text file by means of replacing or not replacing the existing words of the file as follows:

HIDE_SYNONYM$_{\text{DICTIONARY}}$ *(S, T)* → *COVER.*

The cover file COVER is then sent to the receiver. During the extraction of the message, described in Figure 9.9, the receiver has to consider both the text file that is the original and the generated cover file to check the hidden entries. The process reads and compares the words of both the files, and if there is an entry of the word in the maintained dictionary, it checks for any change in the entry. If there is no change in the cover file for the particular entry, the process places the code bit zero

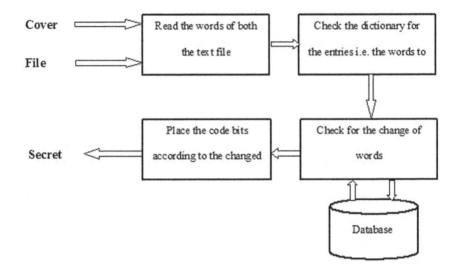

FIGURE 9.9 Extraction process flowchart of synonym replacement method.

for that operation. If it finds any change in the entry, the code bit one is placed. By iterating the process for the message length (concealed as header), the secret message can be extracted from the cover file.

The algorithm EXTRACT_SYNONYM ( ) applies the reverse function of the hiding process and unhides the data from the cover file as follows:

EXTRACT_SYNONYM$_{\text{DICTIONARY}}$ $(T, COVER) \rightarrow S$

EXTRACT_SYNONYM ( )

```
Input: The original Text File T
       Generated cover file COVER
Algorithm:
1. Read the words of both the file T and cover media COVER and
check for the entries of the word in the dictionary.
2. If there an entry for the word and the word is unchanged
with respect to then text file T, place the code bit = 0
3. Else if there is a change then word has been replaced by
its synonym, place the code bit = 1.
4. Find Length(L).
5. Decompress L number of bits for final extraction of S from
the cover file.
6. End.
```

9.4 ADVANCED TEXT STEGANOGRAPHY TECHNIQUES

The Indian languages in the form of digitized text can be exploited to generate advanced text steganography algorithms which are efficient in terms of both the capacity of hiding and transparency to steganalysis attacks.

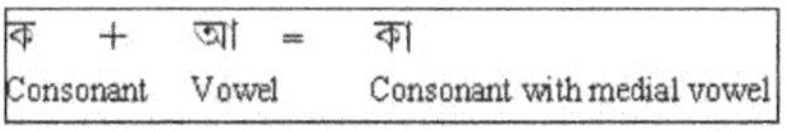

FIGURE 9.10   Consonant with medial vowel ("matra").

9.4.1   Deterministic Feature-Coding Approaches

Feature-coding method hides data by altering or not altering certain features or structures of the characters in a text depending upon the feature-codable properties of the characters present in the particular language. This approach is very effective for hiding large amount of data. Since the writing styles, properties, and features of different languages are mostly different from each other, it is not very easy to analyze the presence of data by using one algorithm for all languages, which thus increases the chance to communicate secretly without drawing any suspicion to the unintended recipients.

Next, we shall demonstrate the use of feature-coding techniques for implementing text steganography in an Indian language, Bengali (ISO 639-1/2/3). The technique works by shifting the "matra", the word used to denote a medial vowel. Writing style also provides for denoting a pure consonant using a specific "matra" so that a syllable may be written as a sequence of generic consonant shapes ending with the final consonant–vowel combination which may be shown using the medial vowel form (Figure 9.10).

9.4.1.1   Algorithm for Steganography by "matra" Shift

One of the major features of the Bengali script is the abundance of vertical lines. The following Bengali characters contain vertical lines:

অ আ খ গ ঘ ঝ ন ণ থ

These characters when added with the "matra" form two parallel lines, as explained in Figure 9.11. In this method, this characteristic is used for steganography (Figure 9.12).

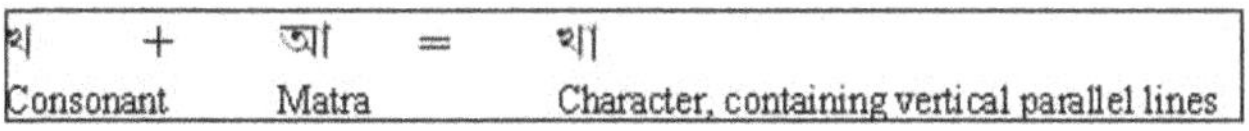

FIGURE 9.11   Character containing vertical parallel lines.

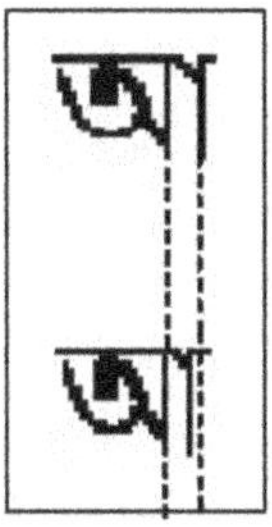

FIGURE 9.12   Horizontal shift (left) of the "matra" of the Bengali character containing vertical parallel lines.

HIDE_MATRASHIFT( ) and **EXTRACT_MATRASHIFT( )** denote the pseudocodes for hiding and extraction, respectively.

HIDE_MATRASHIFT( )

```
1. A[]= Compress(S) //S is the secret information
2. n = Length(A)
3. For i = 1 : n
4.      read A[i]
5.      read the Bengali text file
6.      Search for character with two parallel lines.
7.      If A[i] = 0 then,
               No change occurs to the character chosen
        Else
               Apply Left_Shift_Matra(chosen character)
8. End For
9. For the remaining characters of the text, apply random
changes
10. End
```

As for example:

Step 1: Let us consider a secret text message "X", a string of character. This step compresses the message, converts it into its equivalent binary string, and stores it in an array A.

Step 2: In this step, the length of the bit string is stored in a variable n.

Step 3–Step 8: For 1 to n (the length of the message), first, the algorithm reads the element of A one by one. Then by scanning the characters of the Bengali text file (cover media, where the message will be hidden), whenever it finds a Bengali character containing parallel lines as explained before, the entry of the array A is checked. If the entry of A is 1 (one), replace the character with a new character where "matra" is shifted a little toward left, as shown in Figure 9.13. If the entry of A is 0 (zero), there is no change.

Step 9: After hiding the message, this step changes some characters randomly so that the reader's attention gets diverted.

EXTRACT_MATRASHIFT( )

```
1. Find the length of the hidden information.
2. Find the characters which have been changed.
3. Arrange the bits sequentially and derive composite
information.
4. Decompress and extract original message.
5. End.
```

To decode the message, the algorithm first finds the length of the hidden information, which is stored in the first eight chosen characters of the cover file, i.e., Bengali text

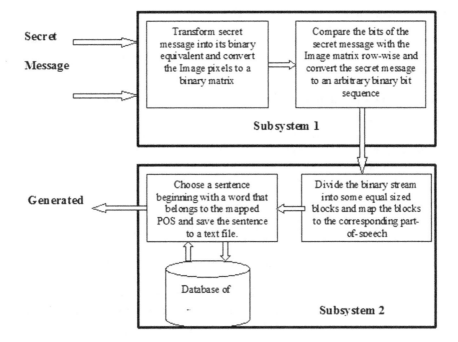

FIGURE 9.13 Hiding process flowchart of part-of-speech mapping.

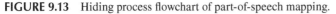

file. Then, for the length of the message (1 to *n*), it scans the Bengali text file searching for the character containing vertical parallel lines. Here, if any changed character (i.e., whose "matra" is shifted toward left) is found, the algorithm considers the corresponding bit as 1; otherwise, the corresponding bit is 0. After getting the bit string, the algorithm decompresses the message and extracts the original information.

Table 9.1 shows the capacity of data hiding of the cover media, collected from different newspapers, using this method.

TABLE 9.1

Capacity Measure of Steganography by Shifting "matra"

Source	Web Address	Genre (one)	Characters	Selected Characters	Percentage (%)
Anandabazar Patrika	http://www.anandabazar.com	Sports, cricket	2262	292	13
Aajkaal	http://www.aajkaal.net	Sports, cricket	3221	476	15
Dainik Sambad	http://www.dainiksambad.com	Movie, gossip	2437	285	11

9.4.2 Non-Deterministic Feature-Coding Approaches

Here, we classify the transformable symbols into several categories and define the event as "sustaining" or remaining in a specific category (same or different) during the transformation of symbols that are transformable.

9.4.2.1 The Steganography Method

This method models the event invocations and represents a bit sequence with the finite state machine (FSM) model. Here, each state is represented as one of the categories of symbols that can undergo transformation. The transition function, δ, is therefore defined as follows:

$$\delta(q,0) = q$$

$$\delta(q,1) = \{p \mid \text{such that } p \in Q \text{ but is not } q\}, \ Q \text{ being the set of states}$$

It is possible to backtrack from the kth invocation of the event to the first, thereby reaching a situation where the outcome of the zeroth invocation can be known, which does not physically comply. As a solution, the first invocation for the outcome transformation of the symbols that can undergo transformation is fixed, from the similar category, and then carried on. The uncertainty in selecting a particular "q" from Q to designate as the first state is resolved by:

- Choosing the first symbol that is transformable
- Finding out the category of the symbol
- Applying the transformation
- Defining this category as "q", or the starting state.

Now, the encoding of zeros and ones proceeds according to the transition function of the FSM.

9.5 LINGUISTIC APPROACHES

With respect to the security concern, the existing linguistic text steganography approaches such as synonym replacement, syntactical text steganography, and Markov chain-based method are not fail-safe. For example, the synonym replacement technique may sometimes replace the synonym of the word which does not agree with the proper language usage or the style of the content. With NICETEXT, though it produces sentences that are correct syntactically by the use of probabilistic context-free grammars (PCFGs), the output text is almost always a set of sentences that are non-grammatical or have semantic anomalies. Markov chain-based approaches may be better, but it may sometimes produce outputs analogous to NICETEXT. By observing the pitfalls, to overcome these, without the replacement of the original word or attempting to create PCFGs, some generalized approaches for steganography through Indian languages can be developed.

Since Indian languages have flexible grammar structure, in one of the approaches, the secret message is hidden in the text by creating meaningful sentences after mapping the data bits to parts-of-speech. In a separate approach, the information is concealed in the text file by changing, without distorting the semantic meaning, the phrase structure of the interrogative sentences.

9.5.1 PART-OF-SPEECH MAPPING OF TEXT STEGANOGRAPHY

To ensure more security, the encoding or hiding system of this method is divided into two subsystems: Subsystem 1 and Subsystem 2, respectively. The Subsystem 1 converts the secret message into an arbitrary binary stream with respect to the original message. The Subsystem 2 takes the converted bit stream as the input and encodes by forming some meaningful yet innocent-appearing sentences. The method takes the advantages to apply the system on Indian languages because:

   i. The flexible use of grammar in sentence or phrase forming draws less attention.
   ii. In Indian languages, unlike in English, personal pronoun shows three points of formalities in second person and two points of formalities in the third person. This property is used to generate large numbers of different sentences devoid of any suspicion.

9.5.2 AN ALGORITHMIC IMPLEMENTATION

The process of data hiding, shown in Figure 9.13, considers a secret message and an image file as the system input. After being processed through the subsystems, the message is concealed within a cover file. The cover file is a collection of innocent sentences.

The following pseudocode presents the algorithm for data hiding by part-of-speech mapping method:

HIDE_POS ( )

```
Input: I, an image of dimension m × n,
       S, the secret message
       K, the number of to be encoded bits
Algorithm:
1. For every pixel Pi in I, IM[i] = Pi mod 2.
2. Let SM[] = Bin(S), where Bin converts S in to binary
equivalent.
3. Let L[] = Bin(Length(SM))
4. For count= 1:7
       CS[count] = L[count] // CS[ ] stores the converted
                                secret message //
5. End For
6. For j = 8 : (Length(CS)+7)
```

```
7.    If (IM[j-7] = = SM[j-7]) then
8.                    CS[j]=1
9.    Else CS[j] = 0.
10. End For
11. Set CS[ ] = Pad(CS, 0 or 1) such that Length(CS) mod K =0.
12. For i = 0: ((length(CS)/K) -1)
13.             Set M = 0
14.        For j = 0 to K - 1
15.            M = M + pow (2, j) * CS[K*i + j+1]
16.        End For
17. N = Random[1,2]
18. Select a word 'W' from the list under POS found in
RSPOS[m, n]
19. Create the cover file COVER by selecting a logically
correct sentence that starts with 'W'.
20. End For
21. End.
```

For extracting the message, as shown in Figure 9.14, consider the image I used for encoding and the generated file COVER as the inputs to the extraction logic.

9.6 DYNAMIC PROGRAMMING AND GREEDY APPROACHES

Considering the availability of more characters and flexible grammar structure of Indian languages, the dynamic programming approaches hide the secret message in the text by creating meaningful sentences after finding the longest common subsequence of two binary strings. One of the strings is the secret message, and the other may be any binary string. The collection of these created sentences is used as the cover media for the mentioned steganography technique. Extraction is performed by applying the reverse of the said mechanism on the cover file.

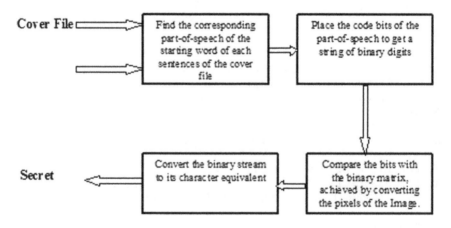

FIGURE 9.14 Extraction flow in POS mapping method.

In greedy approach, the data is hidden into an innocent cover file, containing Indian texts, considering the properties of a sentence such as the number of words, the number of characters, the number of vowels and using the presence of redundant feature-codable character in Indian languages. The approach finds the sentences of the cover file in a greedy manner such that the data can be hidden within the sentences without drawing any suspicion to the readers.

9.7   COMPARATIVE ANALYSIS OF THE TEXT STEGANOGRAPHY TECHNIQUES

In order to cater the readers with a detailed integrated outlook on the different text steganography techniques and their characteristics in terms of a multi-parameter matrix, a comparison grid is presented in Table 9.2.

9.8   APPLICABILITY OF TEXT STEGANOGRAPHY IN INTERNET OF THINGS (IOT)

Steganography is a powerful covert communication tool and therefore has the potential for application in modern communication models such as the Internet of Things, commonly known as the IoT. The IoT is defined as an interconnected network of devices, vehicles, softwares, and other electronic items that facilitate data exchange. In a typical IoT framework, devices can interconnectedly share messages or data in any form to achieve a pre-defined task. Numerous research efforts have been put in Manogaran et al. (2017), Sun et al. (2017), Chervyakov et al. (2019), Raza et al. (2013), Vucini et al. (2015), Yang et al. (2017), Bairagi et al. (2016), Huang et al. (2018), and Shanableh (2012) that enunciate the use of steganography protocols in the IoT domain. Text data is one of the most common forms of data that might be transferred over an IoT network. This paves the way for the development of text steganography techniques that can be used to hide data within device-to-device messages. Such hidden information can be used for various purposes such as control enforcement, device identity purposes, security breach tracking, data breach tracking. At the same time, devices on an IoT network can use embedded text steganography protocols to communicate with other devices covertly as and when required.

9.8.1   ENFORCING IoT SECURITY THROUGH TEXT STEGANOGRAPHY

Despite the popularity of the IoT devices, the Internet of Things paradigm still incurs security challenges such as maintaining the integrity of the ever-increasing number of devices on the IoT network. Text steganography protocols can be used to enforce integrity among the participating devices or nodes by sending covert status messages within the nodes such that any node, if subjected to any unauthorized alteration, would be discoverable to the other nodes and hence can be fixed. Similarly, IoT networks are also prone to data breaches and unauthorized data manipulations.

TABLE 9.2

Comparison among Different Text Steganography Techniques

Methods	Data Hiding Capacity	Complexity	Requirement of Knowledge Power on the Language		Applicable on Printed Documents	Attacks		
			Encoding	Decoding		Retyping	First-Order Statistics	Semantic Checking
Specific characters of words	H	H	Y	N	Y	N	N	Y
Line shift	L	H	N	N	Y	NA	N	N
Word shift	VH	LS	N	N	Y	Y	Y	N
Feature coding in English	H	LS	N	N	Y	Y	Y	N
Adding open space	VH	LS	N	N	N	Y	Y	N
Creating spam texts	H	LS	Y	Y	N	Y	Y	NA
NICETEXT	H	H	Y	N	Y	N	N	Y
Synonym replacement	H	LS	Y	N	Y	N	Y	Y
Markov chain-based	H	H	Y	N	Y	N	N	Y
Feature coding in Indian languages	VH	LS	N	N	Y	Y	Y	N
POS mapping	H	LS	Y	N	Y	N	N	N
Interrogative sentences	L	LS	Y	N	Y	N	N	N
Dynamic programming	H	H	Little	N	N	N	N	N
Greedy approach	VH	LS	N	N	Partially	Y	Y	N

Note: H: high; VH: very high; L: low; LS: less; Y: yes; N: no; NA: not applicable

Such security vulnerabilities can be taken care of by applying text steganography techniques for node-specific watermarking of textual device information or user data over an IoT network. Such an approach would ensure that the transmitted data, if maliciously altered, can be detected for authenticity and data breaches can be contained. Node-specific watermarking through text steganography can also help administrators to deal with security issues arising out of BOTNETs in the IoT domain.

9.9 CONCLUSION

In the era of digital communication, steganography tools can be beneficial for covert communication over potentially unsafe channels and also in situations where protocols prohibit transmission of secret messages between parties. Text documents are the most widely available media objects and can be used suitably as steganography covers that can offer a wide range of scope toward the development of advanced steganography techniques with text documents. The large variation in Indian languages in terms of scripts and their availability in the Unicode renders it a good choice of covers for the development of text steganography techniques. This chapter introduces steganography, its principles, and the domain of text steganography. Also, it elaborates the generic techniques that are applied for implementing text steganography and at the same time it puts light on the advanced techniques that pave the path for the new generation of text steganography. This chapter also provides a detailed parametric comparison chart of the different text steganography techniques, allowing easy comparison and requirement-based selection of the techniques.

REFERENCES

Alattar, A. M. and Alattar, O. M. 2004. Watermarking electronic text documents containing justified paragraphs and irregular line spacing. *Proceeding SPIE – Volume 5306, Security, Steganography & Watermarking of Multimedia Contents VI*, San Jose, CA, 685–695.

Ali, A. E. 2010. A new text steganography method by using non-printing unicode characters. *Eng. & Tech. J.* 28(1):72–83.

Arnold, M., Schmucker, M. and Wolthusen, S. D. 2003. *Techniques and Applications of Digital Watermarking and Content Protection*, Artech House, Norwood, MA.

Artz, D. 2001. Digital steganography: Hiding data within data. *Proc. IEEE Internet Comput.* 5:75–80.

Atallah, M. J., Raskin, V., Crogan, M., Kerschbaum, F., Mohamed, D. and Naik, S. 2001. Natural language watermarking: Design, analysis, and a proof-of-concept implementation. *Information Hiding: Fourth International Workshop, Lecture Notes in Computer Science*, Springer, Berlin, Germany, vol. 2137, 185–199.

Atallah, M. J., Raskin, V., Hempelmann, C. F. Kerschbaum, F., Mohamed, D. and Naik, S. 2002. Natural language watermarking and tamperproofing. *Information Hiding: Fifth International Workshop, Lecture Notes in Computer Science*, Springer, Berlin, Germany, vol. 2578, 196–212.

Bairagi, A. K., Khondoker, R. and Islam, R. 2016. An efficient steganographic approach for protecting communication in the Internet of Things (IoT) critical infrastructures. *Inf. Secur. J. Glob. Persp.* 25(4–6):197–212.

Bergmair, R. 2004. Towards linguistic steganography: A systematic investigation of approaches, systems and issues. Final year thesis handed in partial fulfillment of the degree requirements for the degree "B.Sc. (Hons.) in Computer Studies" to the University of Derby.

Bolshakov, I. A. and Gelbukh, A. 2004. Synonymous paraphrasing using WordNet and Internet. Natural Language Processing and Information Systems, *9th International Conference on Applications of Natural Language to Information Systems, NLDB 2004, Lecture Notes in Computer Science*, Springer, Berlin, Germany, vol. 3136, 312–323.

Brassil, J. T., Low, S., Maxemchuk, N. F., O'Gorman, L. 1995. Electronic marking and identification techniques to discourage document copying. *IEEE J. Sel. Areas Commun.* 13(8):1495–1504.

Cachin, C. 1998. An information-theoretic model for steganography. *Proceeding of the 2nd Information Hiding Workshop*, Springer-Verlag, Portland, OR, 306–318.

Calvo, H. and Bolshakov, I. A. 2004. *Using selectional preferences for extending a synonymous paraphrasing method in steganography.* Avances en Ciencias de la Computacion e Ingenieria de Computo – CIC'2004: XIII Congreso Internacional de Computacion, Mexico, 231–242.

Chandramouli, R., Kharrazi, M. and Memon, N. 2004. Image steganography and steganalysis: Concepts and practice. *Proceeding IWDW and LNCS*, Seoul, Korea, 35–49.

Chapman, M. T. and Davida, G. I. 1997. Hiding the hidden: A software system for concealing ciphertext as innocuous text. *Information and Communications Security. First International Conference, Lecture Notes in Computer Science*, Springer, Beijing, China, vol. 1334.

Chapman, M. T. and Davida, G. I. 2002. Plausible deniability using automated linguistic steganography. *Infrastructure Security: International Conference, Lecture Notes in Computer Science*, Springer, Berlin, vol. 2437, 276–287.

Chapman, M.T., Davida, G.I., and Marc Rennhard. 2001. A practical and effective approach to large-scale automated linguistic steganography. *Information Security: Fourth International Conference, Lecture Notes in Computer Science.* Springer, Malaga, Spain, vol. 2200, 156ff.

Chervyakov, N., Babenko, M., Tchernykh, A., Kucherov, N., Lopez, V.M. and Cortés-Mendoza, J. M. 2019. AR-RRNS: Configurable reliable distributed data storage systems for Internet of Things to ensure security. *Future Gener. Comput. Syst.* 92:1080–1092.

Grothoff, C., Grothoff, K., Alkhutova, L.,Stutsman, R. and Atallah, M. 2005a. Translation-based steganography. *Proceedings of Information Hiding Workshop (IH 2005)*, Springer, Barcelona, Spain, 213–233.

Grothoff, C., Grothoff, K., Alkhutova, L., Stutsman, R. and Atallah, M. 2005b. *Translation-Based Steganography. Technical Report TR 2005-39*, Purdue CERIAS, West Lafayette, IN.

Herodotus. 440 BC. *The Histories*, Penguin Books, London, 1996. Translated by Aubrey de Sélincourt.

Huang, C. T., Tsai, M. Y., Lin, L. C., Wang, W.J., and Wang, S.J. 2018. VQ-based data hiding in IoT networks using two-level encoding with adaptive pixel replacements. *J. Supercomput.* 74(9):4295–4314.

Huang, D. and Yan, H. 2001. Interword distance changes represented by sine waves for watermarking text images. *IEEE Trans. Circ. Syst. Video Technol.* 11(12):1237–1245.

Igor, A. and Bolshakov, I.A. 2004. A method of linguistic steganography based on collocationally-verified synonymy. *Information Hiding: 6th International Workshop, Toronto, Canada, Lecture Notes in Computer Science*, vol. 3200, 180–191.

Johnson, N. F. Duric, Z. and Jajodia, S. 2001. *Information Hiding: Steganography and Watermarking: Attacks and Countermeasures*, Kluwer Academic, Norwell, MA.

Johnson, N. F. and Jajodia, S. 1998. Steganalysis: The investigation of hiding information. *Proceeding Information Technology Conference*, IEEE, Syracuse, NY, 113–116.

Kahn, D. 1996. Codebreakers: The Story of Secret Writing. Revised ed., Scribner, New York, NY.

Khairullah, M. 2009. A novel text steganography system using font color of the invisible characters in microsoft word documents. *Second International Conference on Computer and Electrical Engineering ICCEE '09*, Dubai, United Arab Emirates, vol. 1, 482–484.

Kim, Y., Moon, K. and Oh, I. 2003. A text watermarking algorithm based on word classification and inter-word space statistics. *Proceeding Seventh International Conference on Document Analysis and Recognition (ICDAR '03)*, Edinburgh, Scotland, 775–779.

Low, S. H., Maxemchuk, N. F., Brassil, J. T. and O'Gorman, L. 1995. Document marking and identification using both line and word shifting. *Proceeding of the Fourteenth Annual Joint Conference of the IEEE Computer and Communications Societies (INFOCOM '95)*, Boston, MA, vol. 2, 853–860.

Manogaran, G., Thota, C., Lopez, D. and Sundarasekar, R. 2017. *Big Data Security Intelligence for Healthcare Industry 4.0. Cybersecurity for Industry 4.0*, Springer, Cham, Switzerland, 103–126.

Murphy, B. 2001. Syntactic information hiding in plain text. Master's thesis, Department of Computer Science, Trinity College, Dublin.

Murphy, B. and Vogel, C. 2007a. Statistically constrained shallow text marking: techniques, evaluation paradigm, and results. *Proceedings of the SPIE International Conference on Security, Steganography, and Watermarking of Multimedia Contents*, San Jose, CA.

Murphy, B. and Vogel, C. 2007b. The syntax of concealment: reliable methods for plain text information hiding. *Proceedings of the SPIE International Conference on Security, Steganography, and Watermarking of Multimedia Contents*, San Jose, CA.

Rabah, K. 2004. Steganography – The art of hiding data. *Int. J. Inf. Technol.* 3(3):245–269.

Raza, S., Shafagh, H., Hewage, K., Hummen, R. and Voigt, T. 2013. Lithe: Lightweight secure CoAP for the Internet of Things. *IEEE Sensors J.* 1(10):3711–3720.

Roy, S. and Manasmita, M. 2011. A novel approach to format based text steganography, *ICCCS'11*, Rourkela, Odisha, India.

Shanableh, T. 2012. Data hiding in MPEG video files using multivariate regression and flexible macroblock ordering. *IEEE Trans. Inf. Forens. Secur.* 7(2):455–464.

Shirali-Shahreza, M. and Shirali-Shahreza, M. H. 2008a. An improved version of Persian/Arabic text steganography using "La" word. *6th National Conference on Telecommunication Technologies 2008 and 2008 2nd Malaysia Conference on Photonics*, Putrajaya, Malaysia, 372–376.

Shirali-Shahreza, M. and Shirali-Shahreza, S. 2008b. Persian/Arabic unicode text steganography. *Fourth International Conference on Information Assurance and Security, ISIAS '08*, IEEE, Naples, Italy, 62–66.

Shirali-Shahreza, M. H. and Shirali-Shahreza, M. 2010. Arabic/Persian text steganography utilizing similar letters with different codes. *Arab. J. Sci. Eng.* 35(1B):213–222.

Simmons, G. 1985. The prisoner's problem and the subliminal channel. *Proceeding EUROCRYPT, LNCS*, Springer, Heidelberg, Germany, 364–378.

Stutsman, R., Atallah, M., Grothoff, C. 2006. Lost in just the translation. *Proceedings of the 21st Annual ACM Symposium on Applied Computing (SAC 2006)*, Dijon, France.

Sun, H., Wang, X, Buyya, R. and Su, J. 2017. CloudEyes: Cloud-based malware detection with reversible sketch for resource-constrained Internet of Things (IoT) devices. *Softw. Pract. Exp.* 47(3):421–441.

Taskiran, C.M., Topkara, U., Topkara, M., and Delp, E. J. 2006. Attacks on lexical natural language steganography systems. *Proceedings of the SPIE International Conference on Security, Steganography and Watermarking of Multimedia Contents*, San Jose, CA.

Topkara, M., Riccardi, G., Hakkani-Tur, D. and Atallah, M. J. 2006. Natural language watermarking: Challenges in building a practical system. *Proceedings of the SPIE International Conference on Security, Steganography, and Watermarking of Multimedia Contents*, San Jose, CA.

Topkara, M., Taskiran, C. M. and Delp, E. J. 2005. Natural language watermarking. *Proceedings of the SPIE International Conference on Security, Steganography, and Watermarking of Multimedia Contents*, San Jose, CA, vol. 5681.

Topkara, M., Topkara, U. and Atallah, M. J. 2006a. Words are not enough: Sentence level natural language watermarking. *Proceedings of the ACM Workshop on Content Protection and Security*, Geneva, Switzerland.

Topkara, U., Topkara, M., and Atallah, M.J. 2006b. The hiding virtues of ambiguity: quantifiably resilient watermarking of natural language text through synonym substitutions. *MM&Sec '06: Proceeding of the 8th Workshop on Multimedia and Security*, ACM Press, New York, NY, 164–174.

Vucini, M., Tourancheau, B., Roussau, F., Duda, A., Damon, L. and Guizzeti, R. 2015. OSCAR: Object security architecture for the Internet of Things. *Ad Hoc Netw.* 32:3–16.

Warchalking. 2003. Collaboratively creating a hobo-language for free wireless networking. http://www.warchalking.org/ (Accessed December 2019).

Wayner, P. 1992. Mimic functions. *Cryptologia* 25(3):193–214.

Wayner, P. 1995. Strong theoretical steganography. *Cryptologia* 19(3):285–299.

Wayner, P. 2002. *Disappearing Cryptography: Information Hiding: Steganography & Watermarking*, 2nd Edition, Morgan Kaufmann, San Francisco, CA.

Yang, Y., X. Liu, X., and Deng, R. H. 2017. Lightweight break-glass access control system for healthcare Internet-of-Things. *IEEE Trans. Ind. Inf.* 14(8):3610–3617.

10 IoT as a Means of Strengthening Individual Farmers' Position in Collaboration for Transformation to Sustainable Farming

J.S. Gusc
University of Groningen

S. Jarka
Warsaw University of Life Sciences

I.J. de Zwarte
Blackbox Measurements

CONTENTS

10.1 INTRODUCTION: BACKGROUND AND DRIVING FORCES

New solutions to collaboration toward sustainability are needed for the agri-food sector, covering sustainability in a broader sense, including social, economic, and environmental aspects. While the introduction of the Internet of Things in precision farming enables us to learn about the current status of the operations at the farm and adjust the system [1,2], it has not so far incorporated a broader sustainable perspective. For example, the pollution is at its worst, and the registration without altering the operation or altering only one specific aspect may have adverse consequences for other areas of the operations of the farm. Acting upon that one aspect of the pollution data, probably taking actions to reduce the environmental impact, overlooks other (negative) effects on areas, i.e., social (more pressure on the farmer) or financial (large investment required in the new equipment), of the farm. Connecting IoT with the sustainable development can enable transformation toward smart farm creating value in multiple areas and for multiple stakeholders.

The sustainable development consists of three dimensions – economic, social, and environmental – that are interconnected and balanced. The aim of sustainable development is to obtain the (right) balance between the categories and reconciling conflicts [3]. The model shows three rings – environment, society, and economy – all of equal size with overlapping areas representing each category of sustainable development.

There are multiple attempts to progress sustainable development and reach the "right" balance in the three dimensions. Hopwood, Mellor, and O'Brien [4] grouped the development strategies on a matrix according to the level of importance of the human well-being and equality (the socioeconomic axis) and to the priority of the environment (the environment axis) from low environmental concern through techno-centered to eco-centered. There is a strong ongoing debate around the different views on sustainable development grouped in status quo, reform, and transformation. The debate is beyond the scope of this chapter, but it is worth mentioning that for ingenious sustainability, transformation approaches are needed. The sustainable smart farming concept proposed here can be classified as transformational approach. Necessary is however to understand the interplay of the environmental concerns and socioeconomic well-being on the opportunities to enhance sustainable development. The transformation will work when the roots of the problems are the changes in the economic and the power structures of society [5]. The transformation supporters see the mounting problems for sustainable development as rooted in fundamental features of society and how humans interrelate with the environment. The fundamental change is needed within the society and the way humans interrelate with the environment and each other in order to prevent a crisis. Following up on the position of the transformation supporters, the question arises: How can a fundamental change be accomplished?

Some indigenous change in the power division and decision-making process needs to take place. That is exactly what happens when IoT technology is implemented. In the current study, we show the transformations toward a more sustainable way of operating a farm due to application of IoT. We argue that application of IoT technology/smart farming may be the vehicle for sustainability development.

Further in this chapter, we describe the stakeholders' aspects of farming and illustrate with a micro-case study how an application of IoT on a farm may contribute to sustainability by changing the stakeholder's constellation. By doing so, we respond to the call of Zambon [6] for more empirical evidence and clearance on the application effects of smart farming on economic, social, and environmental dimensions.

10.2 SMART FARMING

The concept of smart farming has its origin in the "precise farming", which is in turn transferred from the digitization process in the industry in general. In the industry, the digitization as a continuous process of convergence of real and virtual world has become the main driver of the innovation and change in most sectors of the economy [7]. After three consecutive industrial revolutions, the fourth industrial revolution takes place, relying on information technologies such as big data, the Internet of Things, and artificial intelligence. As seen in Table 10.1, there are four stages of industrial revolution corresponding to the development of smart farming.

Technology is changing the world, and farming is slowly catching up. According to Paraforos [8], the development of the digitization of agriculture is at an early stage, but the agri-food sector is becoming an active beneficiary of the digital data. Innovative agriculture uses sophisticated technologies such as robots, temperature and moisture sensors, aerial images, and GPS technology. The new technologies open up huge opportunities for agricultural farms in terms of increasing productivity, efficiency, safety, and sustainability. The Internet of everything from automated farm equipment to a wide array of Internet of Things (IoT) sensors that measure soil moisture and drones that keep track of crops has changed the agriculture business.

World Government Summit calls this movement "Agriculture 4.0", which no longer depends on applying water, fertilizers, and pesticides uniformly across the fields; instead, farms use the minimum quantities required and target very specific areas with ideal timing. The key features of IoT (see Table 10.2) jointly the results t of the digital transformation, (i.e., processing large amounts of data using specialized algorithms and programs), enable generating additional value in the agricultural production chain [1]. The challenge is here, however, not only in processing the data, analyzing, visualizing, and modeling, but also and mostly in interpreting it for operational and strategic purposes in farmers' decisions.

TABLE 10.1
The Stages of Industrialization of the Economy and Smart Farming

Stage		Features	Smart Farming [6]
Industry 1.0	–	Mechanization	Product
Industry 2.0	–	Electrification	Smart product
Industry 3.0	–	Digitization	Smart, connected product
Industry 4.0	–	Networking	Systems of system

TABLE 10.2

Key Features of the IoT [6]

Interconnectivity	Being interconnected with global information and communication infrastructures
Object-related services	Providing object-related services within the limits defined by objects (privacy protection and semantic consistency of physical and associated virtual objects)
Heterogeneity	Working and interacting with different hardware platforms and networks
Dynamic changes	For example, connection and/or disconnection, number of devices, and the context in which devices operate (location, speed, quantity of the product, etc.)
High scalability	Practically no limits in the number of devices

However, the application of the newest technology enables precision farming, and it does not make the farms transform to smart farms. Smart farming goes beyond the use of (*big*) data and is defined as farming enhanced by its context-, situation awareness and triggered by real-time events so called "agile" action on suddenly changed operational conditions or other circumstances (e.g., weather or disease alert) [9]. Next to the development of the electronics and the continuous technological progress, and together with the increasing competition in the market, a new stage of management will emerge alongside [10].

The digital transformation takes effect via four levers: digital data, automation, connectivity, and digital customer access [11].

Farms and agricultural operations run technically different, but the tools to enable farmers to make agile decisions and operate in the network of decisions need to be developed. Only after, the digitization and the data treatment of this information can be used for process automation and management at the operational and strategic levels to transform to become smart. Incorporating a sustainable development perspective into precision farming can contribute to this transformation.

10.2.1 SMART FARMING AND SUSTAINABLE DEVELOPMENT

Smart farming using IoT can contribute to sustainable development on all three sustainability dimensions: environmental, economic, and social (Figure 10.1); and its implementation is positively associated with the following conditions [12]:

- socioeconomic; older farmers with better education and experience who are more likely to introduce precision farming methods,
- agri-environment; farmers farming on larger farms and on better soils, being owners,
- socially oriented farmers who search for "more sustainability",
- for information; farmers who positively assess cooperation with advisors, supporting their professional knowledge and skills.

Further, the economic and legitimacy aspects play a significant role in the IoT adoption decisions, such as prices of agricultural produce and energy, labor costs, and compliance with the environmental laws, to name a few [13].

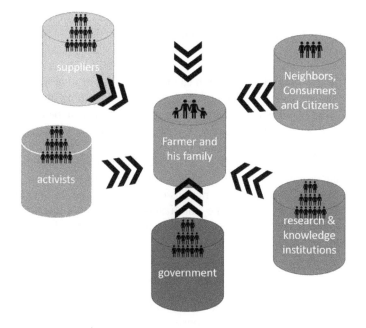

FIGURE 10.1   Stakeholders' pressures on the farmer (see Appendix 1 for a more elaborate overview of the stakeholders).

10.2.1.1   Environmental Dimension

A digital smart farm is more efficient and sustainable than its counterparts of the past [14]. On a smart farm, crops are likely grown using precision agriculture, tractors might be self-driving, the harvest could be determined by digital imagery of the fields, and the farmer is typically working with an agronomist to provide technology know-how. In recent years, significant progress has been made in the field of satellite navigation, variable fertilizer dosage, and yield mapping, i.e., in the area of plant production. In the area of animal monitoring, the use of preventive measures has contributed to reducing the occurrence of diseases and reduced medical treatment [15].

IoT and AI technologies have the potential to transform agricultural practices, paving the way for environmentally safer agricultural methods. Smart monitoring devices and sensors attached to crops monitor their growth in real time, and anomalies can be detected and resolved immediately. Parameters such as hydration, plant nutrition, and diseases are monitored in real time. The data can then be utilized to determine precise irrigation patterns and recommend the best watering cycles. A better variety of crops having high nutritional content can be farmed using AI, which can also lower the use of harmful pesticides indirectly and ultimately benefit people's health. For example, the air pollution generated by farming operations can cause chronic and respiratory problems such as asthma, and fatal diseases such as lung cancer. IoT and AI can provide tools to monitor air pollution in real time. The tools can identify sources of air pollution quickly and accurately.

10.2.1.2   Economic Dimension

The introduction of smart farming is primarily conditioned by economic factors that shape the economic calculation of an agricultural enterprise [16]. These include the costs and benefits of necessary investments in new plant and animal production technology [17]. The results of the research carried out by the Institute of Management WULS indicate that enterprises implementing precision farming obtained on average 8%–14% higher (than regular improvement investments) benefits from the smart farming solutions implemented. The large milk production farms showed a reduction in operating costs, primarily due to the reduction of energy usage and feed, while increasing revenues from milk sales. Interestingly, the smart solution enabled an increase in milk production in periods commonly registering lower milk production, in the high-temperature months of late spring and summer. The smart farming introduction allowed for the stabilization of the environmental conditions for milk production throughout the year, which in turn contributed to an increase in the milk yield.

10.2.1.3   Social Dimension

The social dimension of sustainable development has been probably the most difficult and complex to define [18,19]. Bacon et al. [20] provided an in-depth analysis of the concept and came with the definition consisting of seven factors:

1. human health,
2. democracy,
3. work,
4. quality of life and human well-being,
5. equity, justice, and ethics,
6. resiliency and vulnerability,
7. biological and cultural diversity.

Democracy refers to participation (voice and vote), decision making, rural associations/cooperatives, social capital and community cohesion, inequalities in social power, representation, accountability mechanisms, food sovereignty, social movements, governance, and government policy (overlaps with equity and justice). The factors refer to internal/farm employees as well as the external stakeholders.

Taking the farmer's perspective, Zarco-Tejada [21] indicates endogenous factors regarding the farmer himself and attitudes accompanying him in conducting agricultural activity. These factors include the following:

- the individual situation of the agricultural enterprise (farm),
- the situation resulting from the farmer's age and his attitude toward innovation.

The authors point out that the uncertainty associated with investing in precision farming decreases as farmer learns a new solution and gains experience. Further, the farmers who are 100% -owners of their farms, are more willing to finance new solutions as compared to farmers operating the production means and land with borrowed capital [13]. The interplay of situational and personal circumstances will be the key to exploring the effects of smart farming on the social dimension. For example, the impact of the regulatory environment within the social dimensions concerns the permission/license to operate issuing bodies such as government. Governments can play a key part in solving the food scarcity issue. They need to take on a broader and more prominent role than their traditional regulatory and facilitating function.

By challenging the traditional legacy model and pursuing such a program, governments can:

- Ensure food security and reduce dependency on imports
- Become a net exporter not only of products but also of IP and new solutions
- Increase the productivity and support the shift toward an innovation- and knowledge-based economy.

10.3 FARMERS AND SUSTAINABILITY: THE INTERPLAY OF MULTIPLE PRESSURES

The basis function that legitimizes organizations in society is to create the value for the society. Moving to sustainable operations, the way an organization creates value needs to be considered broader than only from single perspective, either focus on financial value or compliance to requirement from a selected stakeholder, i.e., capital provider or policy maker [19]. The common practice among farming companies has been to prioritize pressure and eventually focus on policy makers as they provide or can withdraw license to operate, exercising the compliance pressure on the farmer.

Second, the non-economic impacts on the stakeholders in the environment and society must be integrated, for example, impacts on public environmental goods such as clean water and air, but also social impacts such as job generation and health hazards. Ever since "the triple bottom line" (TBL), sustainable development has consisted of three dimensions: economic, environmental, and social [22], repeated by many others but always sustaining embracing economic development, environmental aspects, and social inclusion [23]. Despite the triple bottom definition, the social dimension is an underdeveloped and somehow neglected element. It is often minimal and present in relation to or as a part of the environmental or economic sustainability: "a better understanding of the concept's social elements is crucial in reconciling the often-competing demands of the society–environment–economy tripartite [24]". However, the business in general tends to overlook social justice or postpone addressing social concerns to the later stages of sustainable development. Social sustainability is a wide-ranging multi-dimensional concept, with the underlying question "what are the social goals of sustainable development?", which is open to a multitude of answers, with no consensus on how these goals are defined [20,25]. Although the boundaries of a category as broad as "social dimensions of sustainable agriculture" can be ambiguous, social scientific approaches should examine the following: (1) the interdependencies of social, economic, and ecological systems; (2) the social processes, patterns, and factors that shape and maintain economic processes; (3) the social structures, values, and hierarchies often assigned to markers of identity such as race, class, and gender; and (4) the processes of decision making, alliance building, and governance, especially in relation to democracy and rights [25].

As Figure 10.1 shows, the current structures in the agricultural sector as well as the stakeholders' patterns all aim at exercising (multiple and simultaneous) pressures on a farmer's person.

In the current situation, the stakeholders seem to distrust each other and do not recognizeothers as accountable partners for collaboration. Collaboration is about connecting forces and therefore also about sharing – sharing knowledge and insights together. At the same time, sharing in the current relationship models is often a struggle across conflicting interests. At this moment, the party in charge of the data creates pressure on the (other) participants of this stakeholder network, with a farmer being often "pushed to the corner". How did that happen? First, the origin of the pressures is probably to be found in the most savage actions of the environmental activists who gained unlawful access to yards and stables and farmers houses. The activists' goal is to influence public opinion in their worldview therefore they spread an image depicting the farmyard as one malicious environment for humans and animals. They probably are not aware of the side-effects of their pressures on the broader stakeholder network.

Next, a citizen is a customer who seems to be less and less aware of where an egg, a glass of milk, a meatball, or its red cabbage comes from and how it has been grown. At the same time, the citizen expects that farmers' products will always be there, 24/7 (such as the availability of strawberries in the winter season in Europe), and that they will be healthy, of high quality, and reliable.

Finally, the relationship with the national and local government sometimes looks more like a fight. In the European Union, some governments seem to rely on not-impartial reports when making implementation decisions of more rigorous laws and regulations and subsequently appear to lock things up in a fixed status-quo (see, e.g., "nitrogen law[1]" and protest[2] in 2018/2019 in the Netherlands). The above pressure comes on top of the existing ones such as the pressure on scaling up the farm to reduce the cost per unit and the dependence on financing from the banks.

With all these pressures being exercised, it is certainly easy to be a farmer; not to mention to cooperate or share data in these constellation of conflicting interests. In fact, farmers nowadays almost literally arm themselves against the future and the outside world.

The recent protest of farmers in Europe, although at first sight related to the single environmental rule imposed on them, has a broader underlying problem and deeper causes.

> (…) from a good future perspective in 2005 by government intervention eventually relocated the farm and ever since constantly thwarted by changing regulations. While throughout all these years we have focused on land-based agriculture, now it seems that what people want to bet on for the future is intensive dairy farming, they must now pull the plug (…)
>
> (…) A farmer does not want to take the helm; the real farmer is already busy enough (…)

Anthropological research among farmers showed that the farmers and citizens are being played off against each other by the government; as a result, a farmer sees no way out [26]. All these pressures together narrow the functioning of a farmer as an entrepreneur and make him feel desolate, in fact certainly blocking the road of transformation toward (more) sustainable agriculture.

10.4 SMART FARMING AS A WAY OF MANAGING THE PRESSURES

Industrial agriculture and the globalized food system have increasingly occluded the relationship between the stakeholders in the food systems, expanding the physical and cognitive distances among producers, consumers, and their supporting environments [27]. An attempt to broach sustainable agriculture, therefore, demands attention to its social–ecological nature, and an understanding that agriculture produces landscapes that are at once social, cultural, and ecological. The increased crop yields and reduced costs are key for farmers to buy into tech, but at the same time the society requires from farmers to answer the questions: "Is it safe? Is it moral? Does it have value?" Smart farming can contribute to creating a stronger legitimacy among the stakeholders by enabling direct relationships with stakeholders. It can reshuffle the current power division in the chain and initiate collaboration on the partnership terms without unbalanced power relationships.

Collaborating through (IoT) connection, sharing knowledge and insights together, which seems unfeasible in the current status quo of the stakeholders' structure, can emerge in the transformation.

10.5 METHODOLOGY: INTERVENTIONIST APPROACH

This exploratory micro-case research aims to illustrate the application of IoT in a farm and its effects in changing the position of a farmer and his farm in the chain of stakeholders' relationships. This micro-case shows the initiating process of transforming it to a smart farm. To explore this change, the study has designed and implemented a tailor-made IoT system for the farm. In order to do so, the research initiated the collaboration with the members of the case farm and the external stakeholders: knowledge institutes, stable equipment suppliers, government, and customers. This form of collaboration in research requires the use of the interventionist research methodology. Interventionist research is defined as a reflective process of progressive problem solving for which the researchers and members of the case organization work together to improve the way organizational issues are addressed [28]. The intervention here was the creation and implementation of the tailor-made IoT prototype. There exist different streams within the interventionist research, for which the stream of action research has been chosen for this study. Retolaza et al. [19] suggest the following three perquisites for reliable action research:

1. The research project must entail a social challenge with potential improvements;
2. The project must proceed progressively in a loop structure comprising planning, action, feedback, and reflection;
3. The project must take place collaboratively, bringing together all the actors involved and, at least progressively, all those affected.

TABLE 10.3

Overview of the Five Quality Factors of Action Research

Quality Factors	How Applied in the Current Research
Proceeds from a praxis of participation	Collaboration with practitioners of the case company by means of visits, presentations, receiving feedback, telephone calls, and mail conversations
Is guided by practitioners' concerns for practicality	Needs expressed by the case farm for the use and construction of the model were taken into account and processed
Is inclusive of stakeholders' ways of knowing	The study tries to be as comprehensive as possible with regard to the stakeholders of the case company
Helps to build capacity for ongoing change efforts	By creating a pilot dashboard setup that will potentially function as a decision support tool, it provides the capacity to implement ongoing change efforts. In addition, the model is designed to be flexible in order to adapt to changes
Engages with the issues considered significant for the social and ecological environment of the researched object	The process of creating a shortlist by evaluating the longlist addressed this aspect. This process identifies the important issues for the study

The setup of this research complies with the prerequisites; it reflects a social challenge and is aimed at bringing an improvement in the sustainable decision-making process. In addition, a progressive loop structure has been performed in collaboration with the members from the case farm and data analysts.

As Table 10.3 shows, the current study translated Bradbury-Huang's [29] five quality factors for performing action research to the current research problem:

10.6 MICRO-CASE IOT SMART FARMING

The main objective of the pilot setup was to let the farmer discover his own way on the farm to monitor and manage the (big) data together with the platform developer; by doing so, to add value to the farm's operations and strengthen his own position in the stakeholders' network; to explore the added value and the potential engagement of the stakeholders; and last but not least, to explore the social process of multiple parties. And it is the transition from the pilot project to continuous communication.

CASE COMPANY
The Poultry Farm
Farmer Mr. Jan Noorlander
Located in Barneveld, the Netherlands
70,000 chickens
Three poultry houses with 20,000 hens per house
With 2 stories
The package of sensor devices installed:
Particulate matter, ammonia, and CO_2
The pilot ran throughout a year, from summer 2018 until summer 2019

The case site was developed in cooperation with the Poultry Expertise Center (PEC Barneveld, the Netherlands). The center cooperates with multiple farmers who educate the students (future farmers) of the PEC at their farms. From the variety of farmers, Jan Noorlander appeared to be the most suitable person to participate in the pilot as he is the kind of person who is willing to give a "silent" try to an early-stage innovation. His enthusiasm already from the early beginning of proposal helped us make this decision. Jan Noorlander is a middle-aged man and has lived his whole life on the farm, as his father and grandfather did, and has always been interested in modern technologies. He believes that next to the large companies supplying supermarket chains, there is enough space in the market for niche players, but it is livable only with multiple sources of income and innovative solutions. Because he does not want to grow bigger in terms of animal numbers, he innovates and researches alternatives. Recently, he used the extra construction permit for building a new chicken barn to convert the barn into the office spaces for rent capacity and so spread the risks of operations by assuring additional a "third leg" of income.

At the beginning of the project, several visits aimed at introduction, getting know each other (building trust), and exploration of the need for sensor data took place. During one of the visits, the idea emerged to use the sensor data for communication with buyers and suppliers, e.g., animal nutrition supplier/seller who promised a feed with lower ammonia emissions. But how would a farmer know what actual impact it might have on his animals and yield? According to the farmer, the chickens may lay fewer or more eggs at different costs. How would he know without measurement? For customers, expectation was more of a "future story", and the farmer does not expect consumers to ask questions directly to him, but he does want supermarket chains or outlets to support their quality claim. Asking questions about innovations makes it possible to reap the benefits. Gathering knowledge is essential and being able to apply is essential, as he said.

10.7   FINDINGS

The farmer expected that the government would be interested in the IoT data, but he was cautious, and exploring the potential of sharing the data with stakeholders, he wanted to be "in control" and not disturbed in optimization of his operations, hence practicing the proactive attitude but safeguarding the data ownership. Based on these interviews, it was decided to install a set of sensors coupled with a dashboard for a year as a test installation. The scheme for the sensor device system is exhibited below in Figure 10.2, and the actual installation is shown in Picture 10.1.

During the pilot, the installation technician and the data analyst visited the farm regularly, approximately once a month, to socialize, drink a cup of coffee, make contact, and build relationship (farmer versus outsider) essential for assuring the quality in action research. The meetings were semi-structured with recurring set of questions to the farmer: What has he been able to do with data? Has the farmer explored the data and looked at it? Were there any technical issues? What are the improvement suggestions? And enough time was planned for question from the farmer's site. During the visits, some small technical adjustments were also made to the system. For the sensor devices, the technician and data analyst had a real-time

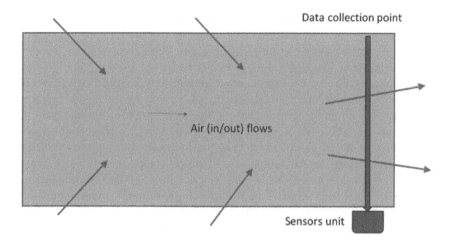

FIGURE 10.2   Animal house scheme for airflows and the sensor device unit (own materials).

PICTURE 10.1   The unit with sensors as installed in the pilot chicken house. (Photo: I.de Zwarte; own materials.)

access online from any spot. The IoT prototype operated without major technical problems. The setup was developed to initiate the transformation.

From the social perspective, more complexity emerged. The transformation to be initiated to let the farmer bundle together, share, and be heard and understood not with physical power but with knowledge, strengths as (collective) effort, and certainly also with the power of real-time data seems to still take place. The farmer in the case was able to continuously monitor, and map his situation in detail, collected data, and observed it when he loged on the device (smartphone/PC). The ownership of data lied at the farmer's hands.

Besides the technical result on the improvement of technical parameters of the farm which have been proven well in other sources [30] and are not the core of the findings here, the farmer did not use the information to the extent we had expected. He collected the data but missed opportunity to manage it within the chain of the relationships.

As Figure 10.3 shows, just at the beginning of the transformations, the stakeholders are a kind of "closed stronghold" by exercising the pressure to pursue each of them his/her own interest while chasing the data. Furthermore, each stakeholder analyzes the data independently from different perspective, and by doing so enhances the frictions and increases the pressures in the relationships.

The IoT platform organization configures and organizes sensors, climate systems, cameras, and peripherals, and collects and processes countless data. Processing and analysis allow for explicating relationships between multiple factors at the same time, for example, outside climate, barn climate, water consumption, electricity, feed, slaughter weight, particulate matter emissions, behavior of your animals, loss, and the number of eggs, but also government and local inhabitants. By combining these data and merging them in one place, the farmer gets relevant insights. But we learned that the farmer needs to be supported with, for example, automatic report function or dashboard solution to be able to create knowledge out of the data himself. The differences between rounds, seasons, barn/chicken houses, or different nutritional ingredients can be analyzed. Based on these relevant data, the farmer can focus on and run informative discussion with any of the stakeholders: veterinary, company coach, or feed supplier. Also, the farmer may decide to grant access to data, for example, in return for compensation (create additional income source).

The platform may become a safe online space where the farmer shares his own results, discusses, or mirrors the neighborhood, other farmers, the country,

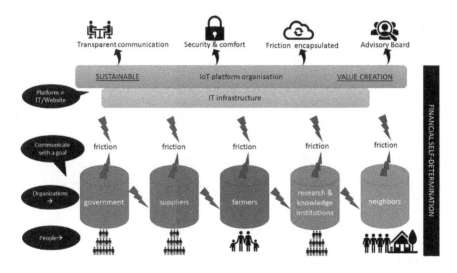

FIGURE 10.3 Stakeholders' structure and relationships with the introduced IoT platform in smart farming.

or the world. The platform as a place where a new future starts with sharing and transparency toward each other resolving existing conflicting interests. The use of IoT technology enables comparing and collaborating farmers with each other in real time at almost no cost.

The results show that using IoT enables the farmer to regain the "being-in-control" position of running his own farm, strengthening multiple dimensions of sustainability.

Sensor data allow for the analysis of peaks and enable farmer to estimate as a professional whether to act or to wait environmental dimensions of sustainability Comparing series with the normal operation level may unveil something important to look at it further. As seen in Figure 10.4, an example the relationship between the peak of particular matter between January 8 and 9 and the down of carbon dioxide at the same time indicates dependency between the two, the causes and consequences of this observed dependency can be explored by farmer self.

If the sensor reads that at certain times it registers more particulate matter during the day when people (stakeholders) are commonly also more outside, the system could adjust the ventilation/fan so that it causes less nuisance for neighbors (it was in the pilot 30% higher during the day than at night). Jan Noorlander looked at the data and saw recognizable patterns, but has not yet discovered useful information for supplier control he was aiming to find. We learned that more-than-clean output data is needed to actively manage the outcomes. During the pilot, we added imaging and made it possible to combine several logins together during the visits.

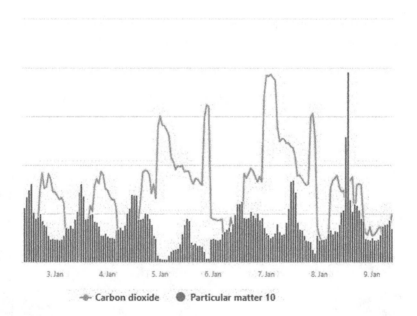

FIGURE 10.4 An example of dashboard indicator of data registration for carbon dioxide and particular matter.

FIGURE 10.5   An example of the dashboard with gauges capturing real-time data by means of IoT (own materials).

Collecting sensor values is not enough to use that output to manage the farm. We decided to introduce improvements in the next phase. In the next steps, the relevant data will be placed on a dashboard with only one login activity as shown in an example of a dashboard in Figure 10.5.

Further, we learned during the pilot that next to the IoT system, a lot of information is already flowing through the farm: electricity, water use, animal weight, egg production, etc. It is mostly analog and located at different places: partly in barn-computer,[3] partly in physical files, or paper reports. The farmer never combined the data in a structured way. Often, these analog data have missing values. Filling out registration of broiler suppliers cannot be completed because not all data is recorded in a barn-computer. Barn-computers are replaced rarely due to a large investment and bothering installation inside the barn (disturbing animal well-being). Barn-computer can record data but cannot install peripheral equipment. At the moment of the pilot, the data with different owners is in separate streams, not united. If it were, colleagues–poultry farmers could identify data to share with each other and thus identify certain signals early, i.e., illness, or avoid epidemics. Combing the streams in a more advanced smart farming system would improve the quality of the management and generate learning outcomes for the farmer [31,32], enhancing the social value of his farm.

10.8   CONCLUSIONS FROM THE MICRO-CASE STUDY

This micro-case study shows some preliminary results that the IoT platform can form an outer shell and bring all data streams together: the data from sensors combined with other data from barn-computer, electricity use, water consumption, and if agreed, data from the other data owners. It transforms the relationships with the stakeholders of the farm and redefines the power relationships. Enhancing the social dimension can create a more balanced sustainable development on all three dimensions, and not only on the environment or financial dimension as precise farming is currently focusing on.

First, determining whether farm/animal house particulate matter emissions rise or fall within or beyond the range is now easy and in real time. Important to the farmer are the real-time access to the data and the immediate learning and ability to do something about it himself [31]. Before the IoT-era and still a common practice to some extent, government has applied a standardized static diligence per animal-stable-system to comply with, while disregarding the actual numbers: like the variation in kg emissions per animal in differing among farmers (this standardization is the compliance rule in NL, B, and DE). Some farmers show emissions far below the standard, but some farmers are far beyond; however, it is irrelevant to the enforcement because of the fixed standards. To illustrate, if a farm exceeds the standard and that is not based on the "actual emissions" but on the "paper-and-tape measures" derived from the static standard, the farmer will be punished with a financial fine or his license to operate will be suspended even if the actual outputs are not exceeding the health norms. This makes the farmers angry because it does not motivate them to use the system properly and sustainably. It seems that the proactive attitude has no recognition and no value. Without IoT, the data has been supplied and distributed by third parties from different sources and the farmer is powerless against those stakeholders. On the other hand, local government enforcement clerks have difficulty in enforcing these standards because they also see differences at different farms, but they cannot act upon them; for example, they use the history of data in decision making.

Second, by applying IoT, a number of measures that are offered from the suppliers can be assessed on their effectiveness and efficiencies. The data allows a harmer to create a stronger position in negotiations about price and type of measures, for example, feed or air cleaning system, with suppliers.

Third, IoT strengthens the position of the farm company in relation to the local government and the local residents. A farmer can show the registration and the history of the measures with a trend line so that the neighborhood and the nearby environment are more involved in the agricultural operations; and their impacts on his farm.

Fourth, health hazard management is feasible. By linking the data output and performing further analyses on the relationships between different data streams (big data), it would potentially be possible to discover, for example, coherence between certain air quality in the sample and germs in the animal.

Fifth, a positive externality of the IoT on the farm is that there are suppliers who want to work together with the farmer to improve the feed so that they can produce top quality together with the farmer. Furthermore, if the farmer decides to share the data with external parties and collaborate, it can generate additional sources of income for the farmer for sharing data with third parties and for suppliers to improve the quality of their product. Potentially more network cooperation between partners could emerge.

Finally, there is a more long-term energy compliance issue; the Dutch government requires farms to provide an energy consumption report. In addition to reporting the energy savings, there is now also an obligation to provide explanation for the use of 50,000 kWh of electricity or above 25,000 m^3 of natural gas (chicken farm (laying hens) with 50,000 used 140,000 KWh per year) meat chick costs per animal 1.3 KWh per year (WUR) as of July 11, 2019; this is mandatory in the Netherlands. Currently, such a report must be compiled manually, which can be done easily and quickly using IoT.

In conclusion, the IoT in smart farming allows for enhanced social sustainability and indirect effects on environmental and economic sustainability at multiple stakeholders. Moreover, IoT with farmers "in control" brings connectivity to their environments, saves resources, and drives down unnecessary costs. It can potentially generate new forms of collaboration.

10.9 FINAL OBSERVATIONS AND FUTURE RESEARCH

The acceptance of the farmers toward implementation of the IoT may be an issue as it depends on several additional factors, such as usability and the identification of best practices. The fact remains that farmers are forced to make a variety of choices that may not be sustainable [33]. Smart farming using IoT as a farmer-centered approach is the opportunity for the concept of smart farming to prove sustainable for the future. The importance of a change in the mind-set of the farmers is crucial to activate an effective and sustainable production system that will last in the long term [10]. There are multiple factors and behaviors explaining the choice of sustainable innovation by a farmer [34], along with the aspect of smart farming adoption as innovation.

Therefore, more practical examples and research articles illustrating the application are needed to create a sound learning base [35]. Currently, non-formal agricultural adult education favored by learners may be fruitfully understood as a proposal for reconfiguring power relationships among farmers, organizations, and the government [36]. These are the avenues for transformations for the future and the potential for international cooperation.

The pilot at the Jan Noorlander farm was successfully completed, and now, the system awaits its full implementation. Currently, the team is working on setting up another pilot, now in Poland. Poland is a high-IoT-adoption potential market. The estimates carried out at the Institute of Management Warsaw University of Life Sciences show that there are about 30 thousand farms in Poland (farms with an area of over 50 ha), for which the use of smart farming will improve the use of agricultural inputs. The farms conduct agricultural activity on 25% of arable land in Poland, but their share in the volume is not more than 3% [37]. It shows they already have the technical, organizational, and intellectual potential to implement and profit from smart farming technologies. Finally, the development of the literature on combing the subjects is a challenge; therefore, more sustainable applications of IoT should be documented in research [9].

Appendix 1

Types of Stakeholders		
Type/Role	Examples	Interest/Role
Farmers and their agents	Landowners, farmworkers, unions, farmers' associations	Victims of exposure; risk management and reduction; potential victims/beneficiaries of risk response (e.g., loss of income)
Government	National-level ministry of agriculture, local municipalities, regional and provincial administration offices	Legitimacy-assuring, license to operate, safeguarding, meeting national and international standards, protecting citizens
Agricultural suppliers and services	Seed suppliers, animal nutrition companies, pesticide manufacturers, fertilizer manufacturers, slaughterhouses, transport companies	Risk management and reduction; potential victims/beneficiaries of risk response
Food distributors and processors	Food wholesalers and retailers, transport companies	Potential victims/beneficiaries of risk response
National/regional health protection agencies	Public health institutions, food standard agencies, occupational health and safety agencies, local/regional health boards and environmental health departments	Risk management and regulation; risk communication
National/regional environmental protection agencies	Ministries of environment, environmental regulatory agencies, local authorities	Risk management and regulation

NOTES

1 https://www.rijksoverheid.nl/onderwerpen/aanpak-stikstof/programma-aanpak-stikstof-achtergrond-en-inhoud.
2 https://www.arc2020.eu/nitrogen-crisis-dutch-farmers-rage/.
3 Barn-computer is a total of hardware and software in animal house usually provided as an element of the animal system equipment.

REFERENCES

1. Muangprathub, J., Boonnam, N., Kajornkasirat S., Lekbangpong, N. Wanichsombat, A. Nillaor, P. 2019. IoT and agriculture data analysis for smart farm. *Computer and Electronics in Agriculture* 156:467–474.
2. Shabadi, L. S., Biradar, H. B. 2008. Design and implementation of IoT based smart security and monitoring for connected smart farming. *International Journal of Computer Applications* 975:8887.

3. Giddings, B., Hopwood, B., O'Brien, G. 2002. Environment, economy and society: Fitting them together into sustainable development. *Sustainable Development* 10(4):187–196.
4. Hopwood, B., Mellor, M., O'Brien, G. 2005. Sustainable development: Mapping different approaches. *Sustainable Development* 13(1):38–52.
5. Rees, W. 1995. Achieving sustainability: Reform or transformation? *Journal of Planning Literature* 9:343–361.
6. Zambon, I., Cecchini, M., Egidi, G., Saporito, M. G., Colantoni, A. 2019. Revolution 4.0: Industry vs. agriculture in a future development for SMEs. *Processes* 7:36.
7. Kosior, K. 2018. Potencjał technologii blockchain w zapewnianiu bezpieczeństwa i jakości żywności. *Food Science Technology Quality* 4:117.
8. Paraforos, D. S., Reutemann, M., Sharipov, G., Werner, R., Griepentrog, H. W. 2017, Total station data assessment using an industrial robotic arm for dynamic 3D in-field positioning with sub-centimetre accuracy. *Computers and Electronics in Agriculture* 136:166–175.
9. Wolfert, S., Ge, L., Verdouw, C., Bogaardt, M. 2017. Big data in smart farming – A review. *Agricultural Systems* 153:69–80.
10. O'Grady, M., O'Hare, G. 2017. Modelling the smart farm. *Information Processing in Agriculture* 4:179–187.
11. Berger, R. 2015. The digital transformation of industry. https://www.rolandberger.com/media/pdf/Roland_Berger_digital_transformation_of_industry_20150315.pdf (accessed 15.12.2019).
12. Tey, Y. S., Brindal, M. (2012). Factors influencing the adoption of precision agricultural technologies: A review for policy implications. *Precision Agriculture* 13(6):713–730.
13. Samborski, S. 2018. *Rolnictwo precyzyjne*. Wydawnictwo Naukowe PWN, Warszawa.
14. Doruchowski, G. 2008. Postęp i nowe koncepcje w rolnictwie precyzyjnym. *Inżynieria Rolnicza* 9:107.
15. Jarka, S. 2019. Food safety in the supply chain using blockchain technology. *Acta Scientiarum Polonorum. Oeconomia* 18:4.
16. Beckman, J., Schimmelpfennig, D. 2015. Determinants of farm income. *Agricultural Finance Review* 75(3):385–402.
17. Schimmelpfennig, D., Ebel, R. 2016. Sequential adoption and cost savings from precision agriculture. *Journal of Agricultural and Resource Economics* 41:1.
18. Dillard, J., Dujon, V., King, M. C. (Eds.). 2008. *Understanding the Social Dimension of Sustainability*. Routledge, New York, NY.
19. Retolaza, J. L., San-Jose, L., Ruíz-Roqueñi, M. 2016. *Social Accounting for Sustainability: Monetizing the Social Value*. Springer, Berlin.
20. Bacon, C. M., Getz, C., Kraus, S., Montenegro, M., Holland, K. 2012. The social dimensions of sustainability and change in diversified farming systems. *Ecology and Society* 17(4):41.
21. Zarco-Tejada, P. J., Hubbard, N., Loudjani, P. 2014. *Precision Agriculture: An Opportunity For EU Farmers -Potential Support with the CAP2014–2020*. Policy Department B: Structural and Cohesion Policies at the EU Parliament, Brussels.
22. Elkington, J. 1997. *Cannibals with Forks; The Triple Bottom Line of 21st Century Business*. Capstone, Oxford.
23. Busco, C., Fiori, G., Frigo, M. L., Riccaboni, A. 2017. Sustainable development goals: Integrating sustainability initiatives with long-term value creation. *Strategic Finance* 99(3):28–38.
24. Vallance, S., Perkins, H. C., Dixon, J. E. 2011. What is social sustainability? A clarification of concepts. *Geoforum* 42(3):342–348.

25. Griessler, E., Littig, B. (2005). Social sustainability: A catchword between political pragmatism and social theory. *International Journal for Sustainable Development* 8(1/2):65–79.

26. van Leeuwen, L. 2018. De hanenbalken. *Zelfmoord op het platteland.* Atlas Contact, 335.

27. Goodman, D., Watts, M. (Eds.). 1997. *Globalising Food: Agrarian Questions and Global Restructuring.* Psychology Press, New York.

28. ter Bogt, H., van Helden, G. 2011. The role of consultant-researchers in the design and implementation process of a programme budget in a local government organization. *Management Accounting Research* 22(1):56–64.

29. Bradbury-Huang, H. (2010). What is good action research? Why the resurgent interest? *Action Research* 8(1):93–109.

30. Winkel, A. 2016. *Particulate Matter Emission from Livestock Houses: Measurement Methods, Emission Levels and Abatement Systems.* Doctoral dissertation, Wageningen University.

31. Fielke, S., Taylor, B., Jakku, E. (2020). Digitalisation of agricultural knowledge and advice networks: A state-of-the-art review. *Agricultural Systems*, 180 (Article 102763) 2020.

32. Vogl, C. R. 2017. Farmers' own research: Organic farmers' experiments in Austria and implications for agricultural innovation systems. *Sustainable Agriculture Research* 6:526.

33. Pilgeram, R. 2011. The only thing that isn't sustainable … is the farmer: Social sustainability and the politics of class among Pacific Northwest farmers engaged in sustainable farming. *Rural Sociology* 76(3):375–393.

34. Adnan, N., Nordin, S. M., Rahman, I., Noor, A. 2018. The effects of knowledge transfer on farmers decision making toward sustainable agriculture practices: In view of green fertilizer technology. *World Journal of Science, Technology and Sustainable Development* 15:98–115.

35. Pretty, J. 1995. Participatory learning for sustainable agriculture. *World Development* 23(8):1247–1263.

36. Grudens-Schuck, N. 2001. Stakeholder effect: A qualitative study of the influence of farm leaders' ideas on a sustainable agriculture education program for adults. *Journal of Agricultural Education* 42. doi: 10.5032/jae.2001.04001.

37. Polish Statistical Yearbook of Agriculture 2019. National population and housing census. http://www.stat.gov.pl.

11 Security in IoT HealthCare

Manorama Mohapatro and Itu Snigdh
Birla Institute of Technology

CONTENTS

11.1 INTRODUCTION: BACKGROUND AND DRIVING FORCES

The IoT has enabled the next-generation technologies that are a rainbow for the whole business world, and uniquely identifiable smart objects are interconnected within Internet. Recently, Internet infrastructure has empowered with enlarged benefits such as connectivity of advanced smart devices that are imparted beyond the scenario of machine-to-machine (M2M) capabilities [1]. Consequentially, IoT represents one of the most attractive healthcare applications [2]. Internet of Things and wireless body area sensor networks (WBAN) essentially collect and transfer data using a wireless or wired network without requiring any significant human assistance. It could therefore cater to many applications in medical fields, such as elderly care, fitness programs, remote health monitoring, and chronic diseases.

The association of physical things connected to IoT healthcare is illustrated in Figure 11.1.

Healthcare has undergone a wide transformation over time. The targeted applications are therapeutic environments that allow real-time monitoring of a large number of patients continuously. The growth of such systems is attributed to its boundless advantages such as better rehabilitation of patients and higher efficiency

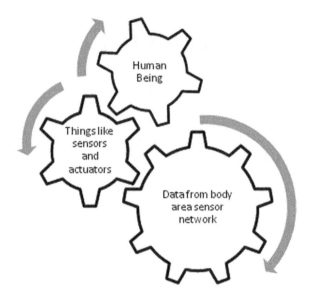

FIGURE 11.1   Things connected with IoT healthcare.

as it largely reduces the workload of medical professionals. The wireless monitoring facility gives patients the necessary mobility and comfort. Such wireless monitoring systems also reduce the healthcare cost and provide an improved quality of life. Now, one does not require a patient to be attached to bulky and uncomfortable gadgets for a prolonged time, thereby enabling off-site monitoring of patients. However, realization of such systems is not as easy in practical scenarios. Some of the challenges are ensuring reliability in communications by removing collisions of sensors signals and external wireless devices interference, requirements to keep the overall cost low, enablement of low power consumption, and user sensitization and customization.

The demand of these devices is primarily attributed to two driving forces: embedded sensors and IoT connectivity, responsible for assembling the data and relaying them forward to other resources for remote monitoring. Medical devices are categorized as follows:

1. **Life-Critical Devices**: These devices are primarily designed to monitor patients. The application requires continuous monitoring and transmission of data which are essential to sustainability of life. These systems are risk-prone in the sense that if they fail, the patient's life is at risk. Examples of such devices are pacemakers and ventilators.
2. **Non-Critical Monitoring Devices**: These IoT healthcare devices are usually responsible for the collection of data and transferring it to a server and they do not monitor for life-critical conditions. Devices such as glucose monitor and blood pressure monitor are some of the examples. These devices may need immediate attention, but not at the same level of urgency as life-critical devices.

3. **IoT Health and Wellness Devices**: Wearable and implanted devices such as smart bands and fit bits belong to this category. These devices encompass continuous monitoring and transmission of data from local servers to personal servers. These are specifically designed for detecting activity and process some set of signals such as accelerometers, pedometers, heartrate, caloric counts [3]. Their success depends on the accuracy and user satisfaction.

This chapter presents a brief outline of the factors that affect the realization of a body area sensor network with an overview of the enabling technologies. We discuss the security concerns in depth with a sample simulation result that shows how network parameters get affected when subjected to a security breach. We also put forth methods that can be adopted for securing a body area sensor network application against major security concerns. This chapter is organized as follows: Section 11.2 explains about the wireless body area sensor networks with enabling technologies. Section 11.3 describes the architecture of wireless body area sensor networks. Section 11.4 deals with security essentials such as data, network, and device security in IoT healthcare. Experiment-based simulation is also described as a part of it. Section 11.5 provides all the existing solutions. Section 11.6 provides conclusions, and Section 11.7 focuses on future work.

11.2 WIRELESS BODY AREA SENSOR NETWORKS

A body sensor network (BSN), commonly referred to as a wireless body area network (WBAN), is the terminology used in the context of analyzing the healthcare system and its probable automation through IoT. WBANs are used to observe functions of a human body and also the impact that the surrounding environment has on it. The node of a WBAN collects critical and proximate information [4]. One of the important applications is medication at home and compliance with treatment by healthcare providers. Different diagnostic and imaging devices, sensors, and medical equipment with intelligence can behave as smart devices or objects as an important part of IoT. Since IoT-based healthcare services are contemplated to increase the quality of life, reduce costs, and enhance the user experience. One of the objectives of BSNs is to reduce the downtime of devices through remote provisioning. Furthermore, IoT-based sensor system can adopt scheduling and provide best timing for reloading the requirements for different types of devices for their continuous and smooth operation [5]. Furthermore, IoT-based systems can also be applied in non-technical or non-medical scenarios such as scheduling and optimal reloading of medical supplies in supply chain management of medical devices and medicines.

WBANs may be classified as follows:

1. **Implant Node**: Nodes placed inside human tissue or just below the skin.
2. **Body Surface Node**: This is either implanted on human body surface or very near (less than 2 cms) the body.
3. **External Node**: These kinds of nodes do not touch human body and are rather placed a few centimeters to or 5 m away from the human body.

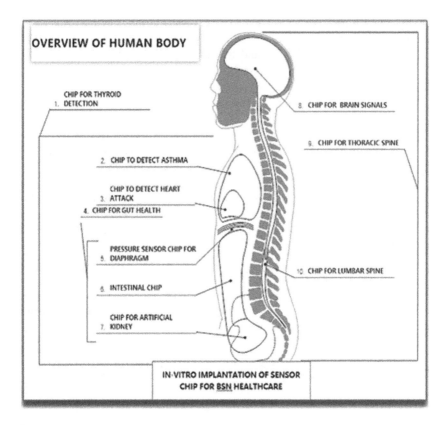

FIGURE 11.2   In vitro implantation of sensor chips in body area sensor networks.

The usage of sensors for human body monitoring is illustrated in Figure 11.2.

The figure shows a wide range of implantable chips/sensor devices [5] and techniques for remotely monitoring the human body. Some of the sensors shown in Figure 11.2 are described below:

1. **Chip for Thyroid Monitoring**: Thyroid carcinoma (TC) is the very common malignant disease of the human endocrine system. Sensory chips are implanted inside the body to monitor the hormone secretion levels of the thyroid gland. The data collected from this chip [6] are a tool for medical staff to diagnose and analyze the curability of the disease.

2. **Chip for the Detection of Asthma**: This chip is particularly designed for detecting unusual respiration patterns found in asthma cases. The chip detects breath from nostrils or airflow by the mouth. Facemask equipped with sensors measures the rate of inhalation and exhalation. The intensity of airflow induce strains that can be detected [6,7]. The chips implanted have been used to detect various asthmatic conditions.

3. **Chip for the Detection of Heart Attack**: This chip is implanted beneath the skin and observes the concentration of various substances in the blood

to help determine the patient's health status. In case of emergency, the medical implant chip does preliminary analysis by collecting data. For this, it uses a patch on the skin enabled with a battery to receive the chip's radio signals for information and for transmitting data through a cell phone, to a doctor [8].

4. **Chip for Intestine**: Gut inflammations may lead to Crohn's disease. Microbial symbionts in the intestine can also be monitored by implanting chips in the intestine. The chip monitors the structural, absorptive, mechanical, pathological, and physiological properties of the human gut [8].

5. **Chip for Monitoring Brain Signals**: Neurons, tissue, and brain are the important things for brain stroke analysis. Metal microelectrodes (electrolyte oxide semiconductor capacitors) are interfaced with neurons to stimulate or record their electrical activity using chip [8].

6. **Sensory Chip for Thoracic and Lumbar Spine Care**: A spinal cord injury is an emergency concern in medical terminology. It may give rise to prolonged acute pain in the neck or back or weakness in limbs. Extreme cases lead to surgeries and neck or back injuries. An in vitro chip is implanted to detect the level of emergency in case of spinal cord injuries [8].

11.2.1 WBAN ENABLING TECHNOLOGIES

For a WBAN application, the most frequently used technologies are Zigbee, 2.4 GHz ISM bands, MICS (Medical Implant Communication Service), and WMTS (wireless medical telemetry services). The WBAN can be simply implemented by adopting MICS to obtain sensory data from the sensors planted within the body and by employing WMTS band on the sink/relay nodes for wireless communication over a long distance.

WBANs are equipped with wearable and implanted sensors [9]. Wearable devices monitor and collect information about one's physiological conditions and activities related to movement. Detection and diagnosis of various neurological, pulmonary, and cardiovascular diseases may be made possible by continuously monitoring the physiological signals using sensors implanted inside the body. Real-time monitoring of a person's movement can be useful in predicting gait, analyzing posture, detection and prediction of fall as well as sleep analysis. Existing IoT systems hold on short-range wireless systems such as Zigbee, WLANs, GSM, and Bluetooth with all its versions (BL, BL4.X, and BL5). LoRa is the one of the newest concepts and methodologies with a good radio network in comparison with other long-range technologies such as NB-IoT, LPWAN, and 5G.4G and 4G LTE provide good quality of connectivity as required in IoT applications. In a WBAN, sensor electronics requires to be small, consume less power, and be able to detect medical signals such as ECG, EEG, PPG, and pulse rate. Owing to the requirements of short range and low data rates, regulated low transmission power suited for wearable gadgets, ultra-wideband (UWB) technology is being used. UWB communication achieves high-bandwidth connections in wireless networking, ensuring low power consumption. This is basically designed for commercial radar systems; it finds applications in consumer electronics and PAN (used in Bluetooth devices). The features of UWB [10], such as short range and low data rate, has made it a popular choice for implementing WBANs, especially in medical and sport monitoring among their other relevant applications. The wearable health monitoring systems consist of various MEMS sensors and electronic equipment, signal processing units, communication modules (both wireless and wired), and actuators [11].

11.2.2 WBAN ARCHITECTURE

WBAN in IOT healthcare [12] is the integral part of a multi-tier telemedicine system. Figure 11.3 illustrates the components of a three-tier WBAN architecture system. Sensor nodes of a WBAN are enabled to sense, sample, and process the physiological data or signals.

Generally, a WBAN comprises of a robust central control unit (CCU) that is remotely worn and many implantable or wearable sensor nodes as shown in Figure 11.4. Figure 11.4 elaborates the significant components of the multi-tier architecture of a WBAN. The sensors basically collect data from lightweight sensor hubs that are utilized to screen the human body capacities surrounded by the environment. Body sensor nodes enable continuous, remote health monitoring which may or may not be non-intrusive in a hostile environment. The communication infrastructure forwards sensory information to servers, which may be stored at local or cloud-based repositories [13]. The medical servers assist in analysis and decision making and comprise the third tier.

Tier 1: Tier 1 specifies the number of wireless sensor nodes or applications from a laptop or a personal computer connected to a local or personal server that is connected to a WBAN. It collects data generated from

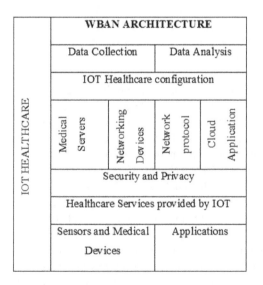

FIGURE 11.3   Elements of WBAN architecture.

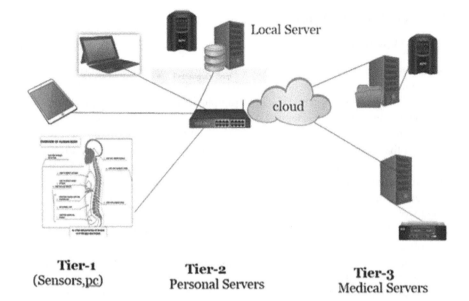

Tier-1
(Sensors,pc)

Tier-2
Personal Servers

Tier-3
Medical Servers

FIGURE 11.4   Three-tier architecture in WBAN.

physiological sensors, user profile, etc. It is equipped with one or more interoperable medical devices (IMDs).Each device is attached to a specific location and programmed to measure a specific parameter or physiological indicator (e.g., ECG and blood pressure). The sensors are integrated into components that amplify and filter data that are collected from sensors and send the measurements to a server or repository as shown in Figure 11.4.

Tier 2: This accomplishes the personal server application which runs on a local server such as a PC or laptop. The personal server is accountable for executing various tasks such as communicating with the user, serving as a gateway, or acting as an interface to the cloud server or a medical server. It performs the necessary network configuration and management tasks being an edge device. This layer comprises of a server where the application runs locally and the processing unit behaves as the gateway or an edge device. The local server (personal server) is responsible for providing an interface for the client, for the medical sensors, and also for the medical server [9]. The network configuration takes care of the following tasks:

- Registration of sensor nodes (type and number of sensors)
- Initialization (such as mode of operation and specifying samples)
- Customization (to the user-specific application)

Tier 3: This layer accesses the medical server through the Internet. In the last tier, multiple servers can run on services that build an ad hoc infrastructure for communication channel to the patient's personal server. Reports are collected from the patient's personal server, and the data are integrated into a medical record [14].

11.3 SECURITY IN WBAN

11.3.1 WHY SECURITY?

Among the many requirements of WBANs, security is the most important component. Security parameters that are important for WBANs are confidentiality, authenticity, availability, and integrity [15]. For example, if we consider a simple application of connected health which contains devices such as an insulin pump, hacking for wrong administration of the dosage could be life-threatening. Similarly, a less basic assault on the refrigeration framework lodging the drug, observed by an IoT framework, can demolish the feasibility or adequacy of the medication because of temperature fluctuations. Another issue is data protection [16], which usually relates to disease information that should not be publicly disclosed. In such cases also, the patient's physiological vital signs are critically monitored, especially if a patient is suffering from a disease, the details of which are not usually appropriate to disclose. Disclosure of such data could cause a patient to suffer humiliation. Any leak in the confidential information [17] would cause him/her psychological stress, sometimes resulting in loss of their jobs or getting insurance protection. Hence, medical data need to be effectively treated as private and the protection becomes a mandatory concern in quality of service (QoS) criteria.

As IoT healthcare technology employs relaying information from one device to another, it is susceptible to security breaches and loss of privacy [18] for the end users. Other than these possible attacks, certain device limitations such as battery life constraints and lightweight computational requirements also affect the security.

Since wearable and implantable sensors are in demand in WBAN, they have the possibility of absolute integration with online social networks to gain popularity [19]. Enough care must be taken to secure identity, ensure protection, identify vulnerabilities, and also inhibit any possibilities for adversaries to mount any attack.

Typical attacks on IoT healthcare systems are the following:

1. **AAA Penetration**: This is caused by backdoor credentials, default name and password enablement options, direct Web access, and man-in-the-middle attacks.
2. **Steal Key Certification**: caused by are due breaking the encryption or by getting the access rights by bypassing the encryption process. This occurs when a user tries to fix a device or certification.
3. **Fake Firmware**: This is usually done by modification of firmware methodologies, including signing in through a stolen key that can be further used to control the device for further penetration.
4. **Data Breach/Ransomware**: The probable sites for this risk are the cloud infrastructure for backend data; it includes stealing the data from repositories and also holding data for ransom.

The three-tier architecture followed by most WBANs, security also needs to be asserted and measures at all levels and interfaces, in view of the basic security-prone [20] areas as illustrated in Figure 11.5. The sensing and network devices such

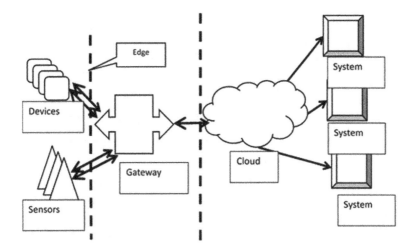

FIGURE 11.5   Multi-level security in WBAN infrastructure.

as gateway could be a part of an already existing infrastructure-based network or purely ad hoc in nature.

Level 1: Internode Domain:

Since the interaction among sensors is limited to the confinement of the patient's body parts, we require a personal server in this level. Sensors which are confined around the body of the patient commence interaction. The personal server's communication signals within the region act as a gateway for transferring the information to the consecutive access point interface (i.e., access point that is present in Tier 2). Hence, encryption techniques such as AES are implemented to ensure that the data traffic generated is well protected. However, we cannot control attacks such as man-in-the-middle, sinkhole, and denial-of-service attacks at this level. This level only ensures device security by means of restricting the cloning of hardware and prefabricating of chips.

Level 2: Internetwork Domain:

The main role here is to bridge the gap between the server and the user via access points (APs). Essentially, information can be easily obtained via various mediums, such as the Internet. Therefore, all the attacks that relate to any wireless network are applicable here as well.

Level 3: Beyond-Network Domain:

This level of communication is ideal for a particular application from the Internet to the medical server (MS). It usually affects a gateway or a data repository. A medical environment database is a particularly important component of communication, as it accommodates the user's medical history and clinical profile that the designs necessarily require to be unique to an application. Therefore, for easy understandability we classify IoT healthcare security under the categories shown in Figure 11. 6.

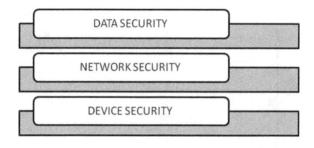

FIGURE 11.6 Security types in IoT healthcare.

TABLE 11.1
Example of Action Table Using Blood Sugar Data [9]

BSN Blood Sugar Data	Action	Response
Blood sugar < 100	No action	Null
Blood sugar ≥ 100	To inform family members	FR:T/F
Blood sugar > 160 and FR:F	To inform local physician	PR:T/F
Blood sugar > 200, FR:F and PR:F	To inform emergency	ER:T/F

FR, family response; PR, physician response; ER, emergency response.

Let us consider an example. Suppose a person suffering from diabetes has to be monitored. The blood sugar level and the course of action that must be taken are shown in Table 11.1.

In a WBAN method, if a local server wants to send regular updates to the WBAN server, the server needs to confirm the local processor's identity via a secure channel. This phase is called the registration phase. The next stage of the anonymous authentication phase is where both the local server and the cloud server authenticate each other before data transmission from the local server to cloud or medical server. Hence, the objective of data security must be ensured by providing data privacy, data integrity, and data freshness.

11.3.2 DATA SECURITY

Data security is one of the major concerns in IoT healthcare [21–25]. The implementation of a ubiquitously functioning healthcare solution has been explored in applications such as Code Blue, Alarm-net, UbiMon, Mobi-care. The following requirements are an integral part of data security:

1. **Data Confidentiality or Privacy**: As explained earlier, owing to the impact of a data leak that affects an individual, WBANs must ensure that any confidential details are not disclosed since the sensory data must be collected and forwarded to a sink node for further processing. This includes

protecting patient's details and location, illness details as well as the prescribed medication from neighboring or external nodes.

2. **Data Integrity**: In a WBAN, data should be received at the destination nodes without alterations or modifications from sink nodes. All methods of alteration/modification of data through any adversary need to be avoided. Manipulation of data is done by inserting other fragments within the packet which can then be forwarded to the collecting nodes causing data loss. In addition to that, the integrity must be maintained with the stored data so that it can never be compromised.

3. **Data Freshness**: The transmission of data from the sensor nodes usually incorporates a key-based authentication and encryption. Therefore, for security both data and keys should be fresh. By freshness, we mean the time a system takes to collect data and apply analytics for processing an action. A data freshness value is set to 40minutes if the above process completes in 40 minutes. It ensures that the current action is not confused with the previously stored results.

4. **Data Authentication**: The identity of an IoT health care device must ensure authenticated data. This authentication is carried out when sensor nodes send the data to the sink or coordinator nodes. Authenticity can sometimes be an extra overhead to the system and is a site for breach of confidentiality between the coordinator and server.

5. **Secure Localization**: WBAN applications may sometimes require information of patient's location. This may be in order to administer the appropriate medical assistance in case of emergency. The estimation algorithms should therefore be secure and avoid data transfusion unknowingly. It should also provide resistance against various security threats and attacks such as impersonation, eavesdropping, replaying.

6. **Availability of WBAN**: The success of IoT healthcare services (either global/local or cloud services) depends on the availability of the WBAN. It means that data storing or transmitting should always be operative.

7. **Anonymity or Untraceability**: When the WBAN application does not require many nodes like usual WBANs, lightweight anonymous mechanisms accomplish security against attacks such as sinkhole, wormhole, or hello flood attack.

8. **Self-Healing Fault Tolerance**: WBANs require protection against accidental failure or energy drain. Collaboration among remaining devices should enable security services even when a fault occurs in the system.

11.3.3 NETWORK SECURITY

While perceiving the overall architecture of IoT healthcare, the complexity of security is to ensure the security and privacy in all three layers of the network are mandatory. Various challenges are faced in the network as the three-layer architecture of WBAN is highly heterogeneous in nature. In addition to that, network vulnerability and limited capacity of multi-node or single node are major causes of IoT network security [26]. Table 11.2 presents the different features and issues of IoT security related to healthcare.

TABLE 11.2

Layer-wise Security Issues in IoT Healthcare

Layers	Features and Issues	Technologies	Attributes	Solutions
Perception layer	Data security with limited computing power and storage space	Cryptographic algorithms in WSNs	**Sensing layer:** Consists of all sensors, RFIDs, and wireless sensor networks(WSN).For example, Google Glass, Fitbit Tracker	Symmetric encryption
	Key distribution, including the distribution of the public key and the secret key	Key management in WSN		Symmetric key algorithm, Group key distribution, Distribution of node master key
	Limitation of power, computing ability, and storage capacity attacks toward routing protocol	Secure routing protocol in WSN		Secure routing protocols designed specifically for WSN
	Limited resources, easy capture of nodes, and unique communication mode	Trust management in WSN		Measurement, evaluation, relationship formalization, formal derivation of trust

(Continued)

TABLE 11.2 (Continued)
Layer-wise Security Issues in IoT Healthcare

Layers	Features and Issues	Technologies	Attributes	Solutions
Transportation layer	Wi-Fi: phishing site, access attacks, malicious AP, and DDoS/DoS, etc.	Access control and network encryption technologies	**Aggregated layer:** Consists of different types of aggregator based on the sensors of sensing layer.	Access control
	Ad hoc: Data security, network routing security	Encryption mechanisms, authentication and key management, symmetric encryption	For example, smartphones, tablets. Processing layer: It consists of server for processing information coming from aggregated layer	Key management, data origin authentication, and data encryption
Application layer	Invalid or insecure data	Middleware, cloud computing, information development platform	**Cloud platform:** All processed data are uploaded in cloud platform, which can be accessed by a large number of users.	Data security protection
	Access control problem			
	Ubiquitous industry, life, environment intelligence	Intelligent logistics, smart home, remote medicine, smart grid, etc.		Access control protocols Environmental monitoring

Other than these issues, the presence of malicious nodes, unsecure communication channel, and physical devices with less security are susceptible to malicious attacks. These possible attacks are as follows [27–30]:

1. **Network Attack**: Eaves dropping, denial-of-service, camouflage, and traffic analysis are some of the network-based attacks in WBANs.
2. **Eavesdropping**: Sometimes various communication channels both wireless and wired networks can be targeted by internal attackers in order to extract data.
3. **Node Capture**: In node capturing, an intruder compromises the entire network by taking control of the transmitting nodes. It is one of the most critical attacks in WBANs.
4. **DoS (Denial-of-Service) or DDoS(Distributed Denial-of-Service)**: IoT can have a number of DoS attacks. These attacks are of traditional type and are based on network bandwidth and exhaust service provider resources. In these attacks, jamming of channels or malicious internal attacks can take control of the infrastructure.
5. **Physical Damage**: This may result from DoS or DDoS threat. This type of attacks ends up with some physical damage or destruction of actual things.
6. **Side-Channel Attacks**: These are one of the concerns for communication network. Fault, power, and timing analyses are part of it.
7. **Cryptanalysis Attack**: Cipher text-only attack, known-plaintext attack, and man-in-the-middle attack come under cryptanalysis attacks.
8. **Software Attack**: Other than the above-mentioned attacks, software attacks such as denial-of-service, virus, Trojan horse, worms are also threats to IoT communication.

11.3.3.1 Case Study

A body sensor network is created to demonstrate an attacking scenario in RPL protocols. The following two scenarios depict the very common attacks of WBAN: DoS attack and sinkhole attack.

Simulation Process: The proposed work is simulated using TETCOS NETSIM. Figures 11.7 and 11.8 show the setup of simulation. Various attacking scenarios are created for the simulation of the complete performance evaluation of the network. The following parameters are observed during the simulation of DoS and sinkhole attacks.

The routing protocol is based on the concept of direct acyclic graph that avoids creation of loops inside the tree constructed by distance vector algorithm. RPL can construct multiple paths back to the same destination and set various alternative routes as and when default routes are not accessible. RPL protocol is intended to target resource-constrained networks in order to command over power, energy, and bandwidth, and this results in high amount of packet loss and a relatively significant error rate [31]. This demonstrates a hospital scenario with two nodes sending data to a Low PAN gateway which communicates to a router later connected to a

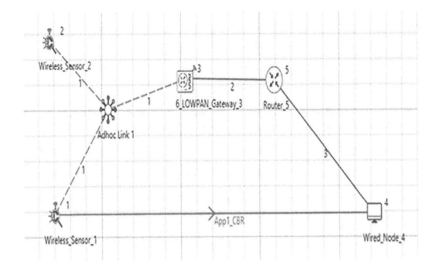

FIGURE 11.7   Scenario before DoS attack.

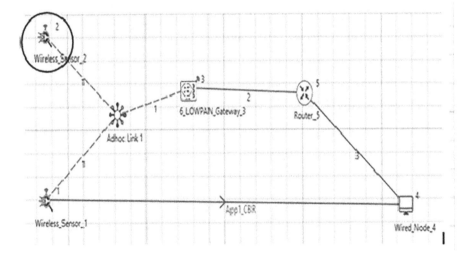

FIGURE 11.8   Scenario after the DoS attack.

wired node. This scenario contains one malicious node which sends a series of SYN flood packets to traffic the network. How throughput and delay are affected by this attack is displayed in Table 11.3 [4].

SCENARIO A:
DoS Attack
SCENARIO B:
Sinkhole Attack

TABLE 11.3

Network Parameters Before and After the Attacks

Simulation in ms	Throughput		Delay		PDR	
			DoS Attack			
RPL	**Without Attack**	**After Attack**	**Without Attack**	**After Attack**	**Without Attack**	**After Attack**
100	0.001067	0	40,031,661	0	547.445	Indefinite
200	0.003139	0	112897807.7	0	186.1042	Indefinite
300	0.002191	0	120545269.8	0	266.5876	Indefinite
			Sinkhole Attack			
100	0.000396	0	4369.126	0	1.0101	Indefinite
200	0.000398	0	4234.57	0	1.0050	Indefinite
300	0.000399	0	4226.406	0	1.0033	Indefinite

In the above NETSIM scenario, there are five wireless sensors creating a wireless body area network with a Low PAN gateway and router with an IoT interface by using routing protocol (RPL). Data that emanated from the sensors are collected at a wired node 3. This scenario explains a situation in data transferring between wireless sensors and a local gateway. The limitation of this scenario is that it can depict only one sink node in its workspace. Now the wired node 3 is a sink or intermediate server for data transmission.

Figures 11.9 and 11.10 demonstrate a sinkhole attack with fifth node as malicious node. This node from the above scenario is stating that data packets are

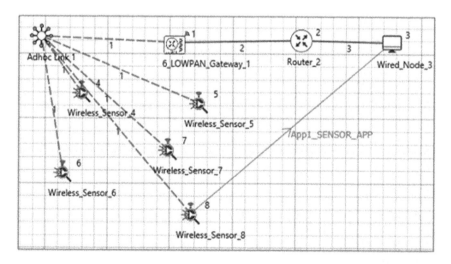

FIGURE 11.9   Scenario before sinkhole attack.

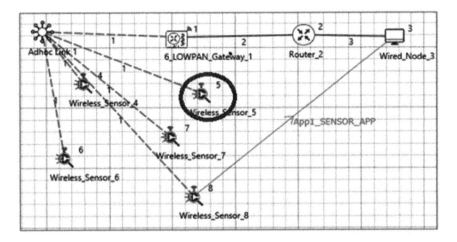

FIGURE 11.10   Scenario after the sinkhole attack.

continuously sent to the destination nodes through DODAG structure formation. But due to the presence of malicious node 5 the packets are not forwarded and dropped. Node 5 is a malicious node, and it produces a low rank and engulfs all data packets (Table 11.3).

Table 11.3 indicates various simulations executed in NETSIM for DoS and sinkhole attacks in RPL protocols. Variations in different network parameters are observed as follows:

Application throughput (in Mbps) = Total payload delivered to destination (bytes)*simulation time (in microseconds)

- **Packet Delivery Ratio**: The ratio between the packets generated in clients originated by the "application layer" CBR sources and the number of packets received by the CBR sink at the final destination.

$$\text{Packet Delivery Ratio} = \frac{\sum \text{Number of packets received}}{\sum \text{Number of packets sent}}$$

- **End-to-End Delay**: End-to-end delay or one-way delay (OWD) refers to the time taken for a packet to be transmitted across a network from source to destination. It is a common term in IP network monitoring and differs from round-trip time (RTT) in that only path in the one direction from source to destination is measured.
- **Result Analysis**: Before the attack, throughput, PDR, and delay have some values. Network parameters simply showing (0) zero values signify the existence of an attack. Hence, an attack can be easily predicted by observing the network parameters constantly.

11.3.4 DEVICE SECURITY

Security threats [28–32] on hardware are categorized as follows:

1. **Hardware Trojan (HT)**: HTs are malicious modifications to the original chip design, which compromise the normal operation of the chip.
2. **Side-Channel Analysis Attacks (SCA)**: In this attack, the hacker [29] analyzes the cryptographic device side-channel signals (e.g., power) to guess the secret key. These attacks can cause denial-of-service and breach of authentication process.
3. **Fault Attacks**: Assault [30] on different physical electronic gadgets, for example USB token, HSM, smart card focuses on the gadget by an outer means like light and voltage. So, this creates blunders prompting a security disappointment in the framework (key recuperation, e-purse balance increase, etc.).
4. **Counterfeit Chips**: For this situation, the forgers [31] basically re-mark or repackage old or mediocre parts and afterward sell them as though they were new and fresh. The main motivation is usually monitory, but problems caused are due to poor reliability.
5. **Reverse Engineering**: Since the chips [32] are usually cheap, the hackers buy the standard chip, get familiarized with the circuitry, and then use reverse engineering to clone the chip.

Installing cloned hardware into networks could enable propelling man-in-the-middle attacks. This hardware would subtly adjust a safer communication way between two systems by bypassing security components such as honesty check, encryption, and endpoint confirmation. The basic framework of network can be disrupted by using software that would reroute information to remote servers. These cloners target network routers or parts of routers. Once a router is cloned, it can take control of the network or redirect communications elsewhere [32]. Fake and recycled segments are also advancing into supply chains. Another worry is the possibility of a vindictive substance separating or trading off restrictive data put away inside the IC (integrated chips). Literature uncovers that there are various distributed assaults effectively removing exclusive data from on-chip flash memory.

11.4 EXISTING SOLUTIONS

Any secure WBAN application must guarantee confidentiality and protection of data. It should enable schemes for data validation and protection of key-based mechanisms. It should also provide trustworthiness and trust setup. Lastly, it should also allow secure group management and aggregation of data. However, implementing all the aforesaid features in a low-level device such as sensors and microcontrollers is extremely difficult right from the design stage. The requirement of any security solution is that the data collected should be guarded from unauthorized access and change of state. For example, the patient-related data should be kept confidential during the storage periods. It should follow encryption and access control procedures to occlude user access and node compromise. It should not allow unauthorized modification during the storage periods. Also, the data should be retrievable in the case of deletion or corruption of data.

IoT security becomes more difficult than IT security due to the following reasons:

1. Enabling security mechanism requires large computing power that the miniature devices do not possess always.
2. Cloud data provide an absolute environment for trial-and-error-based experimentation.
3. Since IoT accommodates heterogeneous devices and layered communication architecture with open interfaces, man-in-the-middle attacks are difficult to manage.
4. IoT solutions are very complex with several attack surfaces and multiple potential vulnerabilities.
5. IoT devices are cheap and easily available, opening grounds for familiarization with hardware, reverse engineering in fabrication and cloning, etc.
6. The device information is stored in the cloud usually, and hence, cloning the device IDs becomes easy.

Since security breaches can occur at any level, i.e., chip, node, edge, cloud, application, for understandability, we need to divide the security solutions at three levels:

1. At the level of product makers and semiconductor companies– making the equipment carefully designed, assembling the equipment securely, guaranteeing secure overhauls, giving firmware fixes and refreshes intermittently, and performing dynamic testing.
2. At the level of solution developers – ensuring secure software development and secure integration.
3. At the level of operator/user– keeping the systems updated, mitigating malware, auditing, ensuring protection of infrastructure, and protecting and safe-guarding credentials.

The solutions should also cater to three aforementioned domains, namely device, network, and data security. The existing security mechanisms are as follows:

1. Cryptography
2. Key management
   a. Trusted server
   b. Key pre-distribution
   c. Self-enforcing
3. Secure routing– avoid injected, malicious node information
4. Resilience to node capture – tamper-resistant hardware and node authentication
5. Trust management – node behavior denoted by data delivery quality
6. Robustness to denial-of-service.

The standard approaches [33,34] ensure that information encoded with a protected key can be decrypted only by the authorized receivers. The use of symmetric key encryption is mostly used for WBANs as public key cryptography is battery intensive for sensor nodes. However, ensuring that the information is encrypted does not protect it from external modifications. A malicious user may modify the internal contents or a few fragments of the packets, thereby reducing the data integrity. The absence of data integrity system is in some cases risky, especially in cases of occurrence of life-critical occasions (when emergency data is altered). Data loss as discussed above can likewise happen because of a vulnerable communication environment. Message authentication codes (MAC) are used for authenticating with the help of a mutual shared key. Perhaps the most ideal way is to suspend the activity of a node that has been assaulted and to replace its functionality with some other node in the system. If there should be an occurrence of fluctuating connectivity, the coordinator associates or disassociates the member nodes in a safe way. In case of device security solutions destructive techniques such as chemical or mechanical polishing are used to detect Hardware Trojans with the help of scanning electron microscopes. Non-ruinous methods observe the chip power and delay, just as system yields to analyze the presence of Hardware Trojans. Investigations to distinguish mismatches between transistors, capacitors, resistors, and even interconnects during assembling can make indistinguishable analog circuit designs to have uniquely various responses which can be utilized to shape the basis of a PUF. These PUFs are employed to detect cloned chips.

For healthcare services provided by WBANs, key agreement for the message authentication is also incorporated. Secure connection among the cooperating nodes that are attached to a patient's body is verified during the initial setup phase of the WBAN. The bootstrapping produces necessary secret keys to secure the subsequent wireless communications. The connected sensor nodes adopt a group device pairing (GDP) scheme if none of the nodes have no earlier shared secrets before they meet. They use authenticated group key agreement protocol where the authenticity of each member node can be visually checked by a human. After initial deployment, different types of secret keys can be generated as and when the demand arises. The GDP scheme supports batch deployment of sensor nodes. This inherently saves the setup time as it allows variable and on-demand node association and dissociation. Moreover, this does not rely on any additional hardware devices and is based on symmetric key cryptography [35]. Currently, anonymization has also been adopted with the help of hashing and clustering of data to protect it from unauthorized access at the cloud and edge devices level. The data, location information, and other critical credentials of the patient are anonymized and then encrypted for maximum safety [33].

11.5 CONCLUSION

This chapter briefly highlighted the causes, impacts, constraints, and existing solutions associate with wireless body area sensor networks. Realizing a fully automated IoT-based infrastructure, fewer healthcare applications are still under research. There has been a lot of research in the field of securing such applications at all the levels.

However, there are still open issues and security breaches such as Mirai and botnets, which impact the IoT community extensively.

11.6   FUTURE WORK

Since IoT (Internet of Things) healthcare has a high impact on our lives, it must be secured from luring targets of cyber-attacks [36]. In order to prevent it from hazardous attacks extensive efforts are made to secure data. Along with lightweight cryptographic solutions, machine learning-based approaches for the prediction of vulnerable attacks for a medical or wearable device are introduced in IoT-based healthcare [37]. Deep learning-based techniques may prevent malware attacks which are difficult to analyze and track. A secured healthcare may provide end-to-end connectivity and affordability [38,39,40]. IoT healthcare with high-end security may improve significant features such as remote medical assistance, integration of devices, and medication management. Future work must focus on better healthcare facilities in order to improve higher patient engagement and decrease errors.

REFERENCES

1. Höller, J., Tsiatsis, V., Mulligan, C., Karnouskos, S., Avesand, S., and Boyle, D. (2014). *From Machine-to-Machine to the Internet of Things: Introduction to a New Age of Intelligence.* Amsterdam, The Netherlands: Elsevier.
2. Pang, Z. (2013, Jan). *Technologies and Architectures of the Internet-of-Things (IoT) for Health and Well-Being,* M.S. thesis. Stockholm, Sweden: Dept. Electron. Comput. Syst., KTH-Roy. Inst. Technol.
3. Amiri, D., et al. (2019). "Context-aware sensing via dynamic programming for edge-assisted wearable systems." *ACM Transactions on Computing for Healthcare,* 1 (2), 1–25. doi: 10.1145/3351286.
4. Islam, S.R., Kwak, D., Kabir, M.H., Hossain, M., and Kwak, K.S. (2015). The internet of things for health care: A comprehensive survey. *IEEE Access,* 3, 678–708.
5. Dinis, H., and Mendes, P.M. (2017 Oct 4). Recent advances on implantable wireless sensor networks. In *Wireless Sensor Networks-Insights and Innovations.* IntechOpen.
6. Majumder, S., Mondal, T., and Deen, M. (2017). "Wearable sensors for remote health monitoring." *Sensors,* 17 (1), 130.
7. Milenković, A., Otto, C., and Jovanov, E. (2006) "Wireless sensor networks for personal health monitoring: Issues and an implementation." *Computer Communications,* 29 (13–14), 2521–2533.
8. Darwish, A., and Hassanien, A.E. (2011). "Wearable and implantable wireless sensor network solutions for healthcare monitoring." *Sensors,* 11(6), 5561–5595.
9. Gope, P., and Hwang, T. (2015). "BSN-care: A secure IoT-based modern healthcare system using body sensor network." *IEEE Sensors Journal,* 16(5), 1368–1376.
10. Nabar, S., Walling, J., and Poovendran, R. (2010). "Minimizing energy consumption in body sensor networks via convex optimization."*2010 International Conference on Body Sensor Networks.* IEEE.
11. Jovanov, E., et al. (2005). "A wireless body area network of intelligent motion sensors for computer assisted physical rehabilitation." *Journal of NeuroEngineering and Rehabilitation,* 2 (1), 1–10.
12. Zhou, W., and Piramuthu, S. (2014). "Security/privacy of wearable fitness tracking IoT devices."*2014 9th Iberian Conference on Information Systems and Technologies (CISTI).* IEEE.

13. Pace, P., et al. (2018). "An edge-based architecture to support efficient applications for healthcare industry 4.0." *IEEE Transactions on Industrial Informatics,* 15 (1), 481–489.

14. Camara, C., Peris-Lopez, P., and Tapiador, J.E. (2015 Jun). "Security and privacy issues in implantable medical devices: A comprehensive survey." *Journal of Biomedical Informatics,* 1 (55), 272–289.

15. Bhushan, B., and Sahoo, G. (2017). "Recent advances in attacks, technical challenges, vulnerabilities and their countermeasures in wireless sensor networks." *Wireless Personal Communications,* 98 (2), 2037–2077. DOI: 10.1007/s11277-017-4962-0.

16. Sharma, M., Tandon, A., Narayan, S., and Bhushan, B. (2017). "Classification and analysis of security attacks in WSNs and IEEE 802.15.4 standards: A survey." *2017 3rd International Conference on Advances in Computing, Communication & Automation (ICACCA) (Fall).* DOI: 10.1109/icaccaf.2017.8344727.

17. Jaitly, S., Malhotra, H., and Bhushan, B. (2017). "Security vulnerabilities and countermeasures against jamming attacks in Wireless Sensor Networks: A survey." *2017 International Conference on Computer, Communications and Electronics (Comptelix).* DOI: 10.1109/comptelix.2017.8004033.

18. Bhushan, B., and Sahoo, G. (2017). "Detection and defense mechanisms against wormhole attacks in wireless sensor networks." *2017 3rd International Conference on Advances in Computing, Communication & Automation (ICACCA) (Fall).* DOI: 10.1109/icaccaf.2017.8344730.

19. Al-Janabi, S., Al-Shourbaji, I., Shojafar, M., and Shamshirband, S. (2017 Jul 1). "Survey of main challenges (security and privacy) in wireless body area networks for healthcare applications." *Egyptian Informatics Journal,* 18 (2), 113–22.

20. Goel, A. K., Rose, A., Gaur, J., and Bhushan, B. (2019). "Attacks, countermeasures and security paradigms in IoT."*2019 2nd International Conference on Intelligent Computing, Instrumentation and Control Technologies (ICICICT).* DOI: 10.1109/icicict46008.2019.8993338.

21. Chakravorty, R. (2006). "A programmable service architecture for mobile medical care." In *Pervasive Computing and Communications Workshops, 2006. Per Com Workshops 2006. Fourth Annual IEEE International Conference on,* 5, IEEE.

22. Malan, D., Fulford-Jones, T., Welsh, M., and Moulton, S. (2004). "Codeblue: An ad hoc sensor network infrastructure for emergency medical care." In *International Workshop on Wearable and Implantable Body Sensor Networks,* vol. 5.

23. Wood, A., Virone, G., Doan, T., Cao, Q., Selavo, L., Wu, Y., Fang, L., He, Z., Lin, S., and Stankovic, J. (2006) *"ALARM-NET: Wireless Sensor Networks for Assisted-Living and Residential Monitoring."* Technical Report 2 University of Virginia Computer Science Department.

24. Khan, R.A., and Pathan, A.-S.K. (2018). "The state-of-the-art wireless body area sensor networks: A survey." *International Journal of Distributed Sensor Networks,* 14 (4): 1550147718768994.

25. Jing, Q., Vasilakos, A.V., Wan, J., Lu, J., and Qiu, D. (2014). "Security of the Internet of Things: Perspectives and challenges." *Wireless Networks,* 20 (8), 2481–2501.

26. Anand, M., Ives, Z., and Lee, I. (2005). "Quantifying eavesdropping vulnerability in sensor networks." *Proceedings of the 2nd International Workshop on Data Management for Sensor Networks.* ACM, in healthcare.

27. AL-mawee, W. (2012). Privacy and security issues in IoT healthcare applications for the disabled users a survey.

28. Deogirikar, J., and Vidhate, A. (2017). "Security attacks in IoT: A survey."*2017 International Conference on I-SMAC (IoT in Social, Mobile, Analytics and Cloud) (I-SMAC).* IEEE.

29. Xu, L., and Wu, F. (2015). "Cryptanalysis and improvement of a user authentication scheme preserving uniqueness and anonymity for connected health care." *Journal of Medical Systems*, 39 (2), 10.
30. Sinha, P., Rai, A. K., and Bhushan, B. (2019). Information security threats and attacks with conceivable counteraction. *2019 2nd International Conference on Intelligent Computing, Instrumentation and Control Technologies (ICICICT)*.
31. Rghiout, A., Khannous, A., and Bouhorma, M. (2014). "Denial-of-service attacks on 6lowpan-RPL networks: Issues and practical solutions." *Journal of Advanced Computer Science & Technology*, 3 (2), 143–153.
32. Liu, Y., et al. (2018)."Flexible, stretchable sensors for wearable health monitoring: Sensing mechanisms, materials, fabrication strategies and features." *Sensors* 18 (2): 645.
33. Li, M., Yu, S., Lou, W., and Ren, K. (2010) "Group device pairing based secure sensor association and key management for body area networks."*2010 Proceedings IEEE INFOCOM*.
34. Arora, A., Kaur, A., Bhushan, B., and Saini, H. (2019). "Security concerns and future trends of Internet of Things." *2019 2nd International Conference on Intelligent Computing, Instrumentation and Control Technologies (ICICICT)*. DOI: 10.1109/icicict46008.2019.8993222.
35. Tehranipoor, M.M., Guin, U., and Bhunia, S. "Invasion of the hardware snatchers cloned electronics pollute the market." *IEEE Spectrum*.
36. Xiao, Liang, et al. (2018). "IoT security techniques based on machine learning: How do IoT devices use AI to enhance security?" *IEEE Signal Processing Magazine*, 35 (5), 41–49.
37. Atlam, H.F., and Wills, G.B. (2019). "An efficient security risk estimation technique for Risk-based access control model for IoT." *Internet of Things*, 6, 100052.
38. Duncan, A., Jiang, L., and Swany, M. (2018, April). Repurposing SoC analog circuitry for additional COTS hardware security. In *2018 IEEE International Symposium on Hardware Oriented Security and Trust (HOST)*, 201–204. IEEE.
39. S. Skorobogatov. (2005). "Data Remanence in Flash memory devices," In *Cryptographic Hardware and Embedded Systems*, Edinburgh, 339–353.
40. Zhang, Z., Wang, H., Vasilakos, A.V., Fang, H. (2012, November). "2018ECG-cryptography and authentication in body area networks." *IEEE Transactions on Information Technology in Biomedicine*, 16 (6), 1070–1078

12 Emerging Wireless Technologies Based on IoT in Healthcare Systems in Poland

E. Stawicka and A. Parlinska
Warsaw University of Life Sciences – SGGW

CONTENTS

12.1 INTRODUCTION

The new generation of health protection will use more and more technological innovations, such as mobile devices, dedicated applications, teleconsultations, and even artificial intelligence or data mining tools. The development of telemedicine is gaining importance; this applies in particular to the area of teleconsultation (where the share of private expenditure is already significant and still growing), as well as telediagnostics and telerehabilitation, which are very promising segments of the healthcare market [1].

Innovative techniques, instruments, tools, and methods enable the use of data values and conducting analytical, diagnostic, and decision-making processes [2]. New digital media and Web-based mobile applications support prevention and health promotion. Modern systems and services operate on the principle of significant user involvement, with an emphasis on education, shaping attitudes, and rationalizing healthcare costs [3]. In Poland, the approach is being developed that artificial intelligence and IoT (Internet of Things) are not a fashion, and there are opportunities to acquire and efficiently store and process large amounts of data (from a marketing point – big data). In addition, the computing power of the processors has increased so much that these data can be effectively analyzed (artificial intelligence renaissance) [4].

The use of modern solutions in medicine and health care is all the more possible because patients in Central and Eastern Europe (CEE) are characterized by a relatively high level of digitization, with the prospect of further dynamic development in this area. In Poland, especially young people are very receptive to new technologies and they absorb them very quickly. About 45% of Poles use or are considering using applications for mobile devices, while in Germany, it is 24% and, in Great Britain, 32% [5].

A group of private providers of medical services, such as LUX MED, Medicover, Enel-Med, and Polmed, has developed particularly well in Poland. They all have plans to develop telemedicine services for private patients. In addition, the average share of private expenditure on health care in CEE countries was around 25% in 2015, while in Western Europe, this level is much lower. The share of private expenditure on health care in CEE countries is still growing, while in most Western European countries, it remains unchanged. In Poland, there is an increased interest in medicine-oriented information systems [6–9].

The biggest changes take place in private facilities where patients pay extra for services or use subscriptions and medical insurance provided by employers. Private healthcare providers must prepare for the ever-increasing demands of these patients and change their preferences. However, the market for modern telemedicine systems and IoT will probably cover the entire healthcare sector, both private and public. The awareness of society increases significantly as to the benefits that telemedicine systems bring. However, conventional methods of providing medical services are still popular with older people.

In Poland, in the case of the public sector financed by a public payer, the presented telemedicine solutions could bring significant benefits. It is about improving access to health care, proper management of patient traffic, access to specialist doctors, and effective rehabilitation. However, these activities are not included in the comprehensive healthcare reform.

The main finding shows that in Poland, emerging wireless technologies based on IoT in healthcare system are aimed at disseminating modern solutions. Activities of the most IoT projects are stopped at the concept stage. Modern solutions such as detailed IoT applications, specialized equipment, techniques, and technologies turn out to be unrealistic to use by stakeholders. Over 60% of the prototypes of intelligent objects fall at an early stage of the enterprise's development. The disadvantage is that IoT activities in the health system are not created holistically and are fragmented. The conditions for the dissemination of solutions in Poland become legal regulations, ensuring data security, technical possibilities, economic justification, acceptance of patients and the medical environment, change of attitude, and awareness of stakeholders. Telemedicine and IoT solutions can have a positive effect if they are accepted by doctors and patients. Education at all stages of cooperation is important.

This chapter consists of six subsections. First, the main problems and findings are presented. The second section presents the methodology process along with research stages and questions. The third section presents synthetic contemporary views on innovative health care in the 21st century, based on a critical review of the literature. The fourth section presents the results of research on patients' behaviors and attitudes on the market of modern medical services. The fifth subsection concerns IoT practices in the Polish healthcare system. The chapter ends with a summary and recommendations.

12.2 METHODOLOGY

The aim of the research is to assess the use of innovative technologies based on IoT in the healthcare system in Poland. The methodology process has four stages and is presented in Figure 12.1.

The first stage of the research was connected with the literature review. Three sixty articles were classified from the article search process at the stage of evaluation of titles. Then, after reviewing the full content, 160 was further eliminated due to context. After a final analysis of the IoT and health services based on the relevance of the full text, 50 articles were considered to fill the gap in the study and were critically assessed.

The second stage concerned the selection of secondary data and own research. The database of the Organisation for Economic Co-operation and Development (OECD), Central Statistical Office (CSO), National Health Fund (NHF), Social Diagnosis, and Ministry of Digital Affairs (MDA) were chosen as the sources of the secondary data. At the same time, the questions of the survey were defined, and finally, interviews were collected. A survey was conducted at the turn of May and September 2019 on the Internet. Using Google Forms, www.google.com/forms, the authors have prepared the electronic version of the survey. The received link was sent to several groups of people: students, employees, retired, and unemployed. Each of the respondents filled the survey on their own. The survey contained closed questions conserving the health condition, with the healthcare system in Poland, and six questions which helped us to characterize the responders by age, gender, education, etc. Of the 303 responders, 73% were female; more than 70% of responders were younger than 49 years; however, the biggest group was represented by the people aged between 18 and 29 years. Most of the responders have a higher education, are employed, and are living in the city with more than 500,000 habitants.

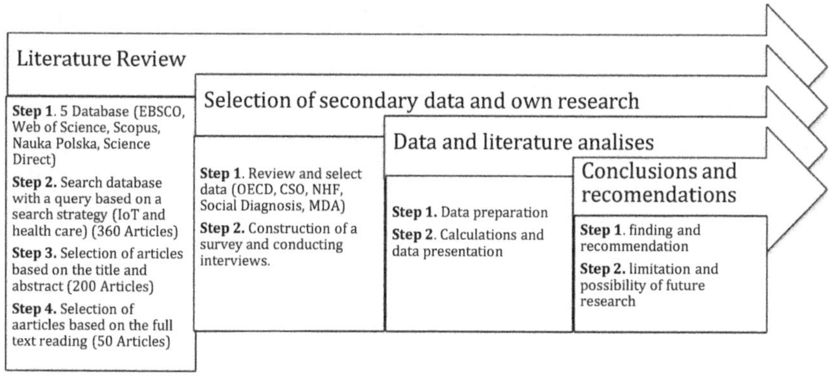

FIGURE 12.1 Research process. (Authors' description.)

The used data were obtained from the literature and from the reports of the Organisation for Economic Co-operation and Development, Ministry of Digital Affairs, PWC Reports, and National Health Fund Development Strategy for 2019–2023.

The analysis was based on selected statistical measures (average, determination and variation coefficients, and Pearson's Chi-square test), which helped to answer the following research questions:

1. What is the structure of the medical service market in Poland?
2. What trends can be observed in the demands and interest of patients in telemedicine?
3. What are the main technologies supporting the development of telemedicine?

The literature review and analysis carried out allowed us to draw conclusions and make recommendations as well as pointed out the limitations of the research.

12.3 INNOVATIVE HEALTH CARE IN THE 21ST CENTURY

One of the key problems in the IoT-based eHealth system is the requirement to have appropriate communication technologies to efficiently share information [10]. This IoT-based eHealth enables people to access emergency services/hospitals and doctors and their relatives to access their health-related data through different applications for immediate and efficient treatment [11].

In Poland, in recent years, there have been ongoing activities consisting in the implementation and dissemination of IoT and artificial intelligence (AI) solutions in medicine and health care. For a long time a slowdown was observed and only development activities were observed, with no implementation of modern technologies and organizational solutions. Currently, in the long term it is recommended to work on the construction of tools to support the collection and classification sets used for learning with supervision (i.e., labelling). Such solutions, often in medicine, require a lot of work and capital, where an expert in this case, a doctor, builds the set of necessary data for a long time. A tool of this type may contain innovations in the field of UI/UX accelerating work as well as AI/ML algorithms trying to propose an annotation that the expert can accept or improve [4].

A constant increase in the demand for medical services, which is the result of an aging society and the decreasing resources of medical staff, poses the challenge of finding a solution that will meet emerging needs. Modern technological and management solutions can be a method to adapt supply to demand.

Innovative IoT activities in the health system also improve efficiency, both financially and socially [12]. The education of patients and healthcare professionals is also becoming an important issue [13].

Despite the enthusiasm of the medical community for telemedicine solutions, costs have often been barriers, especially in the case of financing telemedicine from public funds as well as educational shortcomings [14].

As the European Commission points out,

> thanks to the use of sensors and mobile applications, it enables the collection of a significant amount of medical data, data on physiology, lifestyle and daily activity, as well as data on environmental factors. These data can be used as a basis for shaping medical practice and scientific research. At the same time, thanks to such solutions, patients have easier access to information about their health at any place and time. M-health can also support the provision of high-quality health services and enable more accurate diagnosis and treatment. Such solutions can increase the effectiveness of healthcare, as mobile applications encourage patients to comply with the principles of a healthy lifestyle. This will result in better adaptation of treatment and pharmacological recommendations to the patient's needs. [15,16,32]

In addition, the availability of modern technologies, ease of transmission of information, unlimited range of their spread has an educational aspect, is an important element in the training of medical staff and also raises the level of knowledge and medical awareness of the society [17,18].

The new strategy of the National Health Fund [19] in Poland for the years 2019–2023 guarantees the stability and continuity of actions taken in implementing the assumptions of the European Commission in the field of innovation in modern technological and management solutions. It can be argued that IoT is seen as an opportunity, but these activities are not included in the comprehensive healthcare reform. The SWOT analysis of the National Health Fund strategy in Poland highlighted weaknesses, strengths, opportunities, and threats. The weaknesses include low attractiveness of the NHF as an employer, large staff fluctuation, problems in recruiting specific specialists due to low pay, lack of incentive system, career paths, lack of training, lack of standards of use/targeting training, uneven distribution of human resources, lack of working-time flexibility in the NFH, inefficient internal communication, inefficient IT system, dependence on one IT system supplier. The strengths include the possibility of submitting a request to change the provisions of generally applicable law, training budget, stable employer, partial automation of processes, the opportunity to exchange experiences and good practices between regions. An opportunity was recognized for the development of IoT in health care, such as e-administration and integrated IT systems, and a wide training offer. However, the threats include lack of coherence of actions in the field of computerization in the public sphere [19].

Public confidence in the National Health Fund is at a low level. Research shows that most Poles misjudge the functioning of the NHF. According to CBOS [20] research, in 2018 the NHF was positively assessed by only 31% of respondents. This is a low value both in absolute terms and compared to other public institutions. The social capital refers to qualities such as trust, norms, and relationships that will improve the coordination of planned activities. Building social capital requires effective cooperation of the National Health Fund with patients and other institutions operating in the healthcare system. It requires improving the quality of relationships and building the right climate and cooperation skills. Patients expect not only better availability and quality of health services, but also modern service and innovative solutions. Following the example of its foreign counterparts, the NHF should take the initiative in the field of identification, selection, and implementation of innovations in health care and promote service providers using modern technological and management tools. Patients expect innovative, trustworthy solutions in health care. Only reliable and authentic practices can increase the level of public trust in the National Health Fund. The main goal of the National Health Fund strategy is to increase the level of social trust and to provide a high standard of services. First of all, attention should be paid to improve the service providers and patients, primarily to shorten the waiting time for benefits to standards similar to European countries. In Poland, as part of the main objective, the implementation of preventive programs and an IT platform for prevention is planned. Preventive programs are becoming important, expanding knowledge and public awareness on various issues related to health care. The integration of digital technologies and business processes is also becoming the overriding idea. In the area of information technology, the main goal is to build a new ICT system and to reduce the organization's dependence on external IT solution providers. Projects in the area of IT infrastructure and systems are by their nature long term and require significant investment outlays. There will be activities implemented gradually, as far as possible and as needed. They will concern the adaptation of housing conditions to the expected norms and standards on the one hand and the construction of an integrated IT system, tailored to the structure and new tasks of the organization and challenges of the healthcare system, on the other. In order to gradually introduce innovative solutions in various areas of the healthcare system, the NHF plans to cooperate with, among others, the science sector and relevant institutions, both national and international. The key to successful implementation of the strategy will be concentrated around five main elements, namely:

- communication,
- building a strategic portfolio,
- management control,
- updated strategic initiatives, and
- good practices [19].

The sustained low level of confidence in the National Health Fund results in the dominant role of direct private spending on health. Fee-for-service constitutes about 85% of the total, and the remainder are the so-called structured private expenditure. Structured private expenditure was developed in two forms – as a medical subscription or private health insurance. The former is dominated mainly in Poland and Romania, while insurance plays an important role in Slovenia, Slovakia, and Hungary. In the Central and Eastern Europe region, nearly 60% of patients (over 18 years of age and with revenues over EUR 300 net per month) are ready to use telemedicine solutions, and above all:

- teleconsultation,
- telemonitoring,
- telediagnostics, and
- telerehabilitation [21].

In Poland, forecasts for private expenditure on various types of telemedicine services (including teleconsultations in particular) look promising. This is primarily due to:

- still insufficient public expenditure (financing is a limitation and thus the availability of benefits),
- growing income of residents,
- rapidly growing awareness of the need to care for their own health,
- very wide access to the Internet and new technologies, and
- awareness of the benefits of telemedicine services, such as the possibility of reaching a selected specialist and the possibility of obtaining a second opinion (confirmation of the diagnosis) [9,22].

Activities in the field of telemedicine development in Poland will not be revolutionary. Rather, it should be expected that the infrastructure conditioning the provision of telemedicine services will be gradually expanded, if possible. The choice of the evolutionary path for the development of telemedicine is a consequence of the financial problems of health care in Poland. These problems will certainly not be solved quickly, so it is difficult to expect that the development of telemedicine will be financed to a large extent from budget funds.

Data presented by the National Association of Private Health Care Employers indicate that expenditure on private health care increases by about 20% per year. This means that the private sector is becoming financially stronger and is able to finance telemedicine projects [23,24].

In conclusion, the Polish market of private healthcare services using telemedicine (to a greater or lesser extent) is dominated by five groups of private providers. They all have plans to develop telemedicine services for private patients. However, more and more companies operate in the telemedicine sector focusing on technology rather than the mere provision of services, e.g., equipment manufacturers, software developers, database creators, and diagnostic algorithms. Both large international corporations (Philips and Samsung) and local, smaller companies are present in the market (e.g., Telemedycyna Polska, Comarch, Pro-Plus, Neurosoft, telemedi.co, Telmed, and Medicalgorithmics).

12.4   PATIENT ON THE MARKET OF MODERN MEDICAL SERVICES

PwC research shows that currently patients are open to telemedicine services. They are more and more aware of diseases and have knowledge about diseases, and they visit medical offices not only during diseases, but also prophylactically, educationally for the advice of a dietician or psychologist [5].

Among the many challenges faced by telemedicine, there are a number of successes in the prevention, diagnosis, and monitoring of diabetes. Around 171 million people suffer from diabetes worldwide, and it is estimated that by 2030, there will be an increase to 366 million. In Poland, this number is about 2.5 million, i.e., 6% of the population. The vast majority of patients are over 65 years of age. In this case, simple ICT systems enabling measurement of blood glucose and sending data to a central database and further to a medical center for good interpretation can bring good results for interpretation by medical staff. The information is sent via a mobile phone or computer via wireless networks, a standard telephone modem, or Internet networks. The patient receives feedback in the form of SMS or e-mail, with instructions on how to proceed, and in urgent cases by phone. Patient information is archived and constitutes data. It is a source of knowledge about diabetes, controlling the diet and setting the exercise program [25,26]. In the USA, there are very advanced applications for smartphone, which in addition to continuous monitoring of blood glucose give the ability to record insulin dose, carbohydrate intake, physical activity, weight, arterial pressure, activating reminder alarm, and adjusting insulin pumps [24,27].

Telemedicine is best suited to the elderly. Most chronic diseases are associated with age. Telemedicine enables simplification and acceleration of medical procedures. The patient can have access to not only a doctor, but also specialists from any location at any time of day or night. It contributes quick access to medical services and urgent consultations without leaving home. It is especially convenient for the elderly and disabled. The branches of telemedicine include the increasingly widespread telecardiology, telerehabilitation, telediabetology, telemedicine home care (tele-homecare) [28].

There is a lot to be done regarding the right to use non-personal data. Over 80% of patients in Poland have access to the Internet. This allows the use of modern IoT technologies to transfer knowledge and information. PwC studies show that patients are currently open to telemedicine services. They are more and more aware of diseases and have knowledge about diseases, and they visit medical offices not only during diseases, but also prophylactically, educationally for the advice of a dietician or psychologist [29].

However, the development of telemedicine in Poland depends on systemic solutions aimed at shaping the information society, and in particular qualitative and quantitative progress in Polish telecommunications. This may mean that awaiting the technological development of telecommunications, health care, and medicine in Poland, especially the public, deepens the distance to more developed countries and will offer citizens still insufficient medical care [30]. Despite the access to the Internet, there is a low level of telecommunication device skills among patients, especially people over 60 years of age, to whom telemedicine services are directed to a large extent. Research shows that 59% of people over 60 cannot use a computer and that 94% do not want to learn [16,23].

The current expenditure on health care in Poland fluctuates, but also shows an upward trend (Figure 12.2). In 2017, the total health expenditure amounted to 34,435 million current US$ and 6.54% GDP. More than 92% of Polish households used the medical services financed by National Health Fund (NHF), but more than half of them also used the services financed by their private sources [31]. Furthermore, 7% of households have prepaid medical care package. We cannot assume that households who use medical care package disburden public health-care system. There is no relationship between using public and private healthcare system. There is a constant growth in the number of beneficiaries from health care in each source of financing.

Social Diagnosis: Objective and Subjective Quality of Life in Poland pointed out that most frequently, the institutions where services are financed from private resources have been chosen by households from big cities. Private institutions are more often chosen by urban households (60%) than by rural households (44.7%).

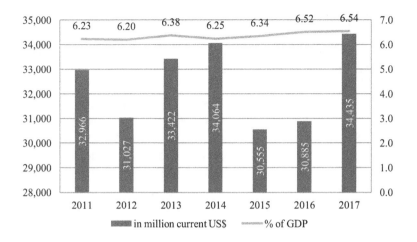

FIGURE 12.2 Current health expenditure in Poland (million current US$ and % GDP). (Authors' own calculation on the basis of data from OECD.)

Out-of-pocket payments for medical services and prepaid medical care package are the most popular in Podkarpackie, Małopolskie, and Mazowieckie and Pomorskie Voivodeships (Table 12.1).

There are a few factors which influence the development of private health insurance market: aging society, societies becoming wealthier, increase in health expenditures, increasing percentage of doctors working in non-public healthcare facilities, growing interest in medical packages and health insurance, and the development of international companies on Polish healthcare market.

Theoretically, public healthcare system in Poland offers a very broad range of services available for patients without any direct payments. Unfortunately, the actual state differs from the legal records. Nowadays, more than 50% of Poles choose private health care which is not financed by the National Health Fund. According to Central Statistical Office (CSO), the main reasons of such situation are as follows:

- long waiting period for medical appointment,
- more qualified specialists in private healthcare institutions, and
- better equipment in private healthcare institutions [32].

According to the analysis of data from OECD (but also mentioned in the Report of CSO), the growth in private expenditures has been observed [26,32]. In 2017, private households' health expenditure amounted to 10,503 million US$. However, the significant growth was observed within voluntary private healthcare payments (Figures 12.3 and 12.4).

Analyzing the changes in voluntary health insurance and household out-of-packet payments during the period 2002–2017, interesting trends written using mathematical models can be observed (Figure 12.5).

According to the polynomial function $y = -42.296x^2 + 1000.4x + 1949.2$ (determination coefficient $R^2 = 0.9056$), household out-of-packet health payments

TABLE 12.1

Percentage of Households Using Different Ways to Finance Their Health Care in 2011 and 2015. (Based on Czapiński, Panek 2015)

Financial Sources	Financed by NHF*		Household Out-of-Pocket Payment		Voluntary Health Care Payment Schemes	
Years	2011 (%)	2015 (%)	2011 (%)	2015 (%)	2011 (%)	2015 (%)
Socioeconomic status						
Employee	90.6	92.0	55.5	59.7	11.1	11.9
Farmers	90.0	92.0	48.7	51.8	-	-
Self-employed	88.5	91.2	71.0	79.2	5.9	6.8
Pensioner	93.8	96.3	41.9	46.1	1.4	0.8
Annuitant	94.4	97.7	28.3	37.3	-	-
Other	82.5	88.2	29.4	28.7	-	-
Residence place						
City with more than 500,000 habitants	89.8	88.9	60.3	65.1	17.3	15.7
City with more than 200–500,000 habitants	90.7	93.6	55.4	57.2	9.3	11.9
City with more than 100–200,000 habitants	92.2	95.9	46.6	54.4	7.1	8.9
City with more than 20–100,000 habitants	92.1	94.3	49.3	53.5	4.5	5.5
City with less than 20,000 habitants	91.1	93.3	44.2	51.9	4.2	3.8
Village	91.5	94.1	44.7	48.8	2.7	2.8
Voivodeship						
Dolnośląskie	91.6	91.2	53.0	59.4	8.0	8.3
Kujawsko–Pomorskie	90.9	95.2	43.3	49.2	2.8	4.5
Lubelskie	92.9	96.4	48.4	52.3	5.2	6.2
Lubuskie	94.2	96.7	56.4	58.3	5.8	8.15
Łódzkie	92.7	94.5	50.2	53.0	2.9	5.0
Małopolskie	92.5	94.9	54.4	55.0	6.3	4.9
Mazowieckie	88.5	89.4	52.6	59.4	13.5	13.2
Opolskie	88.9	97.0	39.7	54.6	2.8	2.3
Podkarpackie	95.1	94.5	56.9	55.4	2.6	3.3
Podlaskie	93.3	92.7	49.5	43.3	2.7	1.5
Pomorskie	91.3	93.4	52.2	55.2	9.9	15.8
Śląskie	91.6	93.6	46.4	53.5	5.7	6.8
Świętokrzyskie	88.9	92.5	47.1	53.2	3.1	2.5
Warmińsko–Mazurskie	89.6	93.5	27.0	30.6	1.9	1.1
Wielkopolskie	91.8	96.5	50.2	59.2	6.7	4.2
Zachodniopomorskie	89.8	89.0	42.8	45.6	5.0	5.1

showed an upward trend until 2013. From 2014, there has been a decline which is associated with a rise in spending on voluntary health insurance. It is also characterized by the polynomial function $y = 14.722x^2 - 123.04x + 276.95$ (determination coefficient $R^2 = 0.9047$) and shows an upward trend throughout the considered period.

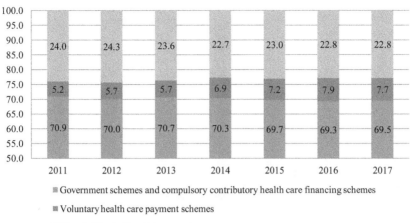

FIGURE 12.3 Structure of the current health expenditure by financing scheme in years 2011–2017 (%). (Authors' own calculation based on the data from OECD.)

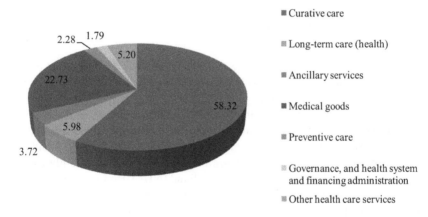

■ Curative care

■ Long-term care (health)

■ Ancillary services

■ Medical goods

■ Preventive care

■ Governance, and health system and financing administration

■ Other health care services

FIGURE 12.4 Structure of the current health expenditure specified by function in 2017 (%). (Authors' own calculation based on the data from OECD.)

In Poland, however, interest in private healthcare and telemetry is developing very quickly. Patients, when choosing a private medical facility, are increasingly looking for something more than the doctor's professionalism and quality of service. What counts for them is trust in the doctor, comfort of the office, a sense of security, and brand strength. Patients are looking for doctors whom they trust and with whom they feel safe. However, it is not enough for today. Patients also want to feel comfortable in the place where the services are provided. They pay attention to how the doctor communicates with them and the surroundings in which they are located. In addition, technologies supporting the development of telemedicine are more often used by the youngest patients. In the literature, many research and analysis results regarding IoT appear in medicine, but the problem is to use the possibilities of IoT in

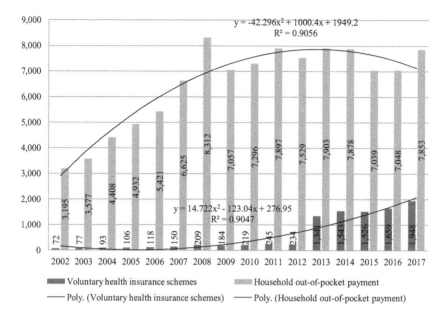

FIGURE 12.5 Health expenditure by financing scheme in years 2011–2017 (%). (Authors' own calculation based on the data from OECD.)

the context of age, gender or due to the patient's nationality [3]. Table 12.2 presents the benefits and barriers to the development of telemedicine and the reactions and attitudes of patients depending on age.

Research shows that patients' interest in private health care is dependent on income. In Poland, additional insurance is associated with a prestigious issue; however, monthly premiums were in the range of up to PLN 100 (Figures 12.6 and 12.7). Also the survey results confirm the observed trends. Responders mainly used the medical services provided by National Health Fund (NHF). However, 98.7% of them use out-of-pocket payments mainly to finance expenses for medicines (65.7%) and medical treatment (22.1%).

A statistical analysis was performed to verify the hypothesis about the distributions independent of the relationship between the amount of premiums paid for private health insurance and gross income. Using the Pearson's Chi-square test, the hypothesis was verified and rejected on the significance level at $\alpha = 0.05$, which showed that the distribution is statistically different for those five income groups. More than 50% of the responders paid for the private health insurance from 0 to 100 PLN, which is connected with the price for the standard health packet offered by private companies (Tables 12.3 and 12.4).

Patients are interested in quick medical assistance, including specialists and avoiding queues. Studies have shown that interviewers declared an extended waiting time (from several days to several months) for visits to specialists, specialist tests, or hospital procedures.

Patients are meticulous about visiting "non-accidental" doctors. They want to visit esteemed and well-known specialists. They convey opinions and comments. They value

TABLE 12.2

Advantages and Barriers That Telemedicine Has Faced

Advantages and Barriers	Attitudes Toward Telemedicine	
	Young People	Elder People
Quick access to diagnostic and specialized medical services, simple operation	Acceptance and eager use of telemedicine	Resistance against the use of new information technologies
Quick access to test results without the need for personal collection	Acceptance and willing use of the computer and mobile phone	Limited ability to operate a computer and a mobile phone
Permanent access to medical records	Eager to use, young people operate computer equipment and smartphone very well	High costs of purchasing the necessary equipment, computer, or smartphone
No need to travel to medical centers	Medical service without leaving home	Lack of knowledge about the availability of telemedicine, resistance to new products
Receiving efficient emergency assistance	Efficient service and quick information	Habit of visiting live, poor use of the telemedicine system
Shortening the length of hospital stay by providing some services using telemedicine system	Acceptance and cooperation with doctors using telemedicine	Poor cooperation, frequent deterioration in performance
Feeling safe and improving the quality of life through constant medical supervision	Acceptance	Acceptance
Educational benefits flowing from the use of telemedicine systems, for example. Telerehabilitation – the patient learns to perform proper exercises alone at home	Acceptance	Acceptance, with a tendency to need personal contact

Source: Authors' description.

individual approach and continuity of care. They like to book medical appointments by phone or Internet without having to confirm in person. Such opportunities are provided by newly created portals of patients, medical companies, or pharmacies. At the same time, individual companies providing medical services provide access to medical data on the Internet or via a phone or a smartphone application [33].

12.5 IoT PRACTICES IN THE HEALTHCARE SYSTEM

Technology in medicine is developing very rapidly. Today, the IT convergence draws a large amount of attention as the next technology for disease prevention and health care in the health and medical service areas as the demand for the improvement in the

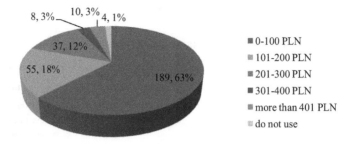

FIGURE 12.6 Structure of answers for the question: How much do you spend on health per month (except NHF health insurance and private health insurance)? (Authors' own calculation based on the data from the survey.)

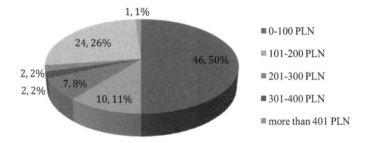

FIGURE 12.7 Structure of answers for the question: How much do you pay per month for private health insurance (i.e., LUX MED and Enel-Med)? (Authors' own calculation based on the data from the survey.)

assessment of quality of life (QoL) of people grows in a smart technology health and wellness society. With an aging global population and an increased human demand for healthy lives, health care and illness prevention have been actively researched in the medical service area [1].

In the era of changing needs and requirements, new technologies are emerging in terms of both equipment and software in medicine. Some technologies are expensive and innovative. However, as telemedicine develops and becomes more widespread, cheaper and affordable solutions appear among patients.

Telemedicine focuses on the exchange of information between the patient and the doctor as between the doctors themselves. It makes it is possible for doctors to examine and diagnose patients located at a place away from them. The widespread use of telemedicine facilitates access to specialized medical care, in particular in areas where access to specialist doctors is difficult, accelerates the diagnosis, and reduces the need for hospitalization and, as a result, the cost of treatment. Telemedicine has a special use in patients with heart disease, diabetes, and asthma. Thanks to the innovative solutions, the sick can stay home and at the same time be under the constant care of a doctor.

The basis for the telemedical market division is the relationship between the persons who take part in the medical procedure. The telemedical market can be

TABLE 12.3
Distribution Matrix of the Relationship between the Amount of Out-of-Pocket Payments and Monthly Gross Income

		Out-of-Pocket Payments (Except NHF Health Insurance and Private Health Insurance)							
		0–100 PLN	101–200 PLN	201–300 PLN	301–400 PLN	More Than 401 PLN	Do Not Use	SUM	%
Monthly gross income	0–2250 PLN	64	10	10	5	3	1	93	30.7
	2250–4500 PLN	61	24	12	0	1	1	99	32.7
	4500–6750 PLN	32	17	6	3	4	0	62	20.5
	6750–9000 PLN	11	2	6	0	2	2	23	7.6
	More than 9000 PLN	21	2	3	0	0	0	26	8.6
	Sum	189	55	37	8	10	4	303	100
	%	62.38	18.15	12.21	2.64	3.30	1.32	100	

Source: Authors' own calculation based on the data from the survey.

TABLE 12.4

Distribution Matrix of the Relationship between the Amount of Premiums Paid for Private Health Insurance and Monthly Gross Income

		Premiums Paid for Private Health Insurance (i.e., LUX MED and Enel-Med)								
		0–100 PLN	101–200 PLN	201–300 PLN	301–400 PLN	More Than 401 PLN	Paid by the Employer	Paid by the Family Member	SUM	%
Monthly gross income	0–2250 PLN	6	0	0	0	0	2	1	9	9.8
	2250–4500 PLN	18	5	2	0	0	9	0	34	37.0
	4500–6750 PLN	13	3	3	0	0	5	0	24	26.1
	6750–9000 PLN	3	1	1	2	2	2	0	11	12.0
	More than 9000 PLN	6	1	1	0	0	6	0	14	15.2
	Sum	46	10	7	2	2	24	1	92	100
	%	50.0	10.9	7.6	2.2	2.2	26.1	1.1	100	

Source: Authors' own calculation based on the data from the survey.

distinguished by two types of services: D2D (doctor-2-doctor) and D2P (doctor-2-patient) (Figure 12.8). The first concerns the doctor–doctor relationship, where telemedicine is used for consultation between doctors. In the doctor–patient relationship, telemonitoring takes place, which provides a constant check of the patient's health. Such solutions are designed for people especially with heart disease, asthma, and diabetes.

The use of mobile phones, smartphone, or computers is the fastest way to access medical services. They provide unlimited consultation not only at the patient–doctor level, but also at the doctor–doctor level. Home stations with remotely connected devices such as weight, blood pressure monitor, ECG, CTG, stethoscope, breath analyzer, mobile telerehabilitation, or telediagnostic devices are becoming more and more common. These systems allow medical services without the need for hospitalization, i.e., effectively and economically. Intelligent electronics, the so-called wearable, e.g., watches with a heart rate monitor or life bracelets, are also used [34]. The software for telemedicine devices, such as applications supporting daily activities, is also changing (shown in Figure 12.10). Modern software supports diagnostics, systems enable remote descriptions, and image viewers.

Artificial intelligence is used, which supports the work of the doctor to respond to errors and help to take a rational path. Data mining has also become a treatment option on historical data. It allows us to change data into information that can be used in treatment.

In Poland, the development of telemedicine is supported by a number of companies providing hardware and software. These include KAMSOFT S.A., Telemedycyna Polska Capital Group, or Comarch. These are companies that support private and state medical institutions as well as provide support for individual clients. Figure 12.9 shows some selected examples in which the equipment and software of these companies are used.

"LESSER POLAND TELE-ANGEL" is a project developed to address the needs of dependent, chronically ill, elderly, disabled people who live alone. The goal is to improve their quality of life and ensure safety. It is organized by the provincial government in cooperation with Caritas of the Kielce Diocese and the European Institute of Regional Development Association from Sucha Beskidzka in cooperation with

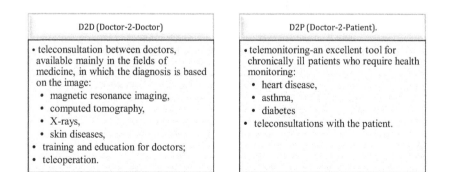

D2D (Doctor-2-Doctor)	D2P (Doctor-2-Patient).

• teleconsultation between doctors, available mainly in the fields of medicine, in which the diagnosis is based on the image:
  • magnetic resonance imaging,
  • computed tomography,
  • X-rays,
  • skin diseases,
• training and education for doctors;
• teleoperation.
	• telemonitoring-an excellent tool for chronically ill patients who require health monitoring:
	  • heart disease,
	  • asthma,
	  • diabetes
	• teleconsultations with the patient.

FIGURE 12.8   Types of services on the telemedicine market. (Authors' description.)

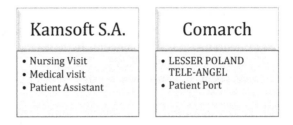

Kamsoft S.A.	Comarch
• Nursing Visit • Medical visit • Patient Assistant	• LESSER POLAND TELE-ANGEL • Patient Port

FIGURE 12.9 Examples of companies that support private and state medical institutions. (Authors' description based on the information from www.comarch.pl/healthcare/aktualnosci/, https://kamsoft.pl. [35])

Comarch. During the period 2018–2020, 10,000 Lesser Poland residents will be covered by the telecare program. Thanks are due to the use of special wristbands, which have an integrated phone, software, and GPS module and allow quick contact with paramedics via the SOS button. The devices provided by Comarch will also allow remote monitoring of the pulse and geolocation of the program's participants. Part of the equipment and software for the implementation was provided by Comarch, an IT systems producer [36].

Telemedicine Poland in partnership with Przychodnia Lekarska Szombierki Sp.z o.o implements the project entitled: Health services for the residents of Bytom using modern technologies. The project is co-financed by the European Social Fund (ESF) under the Regional Operational Program of the Śląskie Voivodeship for 2014–2020. Within the project, the participants can use three or one of the offered services: TELEPHONE CARE "SOS", CARDIOLOGICAL MONITORING, and DIABETOLOGICAL MONITORING. Participants obtain the required equipment and have the possibility to make the test at any time in a day and the possibility of consulting results with a doctor or diabetology expert or of contacting a caregiver [37].

Figure 12.10 presents some of the examples of IoT introduced by the Public Healthcare System.

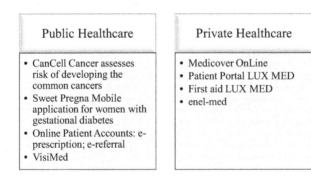

Public Healthcare	Private Healthcare
• CanCell Cancer assesses risk of developing the common cancers • Sweet Pregna Mobile application for women with gestational diabetes • Online Patient Accounts: e-prescription; e-referral • VisiMed	• Medicover OnLine • Patient Portal LUX MED • First aid LUX MED • enel-med

FIGURE 12.10 Examples of IoT application used in private and public health institution. (Authors' description based on the information from www.medicover.pl, www.luxmed.pl, http://aplikacja.enel.pl/, http://cancellcancer.pl, http://sweetpregna.pl, http://pacjent.gov.pl.)

Private medical companies offer their patients online access or mobile application to their accounts, which allows them to:

- order e-prescriptions to chronic medications and, thus, provide with continuous treatment.
- check examination results, both laboratory and imaging ones. The results will be commented by the doctor.
- make an appointment with a particular medical specialist at a chosen time, thanks to the options available in a browser.
- postpone or cancel the appointment if the patient cannot come for any reason.
- send a question to the doctor after the appointment in case of having any additional questions or doubts.
- order medical documentation which will be prepared for you to collect at a center chosen by you.
- look through issued referrals, selecting them by the date of issuance or status "done" or "undone".
- have access to the full history of visits and examination results.

It is worth to stress the application First Aid developed by the LUX MED Group, which helps in an emergency step by step. It contains pictorial and text instructions that will help respond appropriately in various cases, such as choking, traffic accident, poisoning, or hemorrhage. A clear division into categories allows one to quickly reach the actions that should be taken in a specific situation. A flashlight, metronome that sets the rhythm of chest compressions, and the ability to dial an emergency number with one click are additional features to help those in need.

The content of the application was developed by the experts of the LUX MED Rescue Academy – professionally active doctors and paramedics with many years of experience. All content is in accordance with the current guidelines of the European and Polish Resuscitation Council. The application also has a game thanks to which you can easily check your knowledge about the principles of providing first aid [38,39].

In 2017, NHF introduced two applications co-financed by the Norwegian Financial Mechanism:

- The CanCell Cancer application is used to address the needs of healthy people who want to lead a healthy lifestyle through making decisions related to their health daily and thus prevent the incidence of cancer. It is useful to estimate genetic predisposition and assess the risk of developing the most common cancers; monitor diet and physical activity; help to quit smoking; remind them of periodic preventive examinations; help to find the right medical facility; and inform about cancer and its prevention and is available in three languages: Polish, English, and Norwegian [40].
- SweetPregna is a mobile application that, by knowing pre-pregnancy weight, height, and age of pregnancy, will determine caloric demand and

help eat healthy food and properly control glycaemia of pregnant women. The application will also tell them how a balanced diet should look like, which they should not eat so as not to raise their blood sugar level. Women will be able to compose their meals themselves, save their meter measurements, and save it all and give it to their doctor [41].

- Also in 2019, NHF introduced the free application of the Ministry of Health by which an insured person will be able to check health information quickly and safely (their own, their children's, or the person who authorized them to do so). Also it will be possible to find the history of visits (since 2008), paid for by the NHF, and share information about health and history of prescribed medicines with a loved one or doctor [42].

Fog computing in healthcare IoT systems is also becoming very important. Fog computing is considered suitable for the applications that require real-time low and high response times, especially in healthcare applications. All these studies demonstrate that resource sharing provides low latency, better scalability, distributed processing, better security, fault tolerance, and privacy in order to present better fog infrastructure [43].

Digital innovation is a network of heterogeneous elements like an ecosystem that change over time. In consequence, it is increasingly difficult to draw the boundaries of a digital innovation. The landscape of digital innovation is continuously in flux, as the elements evolve through evolutionary patterns such as birth, growth, shrinkage, and death [44].

To sum up, IoT practices in the healthcare system gain importance. Predicting the future of IoT and other digital innovations could be valuable for industry professionals and policy makers. Some recent studies have suggested promising benefits of using social media data and other non-traditional data for forecasting. Future research could explore different methods and techniques of predictive modelling for short-term forecasting of different IoT topics [45–47].

12.6 SUMMARY AND RECOMMENDATIONS

Telemedicine is a good solution to some of the problems in Polish health care. Nevertheless, telemedicine is primarily a benefit for all healthcare professionals. The use of modern technology in diagnostics and research clearly contributes to reducing their costs. The above situation is of particular importance in view of the pressure exerted on the increase in state expenditure on medical services. An advantage of telemedicine is the effective use of the resources of medical personnel, whose deficiencies are already beginning to be felt by patients.

1. An ineffective system of the state health service, lack of sufficient financial resources, and indebtedness are conducive to the development of private healthcare facilities and telemedicine in Poland.
2. A group of private medical service providers, such as LUX MED, Medicover, Enel-Med, and Polmed, is developing very dynamically in Poland.

In addition, the development of telemedicine is supported by a number of companies supplying hardware and software.

3. There is a lack of trust among patients in the state health service. The number of people with private health insurance is increasing. Most often, this type of insurance is usable by people between 30 and 49 years of age, from cities with over 500,000 inhabitants of the Mazowieckie and Pomorskie Voivodeships, with higher education, employed full-time, spending about PLN 100 monthly on additional health insurance, and with an average income from 2250 to 4500 PLN.

4. The conducted research shows that waiting for visits to doctors, especially specialists, is prolonged. Private health insurance shortens the waiting time for specialists. This form of private insurance is often a form of non-wage benefits and applies to employees, where employers pay some or a significant part of the additional insurance.

5. The problem is the low level of health education and trust in telemedicine. Despite the fact that over 80% of patients have access to the Internet, the use of modern IoT technologies in transferring knowledge and information is still low.

6. Modern solutions and IoT are more often used in private healthcare units at the doctor–patient and doctor–doctor levels. The most solutions using IoT are used in the case of teleconsultation, telemonitoring, telediagnostics, and telerehabilitation. The most important applications are applications for telecardiology and prevention, diagnosis, and monitoring of diabetes.

7. There is a noticeable increase in the number of projects using IoT implemented by public medical units at the level of system solutions, e.g., e-prescription, e-exemption, as well as by individual local government units, e.g., project "LESSER POLAND TELE-ANGEL". Noteworthy is the fact that these solutions are introduced with external financial sources, i.e., European Union funds, Norwegian funds, or others.

8. The presented results constitute a preliminary analysis of the use of IoT in health systems. The limitation was the small amount of IoT-related materials in Poland. The literature provides a range of data and information on IoT technical systems in the world, innovations that are very specific to specific countries and have a close connection with economic and social development.

9. A number of barriers and limitations were also noted. Restrictions include the legal issues of solutions using databases or the use of non-personal data, subjective approach of respondents in answering survey questions, and a limited research sample.

The conducted research and analysis contribute to further research on the development of telemedicine in Poland and the possibilities of using modern IoT solutions and innovations in this field. An interesting direction of research would be to examine the interest and use of IoT applications in telemedicine with the consideration of the sex product.

REFERENCES

1. Song C.-W., Jung H., Chung K. (2019). Development of a medical big-data mining process using topic modelling. *Cluster Computing* 22, 1949–1958.
2. Chae B.K. (2019). The evolution of the Internet of things (IoT): A computational text analysis. *Telecommunications Policy* 43, 101848.
3. JungLa H. (2016). A conceptual framework for trajectory-based medical analytics with IoT context. *Journal of Computer and system Sciences* 82, 610–626.
4. Ministry of Digital Affairs (Ministerstwo Cyfryzacji). (2018). Założenia do strategii AI w Polsce, Warszawa, 9 listopada 2018.
5. Raport Pacjent w Świecie Cyfrowym PwC. (2016). www.pwc.pl/pl/publikacje/2016/pacjent-w-swiecie-cyfrowym-raport-pwc.html, access: 10.12.2019.
6. Woźniak J., Lubacz J., Burakowski W. (2009). *Stan obecny i kierunki rozwoju telekomunikacyjnych i teleinformatycznych prac badawczych i wdrożeniowych w Polsce i na świecie. Analiza stanu oraz kierunki rozwoju elektroniki i telekomunikacji.* Warszawa: Komitet Elektroniki i Telekomunikacji PAN.
7. Bujnowska-Fedak M.M., Siejka D., Sapilak B. (2010). Systemy telemedyczne w opiece nad przewlekle chorymi. *Fam MedPrimCareRev* 12 (2), 328–334.
8. Czapiński J., Panek T. (2015, November). Social diagnosis 2015 objective and subjective quality of life in Poland. *Contemporary Economics* 9 (4), 36.
9. Grobelna I., Grobelny M., Bazydło G. (2018). User awareness in IoT security. *A survey of Polish users, AIP Conference Proceedings* Vol. 2040, p. 080002; https://doi.org/10.1063/1.5079136.
10. Faheem M., Gungor V.C. (2018). Energy efficient and QoS-aware routing protocol for wireless sensor network-based smart grid applications in the context of industry 4.0. *Applied Soft Computing* 68, 910–922.
11. Ali R., Qadri Y.A., Zikria Y.B., Umer T., Kim B.S., & Kim S.W. (2019). Q-learning-enabled channel access in next-generation dense wireless networks for IoT based eHealth systems. *EURASIP Journal on Wireless Communications and Networking*, 178 (2019). doi:10.1186/s13638-019-1498-x.
12. Jung H., Chung K. (2016). PHR based life health index mobile service using decision support model. *Wireless Personal Communications* 86 (1), 315–332.
13. Jung H, Chung K. (2016). Knowledge based dietary nutrition recommendation for obese management. *Information Technology Management* 17 (1), 29–42.
14. Dydyt M., Dydyt T. (2018). Determinanty rozwoju innowacji w systemie opieki zdrowotnej. *Zarzadzanie i Finanse [Journal of Management and Finance]* 16, 67–81.
15. Komisja Europejska. (2012) Plan działania w dziedzinie e-zdrowia na lata 2012–2020 – Innowacyjna opieka zdrowotna w XXI wieku, (COM (2012) 736).
16. Trendy w polskiej ochronie zdrowia 2017. (2017). PwC Polska Sp. Z o.o. www. Pwc.pl, access: 19.12.2019.
17. Chojnecka M., Nowak. A. (2016). Telemedycyna na tle polskich regulacji prawnych- szansa czy zagrożenie? *Internetowy Kwartalnik Antymonopolowy i Regulacyjny* 8 (5). 74–81.
18. Lewandowski R., Kożuch A., Sasak J. (2018). *Kontrola zarządcza w placówkach ochrony zdrowia.* Warszawa: Wolters Kluwer.
19. Strategy of the National Healh Fund. (2019). Stategia NFZ na lata 2019-2023. www.nfz.gov.pl, access: 19.12.2019.
20. CBOS. (2018). Jak Polacy oceniają NFZ?, twojezdrowie.rmf24.pl, access 19.12.2019.
21. Raport Pacjent w Świecie Cyfrowym PwC. (2016). www.pwc.pl/pl/publikacje/2016/pacjent-w-swiecie-cyfrowym-raport-pwc.html, access: 10.12.2019.
22. Bujanowska-Fedak M.M, Kumięga P., Sapilak B. (2013). Zastosowanie nowoczesnych systemów telemedycznych w opiece nad ludźmi starszymi. *Family Medicine & Primary Care Review* 15 (3), 441–446.

23. Maziarz W. (2010). Problemy rozwoju telemedycyny w Polsce. *Zeszyty Naukowe Uniwersytetu Szczecińskiego*, Studia Informatica nr 25, 605, 34–44.
24. Sznyk A., Karasek J. (2016). *Innowacyjność w sektorze ochrony zdrowia w Polsce. Wyzwania, bariery, problem i rekomendacje.* Warszawa: Instytut innowacyjna Gospodarka.
25. Bujnowska-Fedak M.M., Puchała E., Steciwko A. (2011). The impact of telehome care on health status and quality of life among patients with diabetes in a primary care setting in Poland. *Telemed e-health* 17(3), 153–160.
26. OECD (2019). State of Health in The EU Poland: Country Health Profile 2019, https://www.oecd.org/health/country-health-profiles-eu.htm, access: 30.01.2020.
27. Tran J., Tran R., White J.R. (2012). Zastosowanie aplikacji na smartfony i glukometrów współpracujących ze smartfonami w leczeniu cukrzycy, przegląd 10 wyróżniających się aplikacji oraz nowego glukometru podłączonego do smartfona. *Diabetol po Dypl* 9 (4), 47–53.
28. Alam H.M., Malik M.I., Khan T., Parady A., Kuusik Y.L., Moullec A. (2018). Survey on the roles of communication technologies in IoT based personalized healthcare applications. *IEEE Access* 6, 36611–36631.
29. Raport Indeks Sprawności Ochrony Zdrowia 2018 PwC (2019). www.pwc.pl/pl/pdf/publikacje/2019, access: 12.12.2019.
30. Unsal A., Stawicka E. (2018). Comparative analyses of health economics indicators in the European Union and Turkey. *Economic Sciences for Agribusiness and Rural Economy: Proceedings of the International Scientific Conference*, Warsaw, 7–8 June 2018. T. 2. Warsaw: Warsaw University of Life Sciences Press, S. 198–202.
31. Parlinska A. (2017). *Konstrukcja i podstawa wymiaru składki na ubezpieczenie zdrowotne, Finanse: instytucje, instrumenty, podmioty, rynki, regulacje (Marian Podstawka)* Wydawnictwo Naukowe PWN, Warsaw pp. 495–499.
32. CSO. (2019). Zdrowie i ochrona zdrowia w 2017. https://stat.gov.pl/obszary-tematyczne/zdrowie/zdrowie/zdrowie-i-ochrona-zdrowia-w-2017-r-,1,8.html, access: 20.12.2019.
33. https://www.medicover.pl/Apps/mobilna/en/index.html; access: 1.12.2019.
34. www.comarch.pl/healthcare/aktualnosci/, access: 19.12.2019.
35. https://kamsoft.pl/, access: 19.12.2019.
36. https://www.malopolska.pl/teleaniol/o-projekcie, access: 20.12.2019.
37. http://teleopieka.bytom.pl/, access: 01.12.2019.
38. https://www.luxmed.pl/dla-pacjentow/o-nas/nasze-aplikacje-mobilne.html, access: 01.12.2019.
39. http://aplikacja.enel.pl/, access: 01.12.2019.
40. http://cancellcancer.pl/, access: 12.12.2019.
41. http://sweetpregna.pl/, access: 12.12. 2019.
42. https://pacjent.gov.pl/internetowe-konto-pacjenta; access; 12.12.2019.
43. Mutlag A.A., Ghani M.K.A., Arunkumar N., Mohammed M.A., Mohd O. (2019). Enabling technologies for fog computing in healthcare IoT systems. *Future Generation Computer Systems* 90, 62–78.
44. Chae B.K. (2019). A general framework for studying the evolution of the digital innovation ecosystem: The case of big data. *International Journal of Information Management* 45, 83–94.
45. Schoen H., Gayo-Avello D., Metaxas P.T. Mustafaraj E., Strohmaier M., Gloor P. (2013). The power of prediction with social media. *Internet Research* 23 (5), 528–543.
46. Kim J.C, Chung K. (2017). Depression index service using knowledge based crowd-sourcing in smart health. *Wireless Personal Communications* 93 (1), 255–268.
47. Yoo H., Chung K. (2017). PHR based diabetes index service model using life behaviour analysis. *Wireless Personal Communications* 93 (1), 161–174.

13 Biometric Monitoring in Healthcare IoT Systems Using Deep Learning

*Shefali Arora, Veenu, M.P.S Bhatia,
and Gurjot Kaur*
Netaji Subhas Institute of Technology

CONTENTS

13.1 INTRODUCTION

With the emerging technology known as the Internet of things, there is an inter-connection between various networks that sense and process a diverse kind of data. Various monitoring systems are quickly emerging, as they assist to measure the essential parameters of an individual or unit. Thus, IoT is useful in preserving a large amount of data and utilizing it in various domains. Figure 13.1 shows the architecture of IoT.

Electronic health care (eHealth) using IoT helps patients, doctors, and nurses to monitor health remotely. It can help to gauge essential parameters or detect chronic diseases.

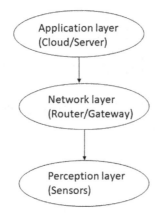

FIGURE 13.1   Architecture of IoT.

Wearable technologies are playing an important role in such healthcare systems. These applications help to reduce the need for physical examination. These wearable technologies could be in the form of armbands [1,2]. Using IoT systems, smart cities can be developed, i.e., an ecosystem with an infrastructure to optimize the usage of various resources in a city. This would help to improve the quality of life of citizens and also monitor important parameters such as traffic. Many sensors such as RFID can be used for smart parking facilities, or to measure the air quality index or the amount of carbon dioxide in the air to monitor pollution [3–5].

IoT is used in autonomous systems and transport vehicles to safely navigate passengers. These vehicles can be equipped with sensors that can gauge the parameters of the engine or track the status of goods that are to be delivered. This is feasible if goods are attached with RFID tags. RFID and NFC provide endless opportunities in the fields of inventory management, supply chain management, etc. IoT is paving the way for automated driving in the future [6]. Social IoT [7], describes a world where humans can form a network in an intelligent manner. Thus, IoT can also be used in applications pertaining to social media. Social IoT helps users to maintain social relationships, track IoT-enabled objects using social media applications, etc. [8]. Figure 13.2 shows some of the popular IoT applications.

In the domain of health care, IoT sensors help in patient-flow monitoring, staffing, and improving the workflow in hospitals. This would help patients to take their doses at the right time. Automatic collection of data using IoT sensors can help in real-time monitoring by measuring blood pressure, heart rate, etc. These IoT-enabled sensors can detect any kind of abnormality and inform the doctor timely. This would further prevent hospitals from overcrowding [9].

The chapter is organized as follows: Section 13.2 explains the related work in the field of health care using IoT. Section 13.3 describes our contribution to the research. Section 13.4 discusses the various machine learning and deep learning algorithms that can be used in IoT frameworks. Section 13.5 describes a case study in which deep learning is used for heartbeat categorization. Section 13.6 concludes the chapter.

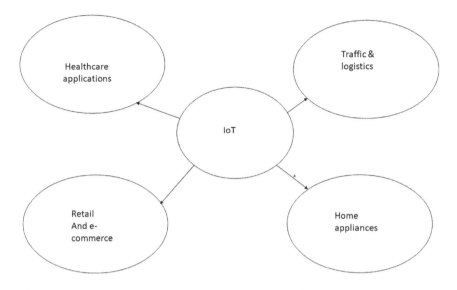

FIGURE 13.2 Popular IoT applications.

13.2 BACKGROUND AND RELATED WORK

Currently, radio frequency (RF)-based technologies help in the transmission of health signal to the gateway. Various devices such as Zigbee, Bluetooth Low Energy (BLE), and IPv6 are considered useful as they are affordable to people. BLE is a noteworthy feature of these because of its low power consumption and robustness to obstacles [10,11]. Some RF-based technologies used in medical devices might be harmful to human health owing to electromagnetic radiations. RF signals can suffer from errors due to inter-channel interference.

Optical wireless communication is being actively worked upon these days. This can help to manage a huge amount of traffic. The optical camera communication has a great potential, in which a camera image sensor is used to receive optical signals from a device. OCC is also a secure technology as it is not affected by the light signal's reflected component [12]. Also, it is less affected by interference as compared to OCC.

PPG-based biometric identification [9] is also being worked upon by researchers. Earlier usage of PPG-based identification focused on data collected from fingers and its Fourier analysis, feature, and classification. Further, the use of deep learning and deep belief networks to monitor PPG signals under ambulant conditions helped to give promising results as compared to traditional techniques. Further, the research challenge would be to update the changing real-time parameters and also ensure the security of the hardware and privacy of the patient[13]. Almotiri et al. [14] used mobile devices to store real-time data collected, on a server, with access to specific clients. Barger et al. [15] developed a smart house facility with sensors so as to track and monitor the movements of a patient. Their primary objective was to check whether the system can outsmart normal behavioral patterns. Chiuchisan et al. [16]

proposed a framework to build smart ICUs and intimate the doctors in case any deviations in his body movements or health are sensed. Dwivedi et al. [17] worked on a framework to secure the information that is transmitted over the IoT systems. The framework is a combination of public key infrastructure and various biometric technologies. Gupta et al. [18] proposed a framework that records ECG and other parameters of a patient using Raspberry Pi. In Ref. [19], the authors used the Intel Galileo development board to collect and retrieve data to and from a database to help doctors and patients. Lopes et al. [20] used IoT in the domain of health care for people who are disabled and would benefit from these technologies.

These monitoring systems implement patches for collection of health signals. These systems would not only gauge parameters but also check for critical conditions in patients. In Ref. [21], the authors made use of wearable technologies to monitor health conditions. While they monitor blood pressure, the authors in Ref. [22] used these to monitor ECG. The system was cost-effective. Researchers used a smartphone to develop a healthcare system for diabetic patients [23]. A remote monitoring system to supervise symptoms of patients developing Alzheimer's was proposed by the authors in Ref. [24], which tracked their patterns and locations. Zigbee was used as a sensor for the same task in Ref. [25]. A wearable device to detect the quality of sleep on the basis of respiration rate was worked upon in Ref. [26].

Lima et al. [27] gave a cost-effective prototype that receives data and displays it in a simple manner. This framework is useful in improving an individual's life. The authors propose an embedded system to measure blood with low coast technology [28]. The framework is hosted by a device, and the results are displayed on a webpage. Healthcare providers can access the Internet and keep a check on the supplied information. Matar et al. [29] proposed a remote body pressure monitoring to determine various body postures. Senthilkumar et al. [30] monitored Sjögren's syndrome by making use of humidity sensors and heart rate sensors. Kitsiou et al. [31] discussed cardiac rehabilitation issues on an m-health platform. Data could be accessed remotely over the cloud[32]. In Ref. [33], the authors developed a cardiovascular disease prediction system that helped people to reduce detected CVD risk factors. Such prediction tools aid doctors in identifying patients who can benefit from such systems offering remote health monitoring [35]. In Ref. [34], the authors presented a pervasive patient health monitoring (PPHM) system integrated with IoT and cloud computing to monitor patients suffering from congestive heart failure.

Patch devices are also used along with wearable technologies to track patients. Thus, patients do not move so that it does not interfere with the device integration. Signals must be transmitted with minimal errors. Low-range power devices must be used to transmit information from one unit to another. Table 13.1 presents the various traits that can be monitored using IoT systems.

Research efforts are being done for automation of wheelchairs that would help disabled people. Various applications are being worked upon for the elderly[43]. A specialized IoT service called Children Health Information (CHI) [44] is used to monitor the emotional, mental, and behavioral symptoms of children in pediatric wards, where an interactive totem may be positioned in a pediatric ward. In Refs. [45,44], the authors described eHealth and IoT policies with regulations in accessing healthcare services. Figure 13.3 shows the flow diagram of an IoT healthcare application.

TABLE 13.1
Role of IoT Sensors in Monitoring Health Traits

Sensor	Purpose	Role of IoT in Recent Applications
ECG sensor	To deliver information about heart rate and rhythm	In various studies [36], optical heart rate sensors [37] give readings of resting heart rate data, or heart rate data when people are working out
Temperature monitoring	To detect body temperature using a sensor unlike the common approach of using a thermometer	An intelligent 3D printed hearable gadget using infrared sensor to monitor temperature based on eardrum [38]
Blood pressure measurement	Wearable cuffless gadget based on photosensors to monitor blood pressure [39]	A framework to monitor blood pressure based on ECG, photoplethysmogram (PPG) from fingertips [40]
BG monitoring	Wearable gadgets	The BG monitoring framework to check BG levels [41] made use of a suitable sensor and designed a front-end interface to display parameters in real time
Asthma monitoring	Heart pulse sensor	Data gathered from the sensor was sent wirelessly to a microcontroller in real time and finally transferred to a remote server [42]

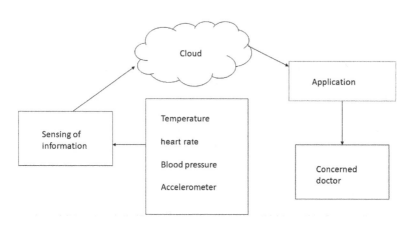

FIGURE 13.3   Flow of an IoT healthcare application.

13.3   CHALLENGES IN HEALTH CARE

The main challenges in the real-time monitoring of patients using IoT are as follows:

- There is no privacy of patient data.
- Timely delivery of accurate response in case of critical conditions is a challenge.

- The IoT-based healthcare applications can be improved using artificial intelligence and neural networks. For example, such an IoT-based application, ambient assisted living (AAL) [46], is being used to assist elderly individuals in their homes.
- Systems may suffer from a lot of bit error rate due to noise and interference from the neighborhood. For intensive care environments, such systems should be secure, demanding low power, within budget, and reliable. Also, the smartphone-based management systems might not be very effective if the patient is far away from the room.
- To analyze large medical data stored in cloud, there are various challenges. If data is misused, it may deteriorate reputation among individuals. Thus, data centers should be capable enough to handle and stream data of a number of subjects, which would be in large volume.
- There might be a risk of cyber-attacks, and thus, implementing security measures is important, keeping in mind the memory limitation and computational power of the system.
- Immediate effort is needed for the standardization of IoT-based healthcare services.

13.4 RESEARCH CONTRIBUTION

1. The usage of and reliance on machine learning algorithms to improve the outcomes of biometric monitoring are dependent on the framework used to support such healthcare platforms.
2. We analyze the methodologies and frameworks being used in healthcare settings, which make use of machine learning to analyze an individual. We look at the various aspects involved in answering fundamental questions with respect to capabilities of improvement in such healthcare outcomes.
3. We explore the use of deep learning methods to improve the performance of such frameworks. The current trend of deep learning is making a vast improvement over existing machine learning algorithms. It can work on datasets containing a variety of data and help to uncover hidden patterns and other useful information, which can help in taking better healthcare decisions.
4. We focus on the application of such a deep learning framework on the ECG Heartbeat Categorization Dataset, which involves analysis and monitoring of various features to identify whether an individual's heartbeat signal is normal or not.

13.5 IOT FRAMEWORKS USING ML ALGORITHMS

Machine learning has simplified and automated tasks that involve the use of big data and cloud computing. Machine learning techniques are being used in a number of fields such as medicine, astronomy, and security.

The authors proposed a framework [46] for monitoring patients with the help of sensor networks. These sensors capture heart rate, temperature and other parameters like blood pressure using Raspberry Pi board. The main stages of this framework are as follows: data collection, storage of data over cloud, and analysis using machine learning algorithms to detect abnormality in a patient's health.

The classification is done using the k-nearest neighbor method. It is a supervised learning technique in which predictions are made by searching through an entire training set of k neighbors. Data points are separated into several classes so as to predict the category of a new instance. The steps are shown in Figure 13.4.

The accuracy achieved using this classifier was 98.02%. Several open-source cloud platforms support data collection using IoT devices like Raspberry Pi board and Raspbian Jessie. This data is uploaded on the cloud and further imported for the application of ML algorithms. For accurate predictions, it is important to train a large amount of data, with data obtained from multiple people. Data should contain samples from both healthy and unhealthy patients.

The authors divide the framework into three parts: one part for monitoring, one for prediction, and one for emergency alert. Data is collected using a sensor and processed using a microcontroller. The analyzed data is stored over cloud and can be retrieved by doctors using Web sites or applications. Encryption can be done to add confidentiality to the system. Using IoT technologies and ML algorithms, the resultant integrated frameworks can help in improving general health. The combination of ML and Semantic Web of Things would result in LDC (less differentiated caregiver). There are various techniques and such ML algorithms available in the market to implement processes with reasonable LDC. These technologies are also described as predictive ML [47].

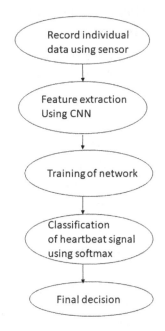

FIGURE 13.4   Steps of KNN classification.

TABLE 13.2
Results Achieved Using ML Algorithms

ML Algorithm	Accuracy in Analyzing Thermal Comfort (in %)
k-nearest neighbors	95.9
Support vector machine	97.9
Naive Bayes	95.3

Implementation of IoT in home healthcare systems, along with ML, makes use of secured hybrid clouds. These devices consist of nanosensors attached to the skin, which help to monitor a patient's important symptoms and perform predictive analytics. Raw data from the patient's database resides in a highly secured cloud platform, and this data can be accessed for previous prescriptions and the use of ML can help to predict threatening life conditions or important drugs.

These days, individuals have a number of technologies to merge IoT and ML, which acts as a practical solution to support day-to-day requirements. Data collected using IoT devices is useful, and efficient data analytics software packages are available along with open-source platforms and cloud technologies.

The authors make use of ML algorithms to perceive thermal conditions of users and analyze heating networks [48]. Thermal comfort perception helps users to interact with the surrounding environment. The results of various ML algorithms to analyze thermal comfort are shown in Table 13.2.

13.6 DEEP LEARNING APPROACHES FOR BIOMETRIC MONITORING

These days, deep learning algorithms are being used in various applications of data analysis, security, health monitoring, e-commerce, etc. These algorithms have an advantage over traditional machine learning algorithms due to the increased number of hidden layers and better capabilities of feature extraction from data in question.

13.6.1 CONVOLUTIONAL NEURAL NETWORKS

Convolutional neural networks (convnets, CNNs) are mainly used for image classification. Originally developed by Geoffery Hinton, they are ideal in detecting objects in different positions in different images. These networks make use of convolution. Given 1D input x and a filter k, this operation is defined as:

$$y[n] = (x*k)[n] \sum_{-\infty}^{\infty} x[m]k[n-m] \qquad (13.1)$$

Convolutional layers are followed by pooling layers and fully connected layers. In convolution, input image matrix is convolved with a filter matrix, which gives a feature map for the following layer. The layers used are as follows: convolutions, pooling layers, and fully connected layer. The dimensions of the filter can be adjusted

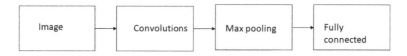

FIGURE 13.5   Architecture of CNN.

to give output maps. This is done by taking the sum of results of element-wise multiplication between some size of input image matrix and filter matrix. Also, multiple filter matrices can be used for convolution, and the results can be concatenated in a feature map. Figure 13.5 gives an example of a convolution.

Next, a pooling layer is used. Max pooling is a widely used type of pooling in which the maximum value is obtained for a given matrix. Thus, the obtained feature map is split into a number of matrices and the maximum value is taken out of each matrix. The final layer or fully connected layer connects the previous units and makes use of some nonlinear activation function (ReLU, tanh, sigmoid, etc.) The softmax or sigmoid function is employed for the final probabilities or scores for classification.

13.6.2   DEEP BELIEF NETWORKS

DBN is a class of deep neural networks with multiple layers of models with directed and undirected edges. There are multiple hidden units in a deep belief network. Layers are connected to each other, but units are not [46]. Figure 13.6 depicts a DBN.

Boltzmann machine is a stochastic recurrent neural network consisting of stochastic binary units and undirected edges between units. Restricted Boltzmann machines constitute a deep belief network, with various layers stacked to each other.

13.6.3   LSTM

LSTM is used to learn sequential data and is better than RNNs, which suffer from the problem of vanishing gradient. Gates are used to decide the information store and the one to forget. The gates used are input gate i_t, output gate o_t, forget gate f_t, and a gate signifying the cell state c_t. The input gate allows a limited amount of information, and the forget gate decides on the information which is not needed further. The cell state gate gives the state of LSTM cell over a period of time, while the output gate focuses on a certain part of information.

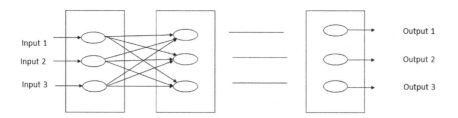

FIGURE 13.6   Architecture of a deep belief network.

$$f_t = \left(W_f \left(h_{t-1}, x_t \right) + b_f \right) \tag{13.2}$$

$$i_t = \sigma \left(W_i \left(h_{t-1}, x_t \right) + b_i \right) \tag{13.3}$$

$$o_t = \left(W_o \left(h_{t-1}, x_t \right) + b_o \right) \tag{13.4}$$

13.7 ECG HEARTBEAT CATEGORIZATION USING DEEP LEARNING

The use of deep networks is implemented on a dataset composed of two constitutions of heartbeat signals taken from the MIT-BIH Arrhythmia Dataset and the PTB Diagnostic ECG Database. The number of samples in both collections is large enough for training a deep neural network. The signals correspond to electrocardiogram (ECG) shapes of heartbeats for the normal case as well as those affected by different arrhythmias and myocardial infarction. Based on more than 100 features, the type of heartbeat is classified into different types. The data has been sampled from 24-hour ECG recordings obtained from different subjects.

The data in both the datasets is summarized in Tables 13.3 and 13.4.

13.7.1 PROPOSED FRAMEWORK

In this case study, classification is done to predict the heartbeat of an individual, i.e., whether it is normal or abnormal. Figure 13.7 shows the process of ECG heartbeat categorization using deep CNN architecture.

TABLE 13.3
Features of the Arrhythmia Dataset

Sample Count	109446
Types	5
Frequency of sampling	125 Hz
Source	PhysioNet's MIT-BIH Arrhythmia Dataset
Classes	['N': 0, 'S': 1, 'V': 2, 'F': 3, 'Q': 4]

TABLE 13.4
Features of the PTB Diagnostic ECG Database

Sample count	14552
Types	2
Frequency of sampling	125 Hz
Source	Data source: PhysioNet's PTB Diagnostic ECG Database

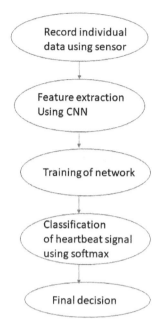

FIGURE 13.7 Flow diagram of the proposed framework.

In the first step, heartbeat is recorded for an individual using a sensor. The signals correspond to electrocardiogram (ECG) shapes of heartbeats for the normal case and the cases affected by different arrhythmias and myocardial infarction. These signals are preprocessed and segmented, with each segment corresponding to a heartbeat. The features of the signal are recorded in a database, along with the type of ECG signal, or categorization of heartbeat which could be normal or affected by types of arrhythmias and myocardial infarction.

Figure 13.8 shows the correlation between features studied to classify the heartbeat, whereas Figure 13.9 shows the recorded heartbeat of an individual selected on a random basis.

13.7.2 TRAINING OF NETWORK FOR FEATURE EXTRACTION

After preprocessing of the dataset, the next step is to train the deep CNN architecture for feature extraction and further classification of ECG signals. Feature extraction is done using a popular pre-trained LeNet model. In the classical LeNet model, the first layer is a convolutional layer with six feature maps or filters having size 5×5 and a stride of one. This is followed by an average pooling layer with filter size 2×2 and a stride of two. The second convolutional layer consists of 16 feature maps having size 5×5. The fourth layer (S4) is again an average pooling layer with filter size 2×2 and a stride of 2. This is followed by two fully connected layers of 120 and 84 feature maps each of size 1×1. The final layer is a softmax layer with the number of classes in the final output, which is 5 in this case study.

FIGURE 13.8   Correlation between features of the MIT-BIH Arrhythmia Dataset.

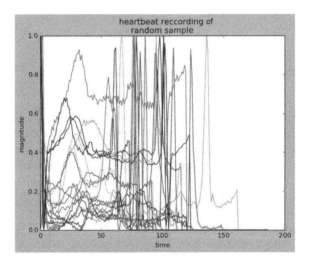

FIGURE 13.9   Heartbeat sample of an individual from the MIT-BIH Arrhythmia Dataset.

In this architecture, we make use of the fully connected layers to extract features from the dataset. The final layer will classify the heartbeat into one of the five classes available. This can be visualized in Table 13.5.

13.7.3   EXPERIMENTAL SETUP

The platform used for categorization of heartbeat is as follows:

- Python 3.3
- TensorFlow (TFLearn)
- Keras

TABLE 13.5
Layers of the LeNet Model

Type of CNN Layer	Number
Fully connected layer FC1	120
Fully connected layer FC2	84
Fully connected layer FC3	5 (corresponding to categories of heartbeat)

13.7.4 PERFORMANCE METRICS

Root Mean Squared Loss: The mean squared loss is used to find the difference between actual values and predicted values.

Accuracy: It is the ratio of the number of correct predictions to the number of inputs.

13.7.4.5 Optimization of Weights

Adam optimization is used to update weights during training as it requires very little hyperparameter tuning and is very fast. It can work on noisy data. It is a faster method of optimizing weights as compared to gradient descent as it makes use of momentum. Thus, it converges faster toward the global minimum. Given velocity V, momentum S, and acceleration dW (where W refers to the weight and B refers to the bias), the algorithm works as shown in Table 13.6.

13.7.4.6 Results

The data has been split into training, validation, and testing part at a ratio of 60:20:20. Once training and validation are done, any unseen record of ECG signal will be categorized into one of the defined classes. The root mean squared loss during training and validation for 100 epochs is 0.132, and the accuracy achieved on the test data is 98.22%. It is observed that the use of deep CNN architecture helps to achieve a superior performance in the categorization of ECG signals as compared to traditional ML algorithms used in IoT frameworks in the existing literature.

Figure 13.10 shows the comparison between the proposed framework and the framework explained in Figure 13.4.

TABLE 13.6
Adam Optimizer algorithm

$$VdW_{corr} = VdW/(1 - \beta_1 T)dW$$
$$VdB_{corr} = VdB/(1 - \beta_1 T)dB$$
$$SdW_{corr} = SdW/(1 - \beta_1 T)dW$$
$$SdB_{corr} = SdB/(1 - \beta_1 T)dB$$
$$W = W - \alpha VdW/\sqrt{SdW} + \varepsilon$$
$$B = B - \alpha SdW/\sqrt{SdW} + \varepsilon$$

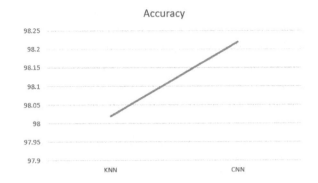

FIGURE 13.10   Comparison between ECG categorization using KNN and CNN.

13.7   CONCLUSION AND FUTURE RESEARCH TRENDS

In this chapter, we discuss the role of health monitoring of human traits such as ECG and temperature using IoT sensors. We review the various aspects researched by the authors using different algorithms. Machine learning is being used to automate such systems, but deep learning is making its mark in this technology. We make use of convolutional neural networks to categorize heartbeat signals of a person. It is a useful and efficient approach, which can be integrated with IoT sensors in the domain of health care. Although time taken is more for training and validation of images, the performance achieved is outstanding and will generalize well on unseen data. Thus, biometric monitoring in healthcare applications can help doctors to indulge in providing services remotely. This can help to reduce overcrowding in hospitals and help individuals to track their health.

Healthcare monitoring is the future of IoT. It is believed that 87% of the healthcare organizations would be making use of IoT applications in the year 2020. Patients are relying on wearable products that transmit data to physicians to monitor vital signs. Remote monitoring is also becoming a fast emerging trend. In the future, IoT applications will offer intelligent services to allow real-time processing; i.e., the devices would perform necessary actions on the patient and send data back to the doctor. Thus, more intelligent IoT devices that can perform autonomous actions would dominate the future of healthcare applications.

REFERENCES

1. Jiang, L., Liu, D.-Y., Yang, B. (2004). Smart home research. In: *Proceedings of 2004 International Conference on Machine Learning and Cybernetics* (IEEE Cat. No. 04EX826), *China*,Vol. 2, pp. 659–663.
2. Bui, N., Zorzi, M. (2011). Health care applications. A solution based on the Internet of things. In: *Proceedings of the 4th International Symposium on Applied Sciences in Biomedical and Communication Technologies, New York, USA*, pp. 131:1–131:5.
3. Biometric Security and Internet of things (IoT). Available from: https://www.researchgate.net/publication/328516514_Biometric_Security_and_Internet_of_Things_IoT [accessed Nov 11 2019].

4. Miorandi, D., Sicari, S., De Pellegrini, F., Chlamtac, I. (2012). Internet of things: Vision, applications and research challenges. *Ad Hoc Netw.* 10(7), 1497–1516.

5. Zanella, A., Bui, N., Castellani, A., Vangelista, L., Zorzi, M. (2014). Internet of things for smart cities. *IEEE Internet Things J.* 1(1), 22–32.

6. Qin, E., Long, Y., Zhang, C., Huang, L. (2013). Cloud computing and the Internet of things: Technology innovation in automobile service. In: *International Conference on Human Interface and the Management of Information, Las Vegas,USA.* Springer, pp. 173–180.

7. Atzori, L., Iera, A., Morabito, G. (2011). SIoT: Giving a social structure to the Internet of things. *IEEE Commun. Lett.* 15(11), 1193–1195.

8. Kleinberg, J. (2008). The convergence of social and technological networks. *Commun. ACM* 51(11), 66.

9. Everson, L., Biswas, D., Panwar, M., Rodopoulos, D., Acharyya, A., Kim, C. H., Van Helleputte, N. (2018). BiometricNet: Deep learning based biometric identification using wrist-worn PPG. In: *2018 IEEE International Symposium on Circuits and Systems (ISCAS),* Florence, Italy, pp. 1–5. doi: 10.1109/iscas.2018.8350983.

10. Dang, L., Piran, M., Han, D., Min, K., Moon, H. (2019). A survey on Internet of things and cloud computing for healthcare. *Electronics* 8(7), 768.

11. Siekkinen, M., Hiienkari, M., Nurminen, J. K., Nieminen, J. (2012). How low energy is bluetooth low energy? Comparative measurements with ZigBee/802.15.4. In: *Proceedings of the IEEE Wireless Communications and Networking Conference Workshops, Paris, France,* pp. 232–237.

12. Hasan, M., Chowdhury, M., Shahjalal, M., Nguyen, V., Jang, Y. (2018). Performance analysis and improvement of optical camera communication. *Appl. Sci.* 8(12), 2527.

13. Ghosh, A. M., Halder, D., Hossain, S. K. A. (2016). Remote health monitoring system through IoT. In: *5th International Conference on Informatics, Electronics and Vision (ICIEV),* Dhaka, 2016, pp. 921–926.

14. Almotiri, S. H., Khan, M. A., Alghamdi, M. A. (2016). Mobile health (m-health) system in the context of IoT. In: *IEEE, International Conference on Future Internet of things and Cloud, Vienna,*pp. 39–42.

15. Barger, T., Brown, D., Alwan, M. (2005). Health-status monitoring through analysis of behavioral patterns. *IEEE Trans. Syst. Man Cybern. Part A: Syst. Hum.* 35(1), 22–27.

16. Chiuchisan, I., Costin, H., Geman, O. (2014). Adopting the Internet of things technologies in health care systems. In: *2014 International Conference and Exposition on Electrical and Power Engineering (EPE), Iasi,* pp. 532–535.

17. Dwivedi, A., Bali, R., Belsis, M., Naguib, R., Every, P., Nassar, N. (2003). Towards a practical healthcare information security model for healthcare institutions. In: *4th International IEEE EMBS Special Topic Conference on Information Technology Applications in Biomedicine, Birmingham, UK,* pp. 114–117.

18. Gupta, M., Patchava, V., Menezes, V. (2015). Healthcare based on IoT using raspberry pi. In: *2015 International Conference on Green Computing and Internet of things (ICGCIoT),Noida,*India pp. 796–799.

19. Gupta, P., Agrawal, D., Chhabra, J., Dhir, P. (2016). IoT based smart healthcare kit. In: *2016 International Conference on Computational Techniques, New Delhi, India,* pp. 237–242.

20. Lopes, N. V., Pinto, F., Furtado, P., Silva, J. (2014). IoT architecture proposal for disabled people. In: *10th International Conference on Wireless and Mobile Computing, Networking and Communications (WiMob), Larnaca, Cyprus,* pp. 152–158.

21. Lin, H., Xu, W., Guan, N., Ji, D., Wei, Y., Yi, W. (2015). Noninvasive and continuous blood pressure monitoring using wearable body sensor networks. *IEEE Intell. Syst.* 30(6), 38–48.

22. Kulkarni, A., Suchetha, M., Kumaravel, N. (2019). IoT based low power wearable ECG monitoring system. *Curr. Sig. Transd. Ther.* 14(1), 68–74.

23. Bui, N., Bressan, N., Zorzi, M. (2012). Interconnection of body area networks to a communications infrastructure. An architectural study. In: *Proceedings of the European Wireless Conference, Poznan, Poland*, pp. 18–20.

24. Cheng, H., Zhuang, W. (2010). Bluetooth-enabled in-home patient monitoring system: Early detection of Alzheimer's disease. *IEEE Wirel. Commun.* 17(1), 74–79.

25. Taub, D. M., Lupton, E., Hinman, R., Leeb, S., Zeisel, J., Blackler, S. (2011). The escort system: A safety monitor for people living with Alzheimer's disease. *IEEE Pervas. Comput.* 10, 68–77. doi: 10.1109/MPRV.2010.44.

26. Milici, S., Lazaro, A., Villarino, R., Girbau, D., Magnarosa, M. (2018). Wireless wearable magnetometer-based sensor for sleep quality monitoring. *IEEE Sens. J.* 18(5), 2145–2152.

27. Rajasekaran, M., Radhakrishnan, S., Subbaraj, P. (2008). Remote monitoring of post-operative patients using wireless sensor networks. *Int. J. Healthcare Technol. Manag.* 9(3), 247.

28. Dantas, E. C. S. (2016). "c" M.S. thesis, Programa de PósGraduação em Engenharia Elétrica, Inst. Federal de Educação, Ciência e Tecnologia da Paraíba, João Pessoa, Brazil.

29. Matar, G., Lina, J., Carrier, J., Riley, A., Kaddoum, G. (2016). Internet of things in sleep monitoring: An application for posture recognition using supervised learning. In: *Proceedings of the 18th International Conference on e-Health Network and Application Services, Munich*, pp. 1–6.

30. Senthilkumar, R., Ponmagal, R., Sujatha, K. (2019). Efficient low power intelligent health care monitoring system using IoT. *IJITEE* 8(12), 1707–1711.

31. Msayib, Y., Gaydecki, P., Callaghan, M., Dale, N. and Ismail, S. (2017). An intelligent remote monitoring system for total knee arthroplasty patients. *J. Med. Syst.* 41(6), 90.

32. Kitsiou, S. et al. (2017). Development of an innovative mHealth platform for remote physical activity monitoring and health coaching of cardiac rehabilitation patients. In: *Proceedings of the IEEE International Conference on Biomedical and Health Informatics*, pp. 133–136.

33. Alshurafa, N., Sideris, C., Pourhomayoun, M., Kalantarian, H., Sarrafzadeh, M., Eastwood, J.-A. (2017). Remote health monitoring outcome success prediction using baseline and first month intervention data. *IEEE J. Biomed. Health Inform.* 21(2), 507–514.

34. Abawajy, J. H., Hassan, M. M. (2017). Federated Internet of things and cloud computing pervasive patient health monitoring system. *IEEE Commun. Mag.*, 55(1), 48–53.

35. Rodrigues, J. J. P. C., De Rezende Segundo, D. B., Junqueira, H. A., Sabino, M. H., Prince, R. M., Al-Muhtadi, J., De Albuquerque, V. H. C. (2018). Enabling technologies for the Internet of Health Things. *IEEE Access* 6, 13129–13141.

36. Yang, Z., Zhou, Q., Lei, L., Zheng, K., Xiang, W. (2016). An IoT-cloud based wearable ECG monitoring system for smart healthcare. *J. Med. Syst.* 40, 286.

37. Bathilde, J. B., Then, Y. L., Chameera, R., Tay, F. S., Zaidel, D. N. A. (2018). Continuous heart rate monitoring system as an IoT edge device. In: *Proceedings of the Sensors Applications Symposium (SAS), Seoul, Korea, 12–14 March 2018*, pp. 1–6.

38. Ota, H., Chao, M., Gao, Y., Wu, E., Tai, L. C., Chen, K., Matsuoka, Y., Iwai, K., Fahad, H. M., Gao, W., et al. (2017). 3d printed "earable" smart devices for real-time detection of core body temperature. *ACS Sens.* 2, 990–997.

39. Xin, Q., Wu, J. (2017). A novel wearable device for continuous, non-invasion blood pressure measurement. *Comput. Biol. Chem.* 69, 134–137.

40. Dinh, A., Luu, L., Cao, T. (2017). Blood pressure measurement using finger ECG and photoplethysmogram for IoT. In: *Proceedings of the International Conference on the Development of Biomedical Engineering in Vietnam*, Ho Chi Minh, Vietnam, 27–29 June 2017, pp. 83–89.

41. Gia, T. N., Ali, M., Dhaou, I. B., Rahmani, A. M., Westerlund, T., Liljeberg, P., Tenhunen, H. (2017). IoT-based continuous glucose monitoring system: A feasibility study. *Procedia Comput. Sci.* 109, 327–334.

42. AL-Jaf, T. G., Al-Hemiary, E. H. Internet of things based cloud smart monitoring for asthma patient. In: *Proceedings of the 1st International Conference on Information Technology (ICoIT '17)*, Erbil, Iraq, 10 April 2017, p. 380.

43. Capodieci, A., Budner, P., Eirich, J., Gloor, P., Mainetti, L. (2018). Dynamically adapting the environment for elderly people through smartwatch-based mood detection. In: Riopelle K., & Gloor P. (eds), *Collaborative Innovation Networks. Studies on Entrepreneurship, Structural Change and Industrial Dynamics* Springer: Berlin/Heidelberg, Germany, pp. 65–73.

44. Ali, B., Awad, A. (2018). Cyber and physical security vulnerability assessment for IoT-based smart homes. *Sensors* 18(3), 817.

45. Dohr, A., Modre-Opsrian, R., Drobics, M., Hayn, D., Schreier, G. (2020). The Internet of things for ambient assisted living. In: *2010 Seventh International Conference on Information Technology: New Generations, Las Vegas, NV*, pp. 804–809.

46. Rakshanasri, S. L., Naren, J., Vithya, G., Akhil, S., Dinesh Kumar, K., Sai Krishna Mohan Gupta, S. (2020). A framework on health smart home using IoT and machine learning for disabled people. *Int. J. Psychosoc. Rehabil.* 24(2), 1–9.

47. Machine learning (ML) and IoT can work together to improve lives. https://opensourceforu.com/2019/10/machine-learning-ml-and-iot-can-work-together-to-improve-lives/.

48. Salamone, F., Belussi, L., Currò, C., Danza, L., Ghellere, M., Guazzi, G., Lenzi, B., Megale, V., Meroni, I. (2018). Application of IoT and machine learning techniques for the assessment of thermal comfort perception. *Energy Procedia* 148, 798–805.

14 Predictive Analysis of Type 2 Diabetes Using Hybrid ML Model and IoT

Abhishek Sharma and Nikhil Sharma
HMR Institute of Technology & Management

Ila Kaushik
Krishna Institute of Engineering & Technology

Santosh Kumar
ITER-SOA

Naghma Khatoon
Usha Martin University

CONTENTS

14.1 INTRODUCTION

Diabetes is an incurable disease that is devastating many lives every day. It is a silent killer. In 2016, diabetes killed 1.6 million people directly. Diabetes is a lifestyle disease that happens due to a lack of awareness about eating habits. It is among the most common diseases in both developed and developing nations. As of 2017, about 425 million people had diabetes in the entire world. Common signs of diabetes are frequent urination, increased thirst, fatigue, unintended weight loss, and also increased hunger. Diabetes increases medical bills. It is a mother of various other harmful diseases. A person with diabetes is at higher risk of developing hypertension, dyslipidemia, cataract, neuropathy, and other harmful diseases. Diabetes doubles the risk of early death. If left untreated, it can cause many complications. Therefore, early detection of diabetes can help patients in controlling their sugar levels.

The American Diabetes Association [1] has classified diabetes into three main categories. The first category is "type 1" diabetes, also recognized as "juvenile diabetes" or "insulin-dependent diabetes mellitus", which happens when a person's body is entirely unable to produce insulin. It requires a person to infuse insulin into his body and is mostly seen in juveniles, but can happen to people of any age-group. The second is "type II" diabetes, also known as "non-insulin-dependent diabetes" or "adult-onset diabetes", which happens when a person's body develops resistance against insulin or has some insulin deficiency. It happens to people who are above the age of 40 and accounts for 90%–95% of all patients with diabetes. The third, "gestational diabetes", is the form that happens to women during gestation because of insulin intolerance. Women who have diabetes during the gestation period are at very high risk of developing "type 2" diabetes later in life.

In the past few years, significant advancements have occurred in the field of sensors, the Internet, and IoT. Due to advanced wirelessly connected sensors, data can now be easily recorded and transferred. IoT has redefined the whole healthcare industry. IoT devices have made hospitals to keep an eye on the health of patients better and deliver better healthcare solutions. Earlier, hospitals used to keep all the records on paper, but now, the healthcare industry is capable of generating treatment and diagnosis data. This data can be beneficial in diagnosing various harmful diseases. In the diagnosis of diabetes, various IoT-enabled sensors can be used to collect the readings of insulin level, glucose level, blood pressure, and other parameters. With the help of various IoT-enabled devices, this data can be stored digitally. This collected data can help doctors to provide better treatment to the patients. This data can act as a key to making data-driven decisions.

Today, machine learning has its applications in various fields, whether it is physics or business. But one field in which machine learning has contributed significantly is medical science. In the field of medical science, machine learning algorithms have helped both doctors and patients. Machine learning is playing a significant role in helping doctors to find the cure of many diseases that were previously thought to be incurable. In the diagnosis of diabetes, the various machine learning algorithms can reduce human labor significantly and help doctors to differentiate between people who have diabetes and healthy people. Diagnosis of diabetes is a very complicated task. With the help of machine learning algorithms, we can significantly minimize the human labor required for this task. Here, the data generated through IoT-enabled devices can be of great help. After proper processing, these symptoms can be employed as features for various modern machine learning algorithms. With the help of machine learning, doctors can quickly identify whether a person has diabetes or not.

In this paper, we are using the Pima Indians Diabetes dataset to simulate the data from multiple IoT-enabled medical sensors and patient reports. In this study, we are proposing a stacking classifier to label a person with diabetes. This proposed model leverages the classification power of various machine learning algorithms. This helps in the precise classification of patients with diabetes. The proposed stacking classifier uses multilayer perceptron (MLP), random forest classifier (RFC), k-nearest neighbors (KNN), and eXtreme Gradient Boosting (XGBoost) as base classifiers and logistic regression as a meta-classifier. This study shows that the stacking classifier outperforms all the base classifiers used.

The rest of the paper is organized as follows: Section 14.2 discusses various research works related to the prediction of "type II" diabetes. Section 14.3 deals with various materials and methods used in this research work. Section 14.4 explains the design and implementation of the proposed model. Section 14.5 analyzes the results of the research, and Section 14.6 provides conclusions and future scope.

14.2 LITERATURE SURVEY

In recent years, notable research has happened about the identification, analysis, and examination of diabetes using machine learning algorithms. This section provides a synopsis of previously done research on machine learning algorithms for the diagnosis of diabetes.

Maniruzzaman et al. [2] worked on the Pima Indians Diabetes dataset and compared four different classifiers, QDA, LDA, GPC, and NB, respectively. Among them, GPC outperformed all other classifiers in accuracy, with a value of 81.97%. Chalabi and Panbechi [3] applied various techniques of data mining for the diagnosis of "type II" diabetes. They used the Pima Indians Diabetes dataset. Among those techniques, the NB algorithm surpassed RBF and J48 with an accuracy of 76.95%. Sabariah et al. [4] used an ensemble of (RF) random forest and regression tree method (CART) classification algorithms for early diagnosis and detection of the second type of diabetes. Gultenb and Ozcift [5] proposed a rotation forest algorithm. They evaluated the effectiveness of the algorithm using the Pima Indians Diabetes dataset.

In Ref. [6], a homogeneity-based algorithm was proposed by Triantaphyllou and Pham to decrease the overgeneralization and overfitting on the Pima Indians Diabetes dataset. In Ref. [7], Diao et al. proposed a Laplacian support vector machine, which is a semi-supervised learning technique to predict diabetes. Li et al. [8] constructed a fuzzy decision tree using BP network. In Ref. [9], Patil et al. proposed a hybrid model for the prediction of "type II" diabetes. In Ref. [10], Polat and Günes proposed another hybrid model that used principal component analysis.

With the help of machine learning methods such as random forest, NB, MLP, modified J48, and random tree, Devi and Shyla [11] analyzed the early detection of diabetes. In this analysis, the modified J48 performed better than the other methods with higher accuracy. Dewangan and Agrawal [12] combined various classification techniques such as random forest, MLP, and C4.5 to create a hybrid model for better performance. Kayaer and Yıldırım [13] used general regression neural networks (GRNN) for early detection of diabetes. They achieved an accuracy of 80.21% on the Pima Indians Diabetes dataset. In their paper, they created GRNN with four layers: an input layer with dimensions of eight for eight features of the dataset, two hidden layers consisting of 32 and 16 neurons, and lastly, an output layer with a single neuron for binary classification.

In Ref. [14], Soltani and Jafarian used PNN (probability neural network) for the detection of diabetes. They used probability neural network consisting of an input layer with dimensions of eight, a hidden layer, and an output layer consisting of two neurons to classify whether a patient has diabetes or not. AlJarullah [15] made use of decision trees and achieved an accuracy of 78.17%. Bozkurt, Yurtay et al. [16] compared various methods for early detection of diabetes using artificial immune system (AIS) and artificial neural networks (ANNs). They achieved an accuracy of 76.0%. In Ref. [17], Chitra and Kumari used the SVM algorithm and achieved an accuracy of 78.00%.

14.3 MATERIAL AND METHODS

The dataset used here is of Pima Indian women, which is by the National Institute of Diabetes and Digestive and Kidney Diseases (NIDDK). This dataset is taken from the UCI repository and contains 768 samples, of which 500 samples are positive – "1" – and 268 samples are negative – "0". Table 14.1 contains eight features, which are as follows.

14.3.1 K-NEAREST NEIGHBOR (KNN) CLASSIFIER

KNN is a nonparametric, unsupervised machine learning algorithm. Nonparametric algorithm means there are no assumptions made on underlying data during training; the model does not summarize the training data. Nonparametric algorithms

TABLE 14.1
Eight Features of the Dataset

Features	Description
Pregnancies	It is the number of pregnancies
SkinThickness	It is the skinfold thickness of the triceps in mm
BloodPressure	It is the diastolic blood pressure level of the subject measured in mm Hg
Glucose	It is the plasma glucose concentration in the 2-hour oral glucose tolerance test
Insulin	It is the insulin level of subject in mu U/ml after 2 hours of fasting
DiabetesPedigreeFunction	It is the diabetes pedigree function
BMI	It is the body mass index of the subject measured in kg/m^2
Age	It is the age of the subject in years

such as KNN [18] are beneficial because, generally, the practical data does not adhere to theoretical assumptions made (e.g., linearly separable, Gaussian mixtures). KNN tries to minimize the intra-cluster distance and maximize the inter-cluster distance [19].

KNN is a lazy machine learning algorithm. It means that there is no generalization done using training data. Therefore, the KNN has a very rapid training phase. Lack of generalization during model training means the KNN model requires all the data used for training the model in the testing phase. There is a trade-off between the quick training phase and the costly testing phase (in terms of both memory and time) [20].

KNN takes data as input and plots it in feature space. Since the points lie in feature space, the Euclidian distance is used to measure the distance between data points and centroids. The value of K is a hyperparameter whose value is manipulated to get the optimum classification of data [21]. For $k = 1$, the algorithm is simply known as nearest neighbors. For K values ≥ 1, it is simply known as k-nearest neighbors. The optimum value of K is determined using the elbow test; when the value of K increases, the distance between the centroid and cluster centroids is reduced [22].

14.3.2 XGBoost (eXtreme Gradient Boosting)

XGBoost is an ensemble machine learning algorithm that merges several machine learning techniques in one predictive model to achieve higher performance. It is a decision tree-based machine learning algorithm that uses a gradient boosting framework. This algorithm has a high bias and low variance [23]. This algorithm is highly scalable and has a faster learning process through parallel and distributed computing. It also offers efficient memory usage. In this algorithm, each base learner learns from its predecessors and aims to reduce the errors of the past tree. These base learners are weak learners in which bias is high and the predictive power is just a little better than random guessing. This high bias is reduced by using sequential decision trees. Each of these weak learners provides some crucial information for making predictions, making the boosting method to produce a strong learner by productively combining all the weak learners [24]. The process in XGBoost is shown in Figure 14.1.

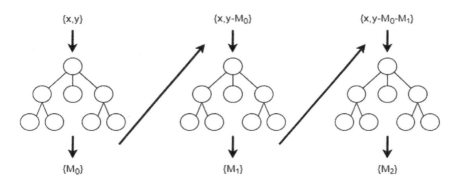

FIGURE 14.1   Sequence of decision trees.

14.3.3 MULTILAYER PERCEPTRON (MLP)

The perceptron is among the most straightforward and simplest artificial neural network architectures. A perceptron is composed of a single layer of threshold logic units (TLUs) [25], with each TLU connected to all the inputs. When all the neurons present in one layer are connected to every neuron present in the previous layer of neurons, the resulting layer is known as the dense layer. But the perceptron has some limitations, which can be eliminated by stacking multiple perceptrons. This stack of multiple perceptrons is known as multilayer perceptron (MLP) [26].

A multilayer perceptron is composed of one input layer, one or more layers of TLUs known as hidden layers, and one final layer of TLU known as the output layer as shown in Figure 14.2 [27]. With the help of the backpropagation algorithm, it can find out how to tweak each connection weight and bias term to reduce the error. Once it is done, MLP just performs gradient descent, and this whole process is repeated until the ANN converges to the solution. The process followed in the training of an MLP is as follows:

*   It takes one mini-batch of instances at a time and passes through the whole training data multiple times. Each pass is known as epoch [28].
*   Each mini-batch is passed to the input layer, which then passes it to the first hidden layer. The algorithm then computes the output of all the neurons present in the first hidden layer. These outputs are then passed to the next hidden layer, and this process goes on until we reach the output layer. This step is known as the forward pass [29].
*   Next, using the loss function, the algorithm calculates the error. Then, it goes through each layer in reverse order to measure the contributions made by each connection to the error. This step is known as the reverse pass [30].
*   Lastly, the algorithm modifies the connection weights to reduce the error. This step is known as the gradient descent step.

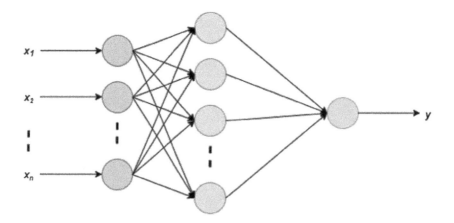

FIGURE 14.2   A multilayer perceptron used for binary classification.

14.3.4 RANDOM FOREST CLASSIFIER (RFC)

A random forest classifier is a decision tree-based ensemble machine learning algorithm. It is a bootstrap aggregation algorithm. In bootstrap aggregation, also known as bagging, we create many random subsamples of the dataset with replacement and pass this data into randomly selected uncorrelated decision trees, also known as base learners [31]. These base learners work in parallel and produce their output. This step is known as bootstrap. The final output of the algorithm is the output determined by max voting on the outputs of all the individual base learners. This step is known as aggregation [32]. Figure 14.3 shows the process of the random forest classifier.

By default, decision trees have high variance and low bias. The high variance of the decision tree is reduced by row sampling and feature sampling of data before feeding it to decision trees and combining multiple decision trees in parallel. The max voting helps algorithm in generalization. The algorithm is known as random forest classifier because a forest of the decision tree is created randomly and used for classification [33].

14.3.5 LOGISTIC REGRESSION (LR)

Logistic regression is a machine learning algorithm, named after the logistic function used at its core [34]. The logistic function is also referred to as the sigmoid function. Logistic function is a curve with "S" shape, similar to tanh(x), but exists only between 0 and 1, whereas tanh(x) exists between 1 and −1. Logistic function can take any real number and map it into a value between 0 and 1, but never exactly 0 or 1 [13].

$$f(x) = \frac{1}{\left(1 + e^{-x}\right)} \tag{14.1}$$

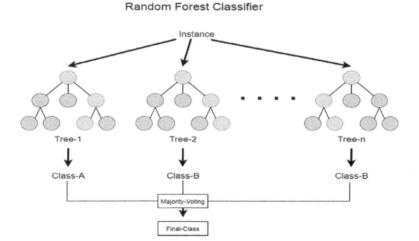

FIGURE 14.3   The process followed in the random forest classifier.

where e is the base of natural algorithm and $f(x)$ is the sigmoid or logistic function.
    An error in logistic regression can be given as:

$$J(\theta) = \frac{1}{m} \sum_{i=1}^{n} \text{Cost}(\hat{y}, y)$$ (14.2)

We can replace the cost function with:

$$\text{Cost}(\hat{y}, y) = \begin{cases} -\log(\hat{y}), \wedge \text{ if } y = 1 \\ -\log(1 - \hat{y}), \wedge \text{ if } y = 0 \end{cases}$$ (14.3)

Now, the resulting equation for error is given by:

$$J(\theta) = \frac{-1}{m} \left\{ \sum_{i=0}^{m} y^i \log(\hat{y}^i) + (1 - y^i) \log(1 - \hat{y}^i) \right\}$$ (14.4)

where $\hat{y}^i$ is the predicted output for the ith row of the dataset, y^i is the actual output for the ith row of the dataset, $J(\theta)$ is the cost function, and m is the number of rows present in the dataset.
    This cost function is reduced with the help of gradient descent. Gradient descent is a technique that uses the derivative of a cost function to change the parameter values, in order to minimize the cost.

14.4   DESIGN AND IMPLEMENTATION

As we have mentioned earlier, our goal is to develop a model that can precisely predict diabetes. For the implementation of this project, we have some machine learning and deep learning libraries such as Keras, TensorFlow, NumPy, Pandas, and Matplotlib. This section describes the approach followed by this paper.

14.1.1   DATA PREPROCESSING

- **Outlier Analysis**: Outliers are the values that deviate from the rest of the values with a large difference. Outliers are the results of error in the data collection process. For the analysis of the outliers, we used boxplots. Boxplots are great for descriptive statistical analysis of the data. The upper and lower extremes of the boxplot are equivalent to 1.5 times the interquartile range. Anything very far from these extremes can be considered as outliers. These outliers are noise in our data, which can reduce the overall accuracy of our model. From Figure 14.4, it can be seen that features such as "Insulin" and "SkinThickness" have outliers, which can be removed from the dataset to reduce the noise.
    After removing the outliers, we have 764 rows left in our dataset.
- **Handling Missing Values**: The count of missing values in the dataset for each attribute is as follows (Table 14.2):

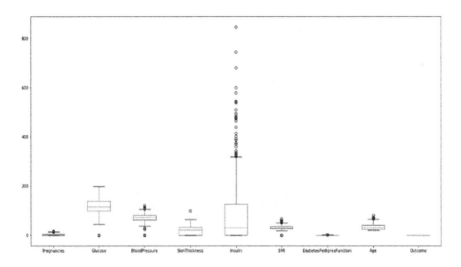

FIGURE 14.4   Boxplot of different features in the dataset.

TABLE 14.2
Count of Missing Values for Each Attribute

Features	Number of Missing Values
Pregnancies	110
SkinThickness	227
BloodPressure	35
Glucose	5
BMI	11
Insulin	374
DiabetesPedigreeFunction	0
Age	0

The number of pregnancies can be zero. But other features with missing values negatively affect our model's performance. Therefore, we replaced missing values in each attribute with the respective attribute's mean.

- **Handling Imbalanced Dataset**: The dataset is imbalanced, which means that it has a higher number of data points belonging to one class. Figure 14.5 shows that our dataset has a higher number of data points that belong to Class 0.

We have 498 cases of "Class 0" and 266 cases of "Class 1". Training on this dataset will make our model better in predicting the cases in which a patient does not have diabetes, but we want our model to be accurate in predicting cases in which a patient has diabetes. We used SMOTE (synthetic minority oversampling technique) to balance our data-set. SMOTE is a statistical technique used for increasing the number of

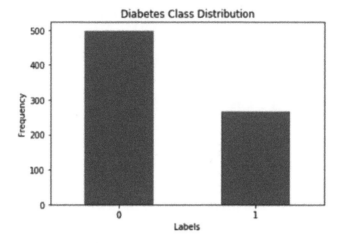

FIGURE 14.5   Class distribution of the imbalanced dataset.

TABLE 14.3
Number of Data Points of Each Class of
Output after Oversampling the Data

Class	Number of Instances
1	474
0	474

---

instances in the dataset in a uniform way. This technique generates new instances from the existing minority instances. Table 14.3 shows the number of data points from each class after oversampling the data.

- **Feature Scaling:** Feature scaling is the process of bringing down the data in features to a standard scale so that the model can be easily trained on it. Feature scaling significantly boosts the performance of the model. Our dataset has some attributes with values ranging from 0 to 100 and others with values from 100 to 1000. We need to perform data normalization to bring our data to one scale. Data normalization is a process in which features are rescaled to the range of 0–1. We used simple feature scaling to normalize the dataset.

In this process, we divided the whole dataset by the maximum value of any feature present in the entire dataset. After performing the above-mentioned preprocessing steps, we split the dataset into three parts: training data (60%), validation data (30%), and testing data (10%).

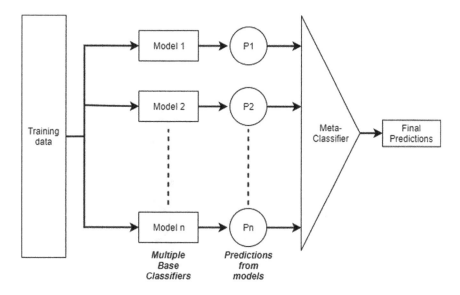

FIGURE 14.6   Schematic of the stacking classifier framework.

14.4.2   PROPOSED MODEL

The proposed model uses the stacking method. Stacking is an ensemble learning technique in which predictions are taken from multiple classifiers, and these predictions are used to train a meta-classifier. This approach gives a better predictive performance compared to a single model. The base-level models, also known as base learners, are trained using the training set, and then, the model at the final level, also known as the meta-model or meta-classifier as in this case, is trained on the outputs of the base-level models. The features present in outputs from the base models are also known as meta-features. The meta-classifier can be trained on either predicted class output labels or probabilities of an output label from the ensemble. Figure 14.6 shows the schematic of the stacking classifier framework.

In this paper, we have used multilayer perceptron, k-nearest neighbors, XGBoost, and random forest classifier at the base level, and logistic regression as a meta-classifier. Each classifier at the base level is trained on the whole training dataset. Then, validation data is used to measure the performance of these classifiers on various metrics. The predictions are taken from these classifiers for both validation and testing data. These predictions include the class labels. Predictions from all the classifiers are combined to form two datasets, one to train the meta-classifier and another to test the performance of the model. Figure 14.7 shows the architecture of the stacking classifier used for the identification of patients with diabetes.

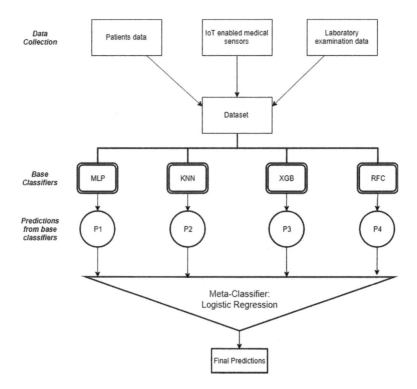

FIGURE 14.7 Architecture of the stacking classifier used for the identification of patients with diabetes.

14.4.3 OUR PROPOSED ALGORITHM

Our Proposed Algorithm:

Input: Base classifiers **B** = multilayer perceptron, k-nearest neighbors, XGBoost, and random forest classifier

Meta-classifier **M** = logistic regression

Dataset **DT** = Pima Indians Diabetes dataset

Dataset preprocessing:
1. Remove outliers from **DT**.
2. Perform replacement of missing values with mean strategy to **DT**.
3. Use SMOTE for oversampling and handling imbalanced dataset **DT**.
4. Use simple feature scaling on **DT** to scale features to common range.

Split dataset DT into following parts:
1. Training set **TR**.
2. Validation set **VA**.
3. Testing set **TE**.

Training base classifiers:
⇒ for each row in **TR** do:
  • for each base classifier in **B** do:
    – Train base classifier.
  • end for
⇒ end for

Generating training and testing datasets for meta-classifier:
⇒ for each row in **VA** do:
  • for each base classifier in **B** do:
    – Predict output labels.
  • end for
⇒ end for
⇒ for each row in **TE** do:
  • for each base classifier in **B** do:
    – Predict output labels.
  • end for
⇒ end for
⇒ Combine predictions from all base classifiers and generate the training dataset **TRM** and testing dataset **VAM** for meta-classifier **M**.

Training meta-classifier M:
⇒ for each row in **TRM** do:
  • Train meta-classifier **M**.
⇒ end for

Testing meta-classifier M:
⇒ for each row in **VAM** do:
  • Test meta-classifier **M**.
⇒ end for

Output: precision, accuracy, recall, and F1-score of the model.

14.4.4 MODEL PERFORMANCE METRICS

Model performance metrics are a set of measurements used for the evaluation of a model's performance. This section elaborates the performance metrics used in this paper to assess the stacking classifier technique.

• **Confusion Matrix**: In case of binary classification problems, confusion matrix is a 2×2 matrix. It is a performance measurement for machine learning classification problem. It consists of four different combinations of actual and predicted values. These four combinations are important properties of any model and act as input to various performance measures. Figure 14.8 shows the confusion matrix for the binary classification problem.
    The four combinations of predicted and actual values are as follows:
    • **True Positive (TP)**: It is the number of instances predicted accurately as a person with diabetes.

Actual

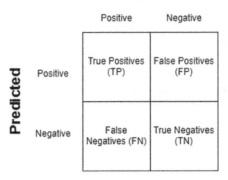

FIGURE 14.8 Confusion matrix for the binary classification.

- **False Negative (FN)**: It is the number of instances predicted incorrectly as a person without diabetes.
- **False Positive (FP)**: It is the number of instances predicted incorrectly as a person with diabetes.
- **True Negative (TN)**: It is the number of instances predicted accurately as a person without diabetes.
- **Performance Evaluation Methods**: The performance evaluation measures used for assessing the performance of the stacking classifier technique are as follows:
  - **Precision**: It is calculated as the ratio of number of accurate positive prophecies to the total number of positive prophecies.

$$\text{Precision} = \frac{\text{True Positive(TP)}}{\text{False Positive(FP)} + \text{True Positive(TP)}} \tag{14.5}$$

  - **Recall**: It is calculated as the ratio of number of accurate positive prophecies to the total number of actual positive instances.

$$\text{Recall} = \frac{\text{True Positive(TP)}}{\text{False Negative(FN)} + \text{True Positive(TP)}} \tag{14.6}$$

  - **_F_1-Score**: It is calculated as the harmonic mean of the recall and precision. Its value **ranges** from 0 to 1, with 0 meaning worst and 1 meaning perfect classification performance.

$$F1\text{-score} = \frac{2 \times \text{Precision} \times \text{Recall}}{(\text{Precision} + \text{Recall})} \tag{14.7}$$

  - **Accuracy**: It is calculated as the ratio of the number of accurate predictions to the total number of predictions made.

$$\text{Accuracy} = \frac{\text{TN} + \text{TP}}{\text{TN} + \text{TP} + \text{FN} + \text{FP}} \tag{14.8}$$

14.5   RESULT ANALYSIS

Performance evaluation methods are applied to the proposed stacking classifier and all the base classifiers to assess their performance. The results of this study have been shown with the help of bar graphs in Figures 14.9 and 14.10. The following remarks can be drawn from Figures 14.9 and 14.10:

- With respect to the $F1$-score measure, the proposed stacking classifier model obtained an $F1$-score of 83.67%. In terms of $F1$-score, it performs better than all the base classifiers individually – RFC (82.47%), MLP (80.26%), KNN (81.18%), and XGBoost (83.04%) – as depicted in Figure 14.9.

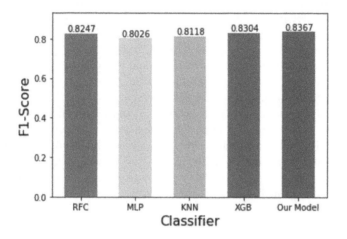

FIGURE 14.9 $F1$-score comparison between the proposed stacking classifier and base classifiers.

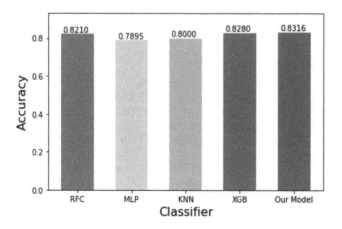

FIGURE 14.10   Accuracy comparison between the proposed stacking classifier and base classifiers.

- The proposed stacking classifier has an accuracy of 83.16%. In terms of accuracy, it performs better than all the base classifiers individually – RFC (82.10%), MLP (78.95%), KNN (80.00%), and XGBoost (82.80%) – as shown in Figure 14.10.

It is visible that the proposed stacking classifier model outperformed all the base classifiers used (i.e., XGBoost, RFC, MLP, and KNN) in terms of all the methods applied to assess the performance of the proposed model.

14.6   CONCLUSION AND FUTURE SCOPE

Gestational diabetes is a dangerous disease. Early detection of gestational diabetes can help doctors to diagnose it in the early stages. With the help of IoT-enabled devices and various sensors, we can now get the proper data for making data-driven decisions. The integration of IoT and machine learning has opened a new door of opportunities for both doctors and patients. In this research work, we proposed a model based on the stacking method. It used multilayer perceptron (MLP), random forest classifier (RFC), k-nearest neighbors (KNN), and XGBoost as base classifiers and logistic regression as a meta-classifier. The experimental outcomes after the evaluation of the proposed technique demonstrate that the model has a higher $F1$-score and accuracy with a significant edge in performance over an individual algorithm. With the day-by-day increasing demand for analysis of medical data, the suggested model can be of great help to research workers and doctors in making much more precise and accurate decisions for patients with diabetes.

In the future, with the latest hospital patient dataset, this model can greatly help doctors to precisely and accurately identify, analyze, and diagnose diabetes and can act as a base for the prediction and diagnosis of other types of diseases. The proposed stacking classifier model performs better than all the other classifiers used, but there is still room for further improvements. There are other ensemble methods that can also be explored to get better results.

REFERENCES

1. American Diabetes Association. "Diagnosis and classification of diabetes mellitus." *Diabetes Care*, 33(Suppl 1), pp. S62–S69, 2010. PMC. Web, 3 May 2018.
2. M. Maniruzzaman, N. Kumar, M. M. Abedin, M. S. Islam, H. S. Suri, A. S. El-Baz, and J. S. Suri. "Comparative approaches for classification of diabetes mellitus data: Machine learning paradigm." *Computing Methods Programs Biomedical*, 152, pp. 23–34, 2017.
3. S. Sa'di, A. Maleki, R. Hashemi, Z. Panbechi, and K. Chalabi. "Comparison of data mining algorithms in the diagnosis of type II diabetes." *International Journal in Foundations of Computer Science & Technology (IJFCST)*, 5(5), pp. 1–12, 2015.
4. M. M. Sabariah, S. A. Hanifa, and M. S. Sa'adah. "Early detection of type II diabetes mellitus with random forest and classification and regression tree (CART)." In *International Conference of Advanced Informatics: Concept, Theory and Application (ICAICTA)*, Bandung, Indonesia, pp. 238–242, 2014.

5. A. Gulten and A. Ozcift. "Classifier ensemble construction with rotation forest to improve medical diagnosis performance of machine learning algorithms." In *Proceedings of the Computer Methods and Programs in Biomedicine*, Bandung, Indonesia pp. 443–451, 2011.

6. H. N. A. Pham and E. Triantaphyllou. "Prediction of diabetes by employing a new data mining approach which balances fitting and generalization." In *Computer and Information Science*, Bandung, Indonesia, pp. 11–26, 2008.

7. J. Wu, Y. B. Diao, M. L. Li, Y. P. Fang, and D. C. Ma. "A semi-supervised learning based method: Laplacian support vector machine used in diabetes disease diagnosis." *Interdisciplinary Sciences Computational Life Sciences*, 1(2), pp. 151–155, 2009.

8. Y. Li, X. Z. Wang, and Q. Hua. "Using BP-network to construct fuzzy decision tree with composite attributes." In *International Conference on Machine Learning and Cybernetics,* Bandung, Indonesia. IEEE, Vol. 3, pp. 1791–1795, 2004.

9. B. M. Patil, R. C., Joshi, and D. Toshniwal. "Hybrid prediction model for Type-2 diabetic patients." *Expert Systems with Applications*, 37(12), pp. 8102–8108, 2010.

10. K. Polat and S. Günes, "An expert system approach based on principal component analysis and adaptive neuro-fuzzy inference system to diagnosis of diabetes disease." *Digital Signal Processing*, 17, pp. 702–710, 2007.

11. M. R. Devi and J. M. Shyla. "Analysis of various data mining techniques to predict diabetes mellitus." *International Journal of Applied Engineering Research*, 11, pp. 727–730, 2016.

12. A. K. Dewangan and P. Agrawal. "Classification of diabetes mellitus using machine learning techniques." *International Journal of Engineering and Applied Sciences*, 2, pp. 145–148, 2015.

13. K. Kayaer and T. Yıldırım. "Medical diagnosis on Pima Indian diabetes using general regression neural networks." In *Proceedings of the International Conference on Artificial Neural Networks and Neural Information Processing (ICANN/ICONIP)*, Bandung, Indonesia, pp. 181–184, 2003.

14. Z. Soltani and A. Jafarian. "A new artificial neural networks approach for diagnosing diabetes disease type II." *International Journal of Advanced Computer Science & Applications* 1(7), pp. 89–94, 2016.

15. A. Jarullah. "Decision tree discovery for the diagnosis of type II diabetes." In *2011 International Conference on Innovations in Information Technology*, Abu Dhabi, pp. 303–307. doi: 10.1109/INNOVATIONS.2011.5893838.

16. M. R. Bozkurt, N. Yurtay, Z. Yilmaz, and C. Sertkaya, "Comparison of different methods for determining diabetes." *Turkish Journal of Electrical Engineering and Computer Sciences*, 22(4), pp. 1044–1055, 2014.

17. V. A. Kumari and R. Chitra. "Classification of diabetes disease using support vector machine." *International Journal of Engineering Research and Applications*, 3(2), pp. 1797–1801, 2013.

18. A. Khamparia, A. Singh, D. Anand, D. Gupta, A. Khanna, N. Arun Kumar, and J. Tan. "A novel deep learning based multi-model ensemble methods for prediction of neuro-muscular disorders." *Neural Computing and Applications*, 2018. doi: 10.1007/s00521-018-3896-0, SCIE (IF 4.6).

19. R. Bansal, S. Kumar, & A. Mahajan. (2017). Diagnosis of diabetes mellitus using PSO and KNN classifier. *2017 International Conference on Computing and Communication Technologies for Smart Nation (IC3TSN)*. doi:10.1109/ic3tsn.2017.8284446.

20. I. L. Cherif and A. Kortebi. "On using eXtreme gradient boosting (XGBoost) machine learning algorithm for home network traffic classification." In *2019 Wireless Days (WD)*, 2019. doi: 10.1109/wd.2019.8734193.

21. A. Khamparia and K. M Singh. "A systematic survey on deep learning architectures and applications." *Expert System*. doi: 10.1111/exsy.12400, Wiley, SCIE (IF 1.5).

22. L. Catarinucci, D. Donno, L. Mainetti, L. Palano, L. Patrono, M.L. Stefanizzi, and L. Tarricone. "An IoT-aware architecture for smart healthcare systems." *IEEE Internet of Things Journal*, 2(6), pp. 515–526, 2015.

23. K. Fawagreh, M. M. Gaber, and E. Elyan. "Random forests: From early developments to recent advancements." *Systems Science & Control Engineering*, 2(1), pp. 602–609, 2014.

24. A. Singh, J. C. Mehta, D. Anand, P. Nath, B. Pandey, and A. Khamparia. "An intelligent hybrid approach for hepatitis disease diagnosis: Combining enhanced k-means clustering and improved ensemble learning." In *Expert Systems*. doi: 10.1111/exsy.12526, 2020, Wiley, SCIE (IF 1.7).

25. J. Sessa and D. Syed. "Techniques to deal with missing data," In *Proceedings of International Conference on Electronic Devices, System and Applications*, Bandung, Indonesia, pp. 1–4, 2017.

26. K. U. Rani, G. N. Ramadevi, and D. Lavanya. "Performance of synthetic minority oversampling technique on imbalanced breast cancer data." In *Proceedings of 3rd International Conference on Computing for Sustainable Global Development*, Bandung, Indonesia,pp. 1623–1627, 2016.

27. UCI Machine Learning Repository. "UCI repository of bioinformatics databases." 2018 [online]. Available: http://www.ics.uci.edu/mlearn/MLRepository.html. [Accessed: 05-Dec-2018].

28. G. Chhabra, V. Vashisht, and J. Ranjan. "A comparison of multiple imputation methods for data with missing values." *Indian Journal of Science and Technology*, 10(19), pp. 1–7, 2017.

29. M. Farahmandian, Y. Lotfi, and L. Maleki, "Data mining algorithms application in diabetes diseases diagnosis: A case study." In Technical Report, MAGNT Research Report, 2015.

30. A. Upadhyay and V. R. Patel. "Comparative study – Prediction of diabetes and heart disease using data mining approaches." *International Journal of Engineering Technology, Management and Applied Sciences*, 4(1), pp. 70–76, 2016.

31. N. H. Cho, J. E. Shaw, S. Karuranga, Y. Huang, J. D. da Rocha Fernandes, A. W. Ohlrogge, and B. Malanda. "IDF diabetes Atlas: Global estimates of diabetes prevalence for 2017 and projections for 2045." *Diabetes Research and Clinical Practice*, 138, pp. 271–281, 2018.

32. A. N. Ramesh, C. Kambhampati, J. R. Monson, and P. J. Drew. "Artificial intelligence in medicine." *Annals of the Royal College of Surgeons of England*, 86(5), pp. 334–338, 2004. PMC. Web, 6 June 2018.

33. M. Maniruzzaman, M. J. Rahman, and M. Almehedihasan. "Accurate diabetes risk stratification using machine learning role of missing value and outliers." *Journal of Medical Systems*, 42(5), p. 92, 2018.

34. V. Vijayan and A. Ravikumar, "Study of data mining algorithms for prediction and diagnosis of diabetes mellitus." *International Journal of Computer Application*, 94, pp. 12–16, 2014.

15 Securing Honey Supply Chain through Blockchain

An Implementation View

Sachin Gupta
MVN University

Bhoomi Gupta
Maharaja Agrasen Institute of Technology

CONTENTS

15.1 INTRODUCTION

There is an emergent need for a platform that secures and streamlines honey supply chain while authenticating the quality of honey delivered from bee farm to table.

15.1.1 FOOD FRAUD

The US Food and Drug Administration (FDA) has defined food fraud in three categories: (1) deliberate adulteration of food with low-priced ingredients, (2) trade of food having forbidden material, and (3) mislabeling of food and ingredients [1]. Food fraud has been rampant, from olive oil that has been cut with cheaper ingredients to seafood being mislabeled as different species and ground coffee contaminated with corn and sawdust.

15.1.2 SUPPLY CHAIN

Food supply chain over the years has become more complex. Manual, paper-based processes, onerous administrative tasks, and increasing regulatory requirements are some of the challenges faced by the food supply chain today. Fragmented and inefficient processes make the data sharing between the food supply partners challenging. Information shared is often incomplete and inconsistent due to the sources being multiple as no one can view the supply chain in its entirety from one end to the other [2]. As consumers are more aware, they want to make informed decisions on what they are buying. Stores carry food products labeled as "organic", "non-GMO", or "farm-raised", but there is no way to find out how the food was grown, stored, inspected, and transported to the grocery store.

  With the increase in food fraud and the complexity in supply chain, besides being transparent in operations, we need accountability and access to known and trusted sources of information. End consumers need to be made aware of the origin and state of food as it moves from farm to table.

15.1.3 Paper Organization

This chapter is organized into five major sections starting with the introduction of the application domain being food supply chain management. Section 15.2 deals with a survey of the existing literature and covers the technical reports about honey-related frauds, and Section 15.3 gives a detailed account of the technical approach to using blockchain for trust among the different stakeholders of the supply chain. The technical architecture of the proposed solution incorporating the blockchain at various platforms along with a risk analysis is elaborated in Section 15.4. The conclusions of the study with future research opportunities are detailed in Section 15.5.

The major contribution of this chapter is a case-study-based demonstration of the applicability of blockchain as a distributed trust management platform for ensuring the elimination of food-supply-related frauds happening in the honey industry. The general applicability of blockchain in similar domains with the requirement of distributed trust can also be envisioned conclusively with this study.

15.2 LITERATURE SURVEY

Due to the decline in honey production in the last 40 years and the growing consumer requirements, this industry in the USA has faced some challenges. An increasing amount of honey must be imported to meet the consumer demand. US beekeepers produced 149.5 million pounds of honey in 2013, 35% less compared to the last 20 years. This decline is due to the combination of many factors – bee's health challenges due to stresses brought on by parasites and insecticides, extreme weather, changes to the habitat where bees forage, and poor nutrition.

15.2.1 Honey Fraud and Supply Chain

A preliminary survey conducted by COLOSS (Honey Bee Research Association) estimated that bee colony loss in the USA was 21.1% over the 2016–2017 winters. Additionally, another COLOSS survey conducted in 29 countries reported an overall colony loss of 11.9% over the 2015–2016 winters. There is no denying that the bee population worldwide is declining, but what is alarming is there is surplus of honey despite the decline in honey production. The reason for the surplus is due to honey adulteration. Honey fraud is committed when it is adulterated with cheap sugar-based substitutes including sugar syrup, dextrose, corn syrup, glucose, flour, molasses, rice syrup, invert sugar, starch, or similar products to make a quick profit from the booming economy associated with honey.

Due to the decline in honey production over the past few decades and the growing consumer demand, the honey industry in the USA has faced some challenges. An increasing amount of honey must be imported to meet the consumer demand. Although there is

> U.S. antidumping duties and quality controls in place to protect U.S. consumers and honey companies from often cheaper and less regulated honey products from abroad. However, some honey brokers and importers illegally circumvent these restrictions,

selling honey to U.S. companies that is of questionable origin and which threatens the U.S. honey industry by undercutting fair market prices and damaging honey's reputation for purity and safety.

For instance, in 2008 the USA imposed China strict antidumping duties for selling honey at artificially low prices. Chinese exporter, however, started using other countries such as Vietnam and Indonesia to avoid the US duties. The rampant malpractices and unauthorized importers cause an economic distress to the ethically honest honey companies as they find it much difficult to sustain with honey sold at such lower prices [3].

15.3 TECHNICAL APPROACH

15.3.1 BLOCKCHAIN TECHNOLOGY

It has been nearly a decade since blockchain-based cryptocurrency bitcoin was released, and the number of cryptocurrencies available over the Internet as of December 22, 2018, is over 1984 and growing. Since its inception blockchain has been referred to as a disruptive technology as it brings benefits such as decentralization, transactions security, persistency, anonymity, and auditability [4]. Blockchain is potentially disruptive and has the potential for ensuring existing technology displacement. It is poised to disrupt the industry as a transforming product that in itself can generate new avenues in the industry, and initial trends with the rise in cryptocurrencies have signaled that blockchain has been disruptive for banking industry and financial services. Cryptocurrency has the potential for shaking the centralized banking system by eliminating the need to pay fees for using credit or debit cards. Recent trends have shown that blockchain is turning out to be a sustaining technology rather than being disruptive and has a huge potential in supply chain, health care, Internet of Things (IoT), education, and public services.

15.3.1.1 Blockchain

The blockchain technology derives its roots from a distributed ledger service that registers every activity as a transaction in the order of appearance. The records are grouped together in the form of blocks and are stored along with hash values that are cryptographically verifiable. Now these blocks are connected together to form a linked record structure called blockchain that is not only a permanent repository of records, but also validated by the community. Each member of the community maintains the individual copy of the records, and yet each transaction update is validated collectively. Figure 15.1 represents a simplified blockchain. Blockchain may record transactions, contracts, information assets, or anything that can be practically stored in digital form. Blockchain records are permanent, tamper-proof, transparent, and searchable. Each new recorded transaction is labeled as a block and is appended to the chain. Initiation, validation, storage, and distribution of each new block are managed by a protocol. Blockchain replaces the need for third-party intermediaries, and the participants of the blockchain ensure the integrity of each recorded transaction/block with the help of complex verifiability algorithms [5].

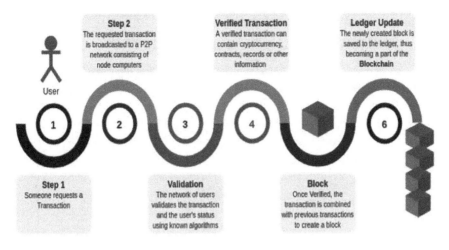

FIGURE 15.1   Blockchain technology workflow.

15.3.1.2   Consensus Algorithms

The basic definition of consensus says that it is a collaborative affirmation of some decision. Technically, we can achieve a consensus by matching the hash-based values of all records stored on the chain. To reach this consensus needs a lot of hard work and resource provisioning of computing power by the participating entities for blockchain maintenance. For this, provision of reward is required for people to contribute resources for maintaining the network. The following are some popular consensus algorithms [6]:

- **Proof of Work (PoW) Mechanism**: The proof of work is one of the most deployed methods for consensus. The participants of the distributed shared transactions running the nodes collect the interrelated transaction blocks which form a blockchain. The proof of work is designed as a hard mathematical puzzle requiring a significant amount of computational resources. There is usually a reward-based mechanism which serves as an incentive for the participants, and at the same time, it keeps the chain harvesting running. The complexity of the mathematical puzzles keeps increasing after each successful addition of block and is also proportional to the number of participants in the network.
- **Proof of Stake** [7]: This consensus mechanism thrives upon the need of expenditure of a native currency by the participating validation node, necessitated by the users' stake in the transaction being carried out. For the Ethereum blockchain, the native coin is ether. To achieve consensus, the network participants spend a tiny amount of ether to exercise their vote.
- **Proof of Activity** [8]: This consensus mechanism drives from the best of the two technologies listed above. A randomly generated number of participants perform work (solve the hard mathematical problem) and sign off on a block using a cryptokey (currency) to make the block official.

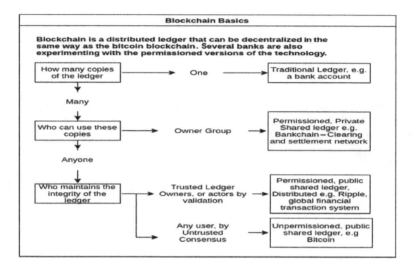

FIGURE 15.2 Blockchain basics.

- **Proof of Capacity** requires miners to allocate a sizable volume of their hard drive to mining, while proof of storage requires miners to allocate and share disk space in a distributed cloud.

There exist other mechanisms like the one used by Ripple (XRP) which rely on social networks for consensus [9]. The value of a member in a social network is determined by the number of unique members/nodes it connects to. This type of mechanism is biased where newcomers need social intelligence to participate. Only authorized member groups reach consensus. This model allows for corporate-style structures to exist with accountability on the blockchain (Figure 15.2).

15.3.2 DATA SOURCE

Food fraud and lack of transparency in the current process calls for a need to secure the honey supply chain by providing a platform to different entities enabling them to share, manage, and authenticate the quality of honey. The proposed platform will be a shared resource where all partners will interact effortlessly with an end goal to provide the much-needed transparency to the end consumer. In this chapter, analysis is gathered from several of sources. Sources include, but are not limited to, the following:

1. **Primary Sources**: Survey/interviews with producers, distributors, retailers, restaurant owners, and end consumers.
2. **Food Fraud Regulatory Data**: US Food and Drug Administration (USDA), US Environmental Protection Agency (USEPA), and COLOSS Honey Bee Research Association.
3. **Blockchain Implementations**: Walmart Pilot Program (Studies conducted by Blockchain Research Institute).

The HoneyBlock software is intended to be made available as dedicated Android and iOS apps and browser-independent Web applications.

The HoneyBlock Web and mobile applications can be accessed by the following users:

15.3.2.1   Producers or Beekeepers

Also called apiarists, the beekeepers operate at different scales. According to the number of colonies they manage, the beekeepers can be categorized into three categories. Small-scale beekeepers having less than 40 colonies, medium-scale keepers with up to 160 colonies, and commercial players with a large scale upward of 160 colonies. Apiarists are generally focused and restrict themselves to the production of honey, while the other tasks are performed by the actors in the distribution chain. The HoneyBlock team would work with third-party RFID tag manufacturers to create tags. These tags will be sent to the producers and distributors used for labelling and sealing the package [10].

15.3.2.2   Distributors

The distribution system in honey supply chain has several actors such as exporters, importers, customs, government authorities, freight forwarders, customs house agent, and transshipment parties.

If the raw honey is processed and converted into new products, the blockchain will be updated to identify and track the new product while retaining the details of the original source.

For example, a lemon honey is prepared by mixing lemon and other ingredients and packed in separate bottles, which will be tracked as a separate contract associated with its own process. The check is mandatory to eliminate adulteration of product along the supply chain [11].

- **Retailers/Restaurant Owners**: They are the entities who order the honey on the HoneyBlock.
- **End Consumers**: The end consumer is the ultimate consumer of honey and has inherent stake in the quality of honey.

15.3.2.3   Third-Party Accreditation Agencies (TrueSourceHoney)

Each customer in the HoneyBlock will undergo accreditation. The accreditation certificate would be issued by TrueSourceHoney [12].

TrueSourceHoney authenticates and validates parties that will be involved as stakeholders along the honey supply chain. The parties will use the smart contract along with the existing notification systems to capture the data as honey moves from hive to home.

15.3.3   Salient Features

HoneyBlock is a decentralized application. The decentralization is handled with backend code running on a peer-to-peer network that is Ethereum in the case of HoneyBlock.

It is different from the normal mobile app with a centralized server deployed for the backend. The DApp uses RestAPI to interact with the Ethereum blockchain. Additionally, Azure Storage (message) Queues and WebJobs are used to enable fault tolerance as the application is decoupled. Besides, SQL and NOSQL data stores will be used, which will interact with Microsoft Azure CDN (Content Delivery Network).

The settlement layer responsible for managing the smart contracts shall act as an interface providing speedy smart contract update. It shall be a dedicated lightweight client of the Ethereum blockchain focused on HoneyBlock contract obligations [13].

15.4 TECHNICAL ARCHITECTURE

See Figure 15.3

15.4.1 SERVICES AVAILABLE ON THE HONEYBLOCK PLATFORM

The honey supply chain process requires multiple documents that need to be shared by several parties. The tracking of these steps is very critical for the success of blockchain. The HoneyBlock platform would update the blockchain as the events are recorded by the parties during the process flow. The tracing process involves smart contract service to track shipment from its beginning in the form of a procurement order submitted by the retailer/restaurant owner to its delivery at the buyer's facility.

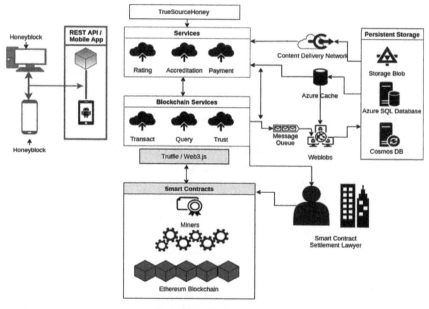

HoneyBlack Technical Architecture

FIGURE 15.3 HoneyBlock technical architecture.

TABLE 15.1
High-Level Summarization of Events on the HoneyBlock

	Documents Required/Action					
Actors	Purchase Order (PO)	Sales Contract	Invoice	Shipping Bill	Certificate of Origin	Letter of Credit
Producer	-	Create	Create	-	Submit	Review
Retailer	Raised PO	Review	Review	-	Submit	Review
Distributor	Inform	-	Inform	Create	-	-

The goal of the smart contract service is to capture all changes to the consignment and provide the users with the ability to view the history. Ethereum blockchain will be used to build the smart contract [14]. The following activities are executed through the process.

- Smart contract creation with end-to-end traceable and verifiable information.
- Generate event alerts on certain events taking place as the honey moves from beekeeper to customer.
- Release of authorization documents and money on the completion of certain events (Table 15.1).

The following micro-services will be available to the actors that can be accessed by the API:

15.4.2 ACCREDITATION SERVICE

The TrueSourceHoney will access the HoneyBlock through the accreditation micro-service. When TrueSourceHoney issues a certificate, it is digitized and stored on the platform. The certificates can be accessed via the HoneyBlock app, thus providing traceability. HoneyBlock would store each customer's certificates in a private blockchain to restrict the capability to authenticate, decentralize, and encrypt outside of the HoneyBlock platform. The platform, however, will provide read access to all users/customers of the HoneyBlock platform. The platform will restrict all customers with missing accreditation or expired certificates from transacting on the platform.

15.4.3 SMART CONTRACT SERVICE

The smart contract on HoneyBlock will undergo the following stages:

1. The retailer or restaurant owner creates an order on HoneyBlock. The order would require sufficient ether tokens in the wallet.
2. The contract entails all the details required to authenticate the order.
3. This triggers the new order notification event for the beekeeper (producer).

4. This is followed by order acceptance and a price estimate by the producer.
5. The distributor is notified when the order is ready to be picked up.
6. The beekeeper seals the shipment with a GPS-enabled RFID tag.
7. The distribution cycle has multiple steps where it must go through packaging, loading, shipping, customs, and intermediate storage. At every stage, the actors scan the tag that in turn invokes a record-keeping transaction on the blockchain.
8. All the parties in the contract can track the shipment from the point it is picked up to the point it is delivered to the entity that placed the order.
9. Once the order is received and acknowledged on the HoneyBlock, the money is transferred to the various parties involved in the contract as per the agreed terms of the contracts.
10. The payment is moved from the wallet of the retailer or restaurant owner to the distributor and producer.
11. In order to speed up the transactions and minimize the time, any calls to the settlement layer must be authenticated by the private key and all double calls will be eliminated by limiting the number of calls between blocks.

Figure 15.4 illustrates the working of smart contracts between two parties.

15.4.4 PAYMENT GATEWAY SERVICE

15.4.4.1 Ether – The Crypto-Fuel for the Ethereum Network

The computational power required to execute transaction on a blockchain needs resources. The resources are quantified as ether and are analogous to "gas" that fuels a vehicle, acting as a currency that powers the Ethereum ecosystem. A small fraction of ether is required to conduct a transaction on the HoneyBlock. As new events in the honey supply cycle occur, new entries are added to the HoneyBlock chain, requiring several small expenditures of ether tokens [15]. The HoneyBlock would add a fee to each transaction, and it would be one of the primary revenue generators for the platform.

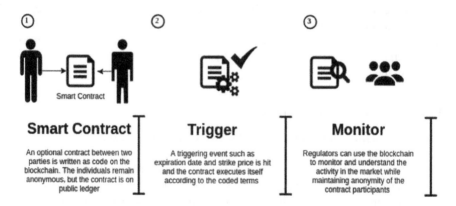

FIGURE 15.4 Smart contracts between two parties.

15.4.4.2 Wallet

The customers (except the end user) will be able to create a wallet. These wallets are synonymous to bank accounts and are used for holding, storing, and transferring ether. Although the Ethereum platform provides the capability to the user to create their own wallet, this will be managed by the HoneyBlock to have more control over parties abiding by the smart contracts. Each customer will be assigned a public and a private key. HoneyBlock would provide customers with their "public key", which is their ether address [16]. The private key is used to unlock the wallet and allow the user to spend the ether that is allocated to that address within the blockchain.

Types of Wallet: The wallets fall into two broad categories: hot storage wallets and cold storage wallets. A hot storage wallet refers to a wallet that is kept online. Hot wallets can be accessed through Internet and hence are prone to hacking. On the other hand, cold storage wallets are offline and its primary purpose is to protect the stored ether against hacking and theft.

15.4.5 RATINGS SERVICE

The different entities from producer to retailer on the HoneyBlock are reviewed and rated on the basis of the product they offer or the services they provide. The reviewer evaluates and assigns them a rating that can be quantified as per predetermined performance metrics. The HoneyBlock platform ratings interface captures quality reviews from all parties on the blockchain, ensuring transparency and immutability.

The main public Ethereum decentralized network which forms the backbone of the HoneyBlock will be linked to a smart contract settlement layer which will be scalable and provide better performance. This in turn would avoid strain on the public Ethereum network and thus reduce the transaction fees and provide scalable architecture. The platform also monitors the smart service, transactions, and actions of different parties and assigns and stores the ratings on the blockchain to ensure the transparency of reviews and prevent fraud [17]. Customers and users can easily check and verify these ratings and the contributing transactions and reviews.

15.4.6 CUSTOMER SEGMENTS

Three distinct customer segments are envisioned:

15.4.6.1 Customer Segment 1

- **End Consumer**: End consumers that buy honey from the retail stores. The end user or the consumer can continuously benefit from a free-of-charge access to the HoneyBlock platform. The consumer can access the platform by downloading the mobile app and scanning a QR code/RFID tag at retail stores to track the history of a product, thus making a conscious decision to buy the product. This customer segment is looking for quality ingredient. They want transparency and reassurance that the quality of the honey they are consuming is exactly what the label says it is. These consumers are willing to pay more for high-quality honey; they would be willing to use the services that will provide transparency in the honey they consume.

The following list of customers will need to go through an accreditation process. Accreditation will be performed by a third-party accreditation firm. A wallet will be setup for each of the partners. The wallet would ensure that the payment is disbursed as per the agreed contract to the partners.

15.4.6.2 Customer Segment 2

- **Restaurant Owner**: Restaurant owners need to design menus that meet customers' different tastes and preferences. They need to accommodate for the list of special dietary restrictions. Customers want to know where their food came from, whether it is organic, and whether it is a high-quality food. Making this information available is food transparency, and it is one of the biggest trends in the food service industry right now. The restaurant owner will utilize the platform for identifying the right set of distributors and producers to ensure the quality and transparency of the ingredients for the end customer, in our case honey. This can be done by subscribing and advertising on the platform to establish the supply chain for its need. This customer unlike the first customer is not subsidized and has to pay for the service.
- **Retailer**: For the retailers, the HoneyBlock can deliver trust in a product, transaction, or the integrity of data effectively and at lower cost. The platform will help retailers collaborate with other partners across the globe by subscribing to the platform. The platform will keep record of every transaction and provide tracking of shipment, thus improving supply chain visibility, ensuring product provenance and authenticity, speeding up transactions, reducing processing fees, and improving the management of networked loyalty programs.

15.4.6.3 Customer Segment 3

- **Distributor**: The platform provides distributers (wholesaler) access to the producers, retailers, and restaurant owners who are looking for high-quality honey. This can be done by subscribing and advertising on the platform to establish the supply chain for its need. The execution of smart contracts will ensure that pre-agreed upon proceeds are paid instantly at the point of delivery [18]. The GEO sensors will help track the location of a shipment along with reducing the amount of paper work by record keeping of every transaction on the HoneyBlock platform.

15.4.6.4 Gains

HoneyBlock Web site and mobile applications shall provide diverse gains such as honey quality assurance, streamlined efficient distribution process, smart contract management, and product provenance and authenticity by providing end-to-end visibility [19]. These shall also provide the ability to track data across supply chain, will speed up transactions, shall provide access to marketing platform, will allow cost cutting, and will reduce food waste. Consumers will receive trusted information on the journey of food, creating a new standard of food quality.

15.4.6.5 Pains

HoneyBlock Web site and mobile applications shall conversely be absorbed with numerous factors such as honey adulteration, food-borne illness, illegal production, complex supply chain, and fractured paper-based systems.

These applications will have more information about how to use the platform. The applications shall encompass social media such as Facebook, Twitter, Instagram campaigns; marketing partners such as Google AdSense and BlockFood; television advertisements in food and health channels; and telemarketing campaigns targeting different customers. The system will be used to rate the retailers/producers/distributors.

15.4.7 RISK ANALYSIS

15.4.7.1 Privacy

The blockchain is an immutable, tamper-resistant transaction, which is visible to all parties within the network, which may not be acceptable for some applications. One of the solutions is to store encrypted data which can lead to different sets of issues if the key is lost, stolen, or published. The problem can be solved by using Zcash on Ethereum which will make it possible to implement anonymous payments [20].

15.4.7.2 Scalability and Latency

Smart contract platforms are still considered unproven in terms of scalability. One of the important characteristics of a blockchain is decentralization. This means that every node on the network must maintain a copy of the ledger [21].

15.4.7.3 External Information

Blockchain stores unalterable sequence of transactions as per the events that take place. If external data sources which are non-deterministic interact with the blockchain, it would create problems with transparency which is the fundamental reason for using blockchain. The problem can be resolved by using trustworthy data services – known as "oracles" – that can execute smart contracts whenever the terms of the contract are met [22].

15.4.8 SOLUTION DIFFERENTIATORS

HoneyBlock platform offers an innovative solution that focuses on providing traceability and transparency in honey supply chain, from bee farm to table. What makes our solution unique is that HoneyBlock platform is solely focused on the honey industry. Our mission is to combat honey fraud by providing a platform where every entity in the honey supply chain can view and track unalterable information at every stage [23]. Additionally, through our partnership with True Source Honey and our rating service, consumers can be assured the honey delivered is ethically sourced and of high quality.

15.5 CONCLUSIONS AND FUTURE WORK

Food fraud in general is a growing problem worldwide that affects country's economy, company's brand image, public health, and consumer confidence. The complex, opaque, and fragmented food supply chain has opened the door to food fraud crime. Likewise, honey is one food product that is at high risk of food fraud. Therefore, to alleviate honey fraud, HoneyBlock is designed to give all participants in the honey supply chain the needed traceability and transparency, from bee farm to table. The application can be extended to different products requiring multi-party trust establishment, and the future may allow for the development feasibility of easy off-the-shelf deployments.

REFERENCES

1. Food Borne Germs and Illnesses; Food Safety (2018). Retrieved from https://www.cdc.gov/foodsafety/foodborne-germs.html.
2. Dani, S. (2015). *Food Supply Chain Management and Logistics: From Farm to Fork*. London: Kogan Page; https://www.koganpage.com/product/food-supply-chain-management-and-logistics-9780749473648.
3. A "food systems thinking" roadmap for policymakers and retailers to save the ecosystem by saving the endangered honey producer from the devastating consequences of honey fraud (2017). www.apimondia.com /docs/honey_white_paper.pdf.
4. Vinay Kumar S., and Gupta, S. (2019). Block chain in supply chain: Journey from disruptive to sustainable. vol. 14, no 2, https://doi.org/10.26782/jmcms.2019.04.00036.
5. Iansiti, M., and Lakhani, K.R. (2017). The truth about blockchain. *Harvard Business Review*, vol. 95, no. 1, pp. 118–127.
6. Mattila, J. (2016). The blockchain phenomenon: The disruptive potential of distributed consensus architectures, ETLA working papers: Elinkeinoelämän Tutkimuslaitos, Research Institute of the Finnish Economy. books.google.com.pk/books?id=StNQnQAACAAJ.
7. King, S., and Nadal, S. (2012). PPCoin: Peer-to-Peer Crypto-Currency with Proof-of-Stake.
8. Bentov, I., Lee, C., Mizrahi, A., and Rosenfeld, M. (2014). Proof of activity: Extending Bitcoin's proof of work via proof of stake [extended abstract]. *ACM Sigmetrics Performance Evaluation Review*, vol. 42, no. 3, pp. 34–37.
9. Schwartz, D., Youngs, N., and Britto, A. (2014). *The Ripple Protocol Consensus Algorithm*. Ripple Labs Inc White Paper, pp. 1–8. Retrieved from ripple.com.
10. Chopra, S., and Meindl, P. (2015). *Supply Chain Management: Strategy, Planning, and Operation* (6th edn.). Pearson Education.
11. Kasireddy, P. (2017). Blockchains don't scale. Not today, at least. But there's hope. Retrieved from https://hackernoon.com/blockchains-dont-scale-not-today-at-least-but-there-s-hope-2cb43946551a.
12. Osterwalder, A. (2012). Alexander Osterwalder: Tools for business model generation. Retrieved from Stanford Technology Ventures Program.
13. Loi, L., Duc-Hiep, C., Hrishi, O., Prateek, S., and Aquinas, H. (2016, October 24–28). Making Smart Contracts Smarter. In Proceedings of the 2016 ACM SIGSAC Conference on Computer and Communications Security (CCS '16). Association for Computing Machinery, New York, NY, USA, pp. 254–269. doi: 10.1145/2976749.2978309.
14. Bambara, J. J., and Allen, P. R. (2018). *Blockchain: A Practical Guide to Developing Business, Law, and Technology Solutions*. New York: McGraw-Hill Education.

15. Ethereum Project; Guides, resources, and tools for developers building on Ethereum; Retrieved from https://ethereum.org/developers/#other-tools.

16. Ethereum Charts & Statistics; Charts history: Ethereum blockchain explorer. (2016). Retrieved from https://etherscan.io.

17. Security and privacy in blockchain environments. (2017). Retrieved from https://www.dotmagazine.online/issues/innovation-in-digital-commerce/what-can-blockchain-do/security-and-privacy-in-blockchain-environments.

18. Wang, S., Ouyang, L., Yuan, Y., Ni, X., Han X., and Wang, F. (2019, November). Blockchain-enabled smart contracts: Architecture, applications, and future trends. *IEEE Transactions on Systems, Man, and Cybernetics: Systems*, vol. 49, no. 11, pp. 2266–2277.

19. Harland, C., Brenchley, R., and Walker, H. (2003). Risk in supply networks. *Journal of Purchasing and Supply Management*, vol. 9 no. 2, pp. 51–62.

20. Malik, A., Gautam, S., Abidin, S., and Bhushan, B. (2019). Blockchain technology-future Of IoT: Including structure, limitations and various possible attacks, *2019 2nd International Conference on Intelligent Computing, Instrumentation and Control Technologies (ICICICT)*, Kannur, Kerala, India, pp. 1100–1104.

21. Ali Syed, T., Alzahrani, A., Jan, S., Siddiqui, M. S., Nadeem, A., and Alghamdi, T. (2019). A comparative analysis of blockchain architecture and its applications: Problems and recommendations. *IEEE Access*, vol. 7, pp. 176838–176869.

22. Sun, Y., Zhang, L., Feng, G., Yang, B., Cao, B., and Imran, M. A. (2019, June). Blockchain-enabled wireless Internet of Things: Performance analysis and optimal communication node deployment. *IEEE Internet of Things Journal*, vol. 6, no. 3, pp. 5791–5802.

23. Sharma, T., Satija, S., and Bhushan, B. (2019). Unifying blockchain and IoT: Security requirements, challenges, applications and future trends. *2019 International Conference on Computing, Communication, and Intelligent Systems (ICCCIS)*, Greater Noida, India, pp. 341–346.

Index